ESTIMATING
in Building Construction

CANADIAN EDITION

ESTIMATING
in Building Construction

CANADIAN EDITION

Frank R. Dagostino

Leslie Feigenbaum
Texas A&M University

Clint Kissoon
George Brown College

PEARSON

Prentice
Hall

Toronto

To my friends, Jim and Lou Ida Marsh, whose help and
guidance has made so much in my life possible.
To Fer and Tash for your support and inspiration.

National Library of Canada Cataloguing in Publication

Dagostino, Frank R.
   Estimating in building construction / Frank R. Dagostino, Leslie
Feigenbaum, Clint Kissoon. — Canadian ed.

Includes index.
ISBN 0-13-039126-3

   1. Building—Estimates. 2. Building—Estimates—Data processing.
I. Feigenbaum, Leslie II. Kissoon, Craig, 1952- III. Title.

TH435.D33 2003   692'.5   C2003-900557-7

ISBN 0-13-039126-3

Vice President, Editorial Director: Michael J. Young
Executive Editor: Dave Ward
Senior Developmental Editor: Lise Dupont
Marketing Manager: Toivo Pajo
Production Editor: Cheryl Jackson
Copy Editor: Rohini Herbert
Proofreader: Lu Cormier
Senior Production Coordinator: Peggy Brown
Formatter: Joan M. Wilson
Art Director: Mary Opper
Interior Design: Alex Li
Cover Design: Alex Li
Cover Image: © Grotesk/Getty/Taxi

1 2 3 4 5   07 06 05 04 03
Printed and bound in Canada

Contents

Preface

This Canadian edition builds on the strong foundation of the sixth U.S. edition. The need for estimators to have a working knowledge of construction materials and methods, and to understand the manual process of quantification is critical prior to the use of any electronic estimating tool, computerized or otherwise. This edition of *Estimating in Building Construction* provides extensive exercises and examples using metric quantification. Although many construction estimating sources, most of which are American, include imperial units, we hope that the presence of this metric-based educational resource will begin to change that trend in Canada. Please note, however, that readers should familiarize themselves with metric to imperial conversions, since imperial measures are still the language of some trades in the Canadian construction industry. This edition also addresses Canadian bidding practices and standards throughout the text. The samples of bid documentation, from invitation to closing, are Canadian and reflect current industry standards and practices. Quantity surveyors and estimators perform extremely important roles at each stage of a construction project. A standardized approach to the measurement of construction work and to the preparation of estimates is fundamental to the increased recognition of these professions and to the contribution they make to a project team.

A new feature in this edition is a list of Web sites (some for vendor and trade associations) at the end of each chapter addressing quantification. These resources provide the most up-to-date information on construction materials and methodology.

This edition features a CD that includes the versatile North American Timberline Precision Estimating Basic Edition software, as well as the spreadsheets that are used throughout the text. These spreadsheets can be used as templates to assist in organizing an estimate. They were created using Excel 2000 and are compatible with other software packages. (See Appendix E for more information.)

This edition also contains a series of drawings at the end of the book—one set for a residential project and one set for a small commercial project. These drawings are actual working drawings and are included to give readers an appreciation of the "digging" for information that is required to prepare an estimate. The needed information may be found on a number of different sheets and must be accumulated and combined to develop an estimate. These drawings are also the basis of the exercises found at the end of some chapters.

Acknowledgments

Particular thanks are due to my colleagues John Wills, Douglas Laporte, Gianni Regina, and William Juranic, all of George Brown College, for their support and encouragement. Thanks also to Robert Loov and to the many reviewers from across Canada, some of whom are listed below, for their comments and suggestions on the Canadian edition manuscript.

Terry D. Bajer, Northern Alberta Institute of Technology

Gregory T. Chayter, College of the North Atlantic

Brian Corrigan, Conestoga College

Gary Kembel, Loyalist College

Chris Makela, British Columbia Institute of Technology

Michael Nauth, Algonquin College

C. (Butch) Petrone, Confederation College

Barry Robbins, College of the North Atlantic

George J.P. Zioteck, Mohawk College

CHAPTER **1**

Introduction to Estimating

After studying this chapter, you will be able to:

1. Describe the type of estimate required at each stage of project development.

2. Identify areas in the construction sector where knowledge of estimating is necessary.

3. Be familiar with the attributes required to be a successful construction estimator.

4. Have knowledge of the contract documents required for bid preparation.

1.1 General Introduction

Building construction estimating is the determination of probable construction costs of any given project. Many items influence and contribute to the cost of a project; and each item must be analyzed, quantified, and priced. Because the estimate is prepared before the actual construction, much study and thought must be put into the bid documents (drawings and specifications). The estimator who can visualize the project from this information and accurately determine its cost will become one of the most important persons within any construction company.

In most instances, it is necessary to submit a competitive bid for the project. The competition in construction bidding is intense, with multiple firms vying for a single project. To stay in business, a contractor must be the lowest qualified bidder on a certain number of projects while maintaining an acceptable profit margin. This profit margin must provide the general contractor an acceptable rate of return and compensate for the risk associated with the project. The estimator must account for site planning, staff organization, scheduling, temporary facilities, material handling systems, safety/environmental issues, and quality control. The ability of the estimator to visualize all the different phases of the construction project becomes a prime ingredient in successful bidding.

The working drawings usually contain information relative to design, location, dimensions, and construction of the project, while the specifications are a written supplement to the drawings and include information pertaining to materials and workmanship. The working drawings and specifications constitute the majority of the contract documents, define the scope of work, and *must* be considered together when preparing an estimate. The two complement each other, and they often overlap in the information they convey. The bid submitted must be based on the scope of work provided, and it is the responsibility of the estimator to account for everything depicted by the drawings and described in the specifications in the submitted bid. Due to the complexity of the drawings and specifications, coupled with the potential costs of an error, the estimator must read and understand everything thoroughly and recheck all items. Initially, he must check the plans and specifications to ensure that they are complete. Then, he can begin the process of quantifying the scope of work presented. Every item included in the estimate must contain as much information as possible. The quantities determined for the estimate will ultimately be used to order and purchase the needed materials. The estimated quantities and their associated projected costs will become the basis of job costing and project cost control.

Estimating the ultimate cost of a project requires the integration of many variables. These variables fall into either *direct field costs* or *indirect field costs*. The indirect field costs are also referred to as general conditions costs in building construction. The direct field costs are the resources of material, labour, equipment, or subcontracted items that are permanently and physically integrated into the building. For example, the labour and materials for the foundation of the building would be a direct field cost. Indirect field costs are the cost for the items that are required to support the field construction efforts. For example, the project site office would be a general conditions cost. In addition, risk factors, such as weather, transportation, soil conditions, labour strikes, material availability, and subcontractor availability, need to be integrated into the estimate. Regardless of the variables involved, the estimator must strive to prepare as accurate an estimate as possible. Since subcontractors or specialty contractors may perform much of the work in the field, the estimator must be able to articulate a scope of work in order for these companies to furnish a price quote. The complexity of an estimate requires organization, estimator's best judgment, complete specialty contractors bids, accurate quantity takeoffs, and accurate records of completed projects.

1.2 Types of Estimates

The stage of project development and the required level of accuracy, together with the amount of information about the project that is available, will dictate the type of estimate that can be prepared.

Detailed Estimate

The detailed estimate includes determination of the quantities and costs of everything required to complete the project. This includes the resources of the contractor's self-directed (own forces) work, subtrades, general expenses, as well as an estimate of profit. To perform this type of estimate, the contractor must have a complete set of contract documents. He must break down each item of the project into its parts for the estimate. Each piece of work that is to be performed by the contractor has a distinct labour requirement that must be estimated. The items that are to be installed by others need to be defined and priced. The contractor must exercise caution to ensure that there is agreement between him and the specialty contractor as to what they are to do and if they are to install or supply and install the items. Additionally, there needs to be an agreement about who is providing support items, such as cranes, scaffolding, and so on.

The detailed estimate must establish the estimated quantities and costs of material, the time required for and the costs of labour, the equipment required and its cost, the items required for overhead and the cost of each item, and the percent of profit desired, considering the investment, the time to complete, and the complexity of the project.

Preliminary Estimates (Volume and Area)

The *volume method* involves computing the volume of the building and multiplying that volume by an assumed cost per cubic metre (foot). Using the *area method*, you compute the area of the building and multiply that area by an assumed cost per square metre (foot). Both methods require skill and experience in adjusting the unit cost to the varying conditions of each project. The amount of information required to produce these types of estimates is much less than with the detailed estimate. For example, a preliminary set of design drawings would have the dimensions for determining the area or volume. These types of estimates are helpful to check whether the project as designed is within the owner's budget; however, they lack accuracy. If the unit price comes from previously completed projects, it is assumed that this project is identical to the completed project. That assumption is clearly not valid in the construction of buildings. Weather conditions, building materials, and systems, as well as design and construction team members, change from project to project, all contributing to the uniqueness of every project. Annual publications, such as R.S. Means' "Yardsticks for Costing," contain a range of unit costs for a wide variety of building types in seven cities in Canada. These guides provide a number of adjustments to compensate for varying building component systems.

Conceptual Estimates

When performing a conceptual estimate, typically, there are no drawings available. What exists is a vague verbal or written description of the project scope. When preparing this type of estimate, the estimator makes assumptions about virtually every aspect of the project. These types of estimates are usually prepared using some unit of measurement that has little to do with actual construction materials or measurements. A typical unit of measurement would be cost per apartment dwelling unit in a multi-family housing project or cost per parking space for a parking garage.

This type of estimate is used early in the design process to check the reality of the owner's wants in terms of their budget. In addition, these estimates are required primarily for controlling the cost of the project during the design and working drawings stages within predetermined cost limits. In order to control costs, estimates for projects are analyzed on an elemental basis by which the cost of each major building element and subelement is identified on a cost per square metre/foot of the gross building area. In addition to providing a means of identifying elements of unusually high cost or potential budget over-runs, elemental cost estimates provide useful reference data when preparing preliminary estimates for other similar projects.

The Canadian Institute of Quantity Surveyors format for elemental cost analysis defines an element as a major component common to most buildings, fulfilling the same function irrespective of its design, specification, or construction. The standardized elemental breakdown used by estimators and quantity surveyors across the country is detailed in the Institute's publication *Elemental Cost Analysis—Method of Measurement and Pricing*.

Contemplated Change Order Estimates

Under the terms of the stipulated lump-sum contract, the consultant must issue a notice of proposed change or contemplated change, together with any revised drawings or specifications, to the contractor to obtain a quote. These changes may include additions, deductions, and alterations to part of the work that may or may not have been completed. These may be due to a change in the owner's requirements, unforeseen conditions, emergencies, or regulatory requirements and may affect the contract price or contract schedule. The estimate for these notices must incorporate any additions or deductions from the original contract documents and include adjustments in contract price and contract time. Unit rates, including levels of profit and overheads, for pricing changes to the work may be established prior to the signing of the contract. Alternatively, the consultant may assess them on a fair and reasonable basis.

1.3 Estimating Opportunities

For anyone who is not aware of the many opportunities in the estimating field, this section will review some of the areas in which knowledge of estimating is necessary. Generally, knowledge of the procedures for estimating is required by almost everyone involved in or associated with the field of construction. From the estimator, who may be involved solely with the estimating of quantities of materials and pricing of the project, to the carpenter, who must order the material required to build the framing for a home, this knowledge is needed to do the best job possible at the most competitive cost. Others involved include the project designer, architectural technologists, engineers, contractors, subcontractors, material suppliers, and technical sales representatives. In the following sections, a few of the estimating opportunities are described.

Architectural Offices. The architectural office will require estimates to plan and control the cost in the four stages of the pre-tender period: program or concept stage (based on costs per unit of gross floor area), schematics stage (costs of major elements per square metre or square foot), design development stage (setting a cost target for all the components of the building), and the contract document stage (trade-by-trade breakdown of costs to evaluate the lowest acceptable bid).

In large and/or collaborative design offices, a professional quantity surveyor or an estimator hired primarily to do all required estimates might do the estimating. In many offices, the senior architectural technologist or the head or lead architect may do the estimating or, perhaps, someone else in the office that has developed the required estimating skills. There are also estimating services or cost consultants who perform estimates on a for-fee basis.

Engineering Offices. The engineering offices involved in the design of building construction projects include civil, structural, mechanical (plumbing, heating, air conditioning), electrical, and soil analysis. All these engineering design phases require preliminary estimates; estimates while the drawings are being prepared; and final estimates as the drawings are completed.

General Contractors. Typically, the general contractor makes *detailed* estimates that are used to determine what the company will charge to do the work required. The estimator will have to "take off" the quantities (amounts) of each material, determine the cost to furnish (procure and ship to the site) and install each material in the project, assemble the bids (prices) of subcontractors, as well as determine all the costs of insurance, permits, office staff, and so on. In smaller companies, one person may do the estimating, whereas in larger companies, several people may work to negotiate a final price with an owner or to provide a competitive bid. Many times, the contractor's business involves providing assistance to the owners, beginning with the planning stage and continuing through the actual construction of the project (commonly called *design-build* contractors). In this type of business, the estimators will also provide preliminary estimates and then update them periodically until a final price is set.

Estimating with Quantities Provided. Estimating for projects with a schedule of quantities involves reviewing the specifications for the contract and material requirements, reviewing the drawings for the type of construction used, and assembling the materials used. The estimator will spend part of the time getting prices from subcontractors and material suppliers and the rest of the time deciding on how the work may be most economically accomplished.

Subcontractors. Subcontractors may be individuals, companies, or corporations hired by the general contractor to do a particular portion of the work on the project. Subcontractors are available for all the different types of work required to build any project and include excavation, concrete, masonry (block, brick, stone), interior partitions, drywall, acoustical ceilings, painting, steel and precast concrete, erection, windows and metal and glass curtain walls, roofing, flooring (resilient, ceramic, and quarry tile, carpeting, wood, terrazzo), and interior wall finishes, such as wallpapering, wood panelling, and sprayed-on finishes. The list continues to include all materials, equipment, and finishes required.

The use of subcontractors to perform all the work on the project is becoming an acceptable model in building construction. The advantage of this model is that the general contractor can distribute the risk associated with the project to a number of different entities. In addition, personnel of the subcontractors' trade perform the same type of work on a repetitive basis and are, therefore, quasi-experts in their niche. However, the general contractor relinquishes a substantial amount of control over the project when this method is employed. The more that the contractor subcontracts out, the more the field operation becomes involved in coordination, rather than direct supervision, of the various tasks of the trade personnel.

The subcontractor carefully checks the drawings and specifications and submits a price to the construction companies that will be bidding on the project. The price given may be a unit or a lump-sum price. If a subcontractor's bid is presented as a charge per unit, then it is a *unit price* (such as: per square metre/foot, per block, per thousand brick, per cubic metre/yard of concrete) bid. For example, the bid might be $27.25 per linear metre (m) of concrete curbing. Even with unit price bids, the subcontractors needs to perform a quantity takeoff so that they can have an idea of what is involved in the project, at what stages they will be needed, how long it will take to complete their work, and how many workers and how much equipment will be required. The subcontractor needs the completed estimate to determine the reasonable amount for overhead and profit. Typically, as the quantity of work increases, the associated unit cost of jobsite overhead decreases. For example, the cost of mobilization for a 100 linear metres of curb project is $3,000 or $30 per linear metre; if the quantity had been 300 linear metres, it would have been $10 per linear metre. The subcontractor would not know how much to add to the direct field cost unit price for overhead unless a quantity takeoff had been performed. If the subcontractor submits a lump-sum bid, then he or she is proposing to install, or furnish and install, a portion of work: For example, the bid might state "agrees to furnish and install all type I concrete curbing for a sum of $23,267."

Each subcontractor will need someone (or several people) to check specifications, review the drawings, determine the quantities required, and put the proposal together. It may be a full-time estimating position or part of the duties assumed, perhaps in addition to purchasing materials, helping schedule projects, working on required shop drawings, or marketing.

Material Suppliers. Suppliers submit price quotes to the contractors (and/or subcontractors) to supply the materials required for the construction of the project. Virtually every material used in the project will be estimated and multiple price quotes sought. Estimators will have to check the specifications and drawings to be certain that the materials offered will meet all the requirements of the contract and required delivery dates.

Manufacturers' Technical Sales Representatives. Manufacturer's representatives represent certain material or product suppliers or manufacturers. They spend part of their time visiting contractors, architects, engineers, subcontractors, owners, and developers to be certain that they are aware of the availability, uses, and approximate costs of the materials. In a sense, they are salespeople, but their services and the expertise they develop in their product lines make good manufacturers' representatives welcome as needed sources of information concerning the materials and products they represent. Representatives may work for one company, or they may represent two or more.

Manufacturers' representatives will carefully check the specifications and drawings to ensure that their materials meet all requirements. If some aspect of the specifications or drawings tends to exclude their product or if they feel there may be a mistake or misunderstanding in these documents, they may call the designers and discuss it with them. In addition, many times, they will be involved in working up various cost analyses of the materials or products installed, devising new uses for the materials and alternative construction techniques, and even the development of new products.

Project Management. Project management companies specialize in providing professional assistance in planning the construction of a project and keeping accurate and updated information about the financial status of the project. Owners who are coordinating large projects often hire such companies. Among the various types of owners are private individuals, companies, corporations, government agencies (such as public works and engineering departments), and various public utility companies.

Both project management firms and the staff of the owner being represented must be knowledgeable in the estimating and scheduling of a project.

Government. When a government agency is involved in any phase of construction, personnel with experience in construction and estimating are required. These are municipal, provincial, federal, and other nationwide agencies, including those involved in highways, roads, sewage treatment, schools, courthouses, nursing homes, hospitals, and single- and multiple-family dwellings that are financed or qualifying for financing by the government.

Employees may be involved in preparing or assisting to prepare preliminary and final estimates; reviewing estimates from architects, engineers, and contractors; the design and drawing of the project; and preparation of the specifications.

Professional Quantity Surveyors. Professional quantity surveyors may be hired to prepare a detailed *Schedule of Quantities* for contractors bidding on a project. Such individuals (designated by the Canadian Institute of Quantity Surveyors) or firms often provide a broad range of estimating services, including cost control throughout the project, and must possess all the attributes of estimators employed directly by contractors.

Freelance Estimators. Freelance estimators will do a material takeoff of a portion of or an entire project for anyone who may want a job done. This estimator may work for the owner, architect, engineer, contractor, subcontractor, material supplier, or manufacturer. In some areas, the estimator will do a material takeoff of a project being competitively bid for and then sell the quantity list to one or more contractors who intend to submit a bid on the project.

Many times, a talented individual has a combined drafting and estimating business. Part of the drafting business may include preparing shop drawings (drawings that show sizes of materials and installation details) for subcontractors, material suppliers, and manufacturers' representatives.

Residential Construction. Estimators are also required for the contractors, material suppliers, manufacturers' representatives, and most of the subcontractors involved in residential construction. From the designer/architectural technologist, who plans the house and produces design drawings, to the carpenters, who put up the rough framing, and the roofers, who install the roofing material, knowledge of estimating is necessary.

The designer/architectural technologist should plan and produce the house drawings using standard material sizes, when possible (or being aware of it when they are not using standard sizes). In addition, they will need to give preliminary and final estimates to the owner. Workers need to have a basic knowledge of estimating so that they can be certain that adequate material has been ordered and will be delivered by the time it is needed.

Computer Software. The use of computers throughout the world of construction offers many different types of opportunities to the estimator. Job opportunities in all the areas mentioned earlier will be centred on the ability to understand, use, and manipulate computer software. The software available today can integrate the construction drawings, estimating, bidding, purchasing, and management controls of a project.

1.4 The Estimator

Most estimators begin their career doing quantity takeoff; as they develop experience and judgment, they develop into estimators. A list of the abilities most important to the success of an estimator follows, but it should be more than simply read through. Any weaknesses affect the estimator's ability to produce complete and accurate estimates. If individuals lack any of these abilities, they must (1) be able to admit it, and (2) begin to acquire the skills they lack. Those with construction experience, who are subsequently trained as estimators, are often most successful in this field.

To be able to do quantity takeoffs, the estimator:

1. Must be able to read and interpret design and working drawings.

2. Must have knowledge of mathematics and a keen understanding of geometry. Most measurements and computations are made using linear, area, and volume concepts. The quantities are usually multiplied by a unit price to calculate costs.

3. Must have the patience and ability to do careful, methodological, and thorough work.

To be a successful estimator, an individual must go a step further. He or she:

1. Must have an eye for detail and be able, from looking at the drawings, to visualize the project through its various phases of construction. In addition, an estimator must be able to foresee problems, such as difficulty in the placement of equipment or material storage, then develop a solution and determine its estimated cost.

2. Must have enough construction experience to possess a good knowledge of job conditions, including methods of handling materials on the job, the most economical methods of construction, and labour productivity. With this experience, the estimator will be able to visualize the construction of the project and, thus, get the most accurate estimate on paper.

3. Must have sufficient knowledge of labour operations and productivity in order to convert them into costs on a project. The estimator must understand how much work can be accomplished under given conditions by given trades. Experience in construction and a study of projects that have been completed are required to develop this ability.

4. Must have the ability to keep a database of information on costs of all kinds, including those of labour, material, overhead, and equipment, as well as a knowledge of the availability of all the required items.

5. Must be computer literate and know how to manipulate and build various databases and use spreadsheet programs.

6. Must be able to meet bid deadlines and still remain calm. Even in the rush of last-minute phone calls and the competitive feeling that seems to electrify the atmosphere just before the bids are due, estimators must "keep their cool."

7. Must be able to deal with a number of bids in various stages of the bidding process.

People cannot be taught experience and judgment, but they can be taught an acceptable method of preparing an estimate: items to include in the estimate, calculations required, and how to make them. They can also be warned against possible errors and alerted to certain problems and dangers, but the practical experience and good judgment required cannot be taught and must be aquired over time.

How closely the estimated cost will agree with the actual cost will depend, to a large extent, on the estimators' skill and judgment. Their skill enables them to use accurate estimating methods, while their judgment enables them to visualize the construction of the project throughout the stages of construction.

1.5 ▮ Schedule of Quantities

A Schedule of Quantities is used mainly in civil engineering projects, where the work is repetitive and the actual quantities are indeterminate. The estimated quantities of materials required on the project are determined by a professional quantity surveyor or engineer and provided to the interested bidders on the project. Figure 1.1 is an example of the quantities that would be provided by a quantity surveyor or engineer.

No.	Quantity	Unit	Item
02525	715	m	Curb, type A
02525	75	m	Curb, type B
04220	12,500	each	100 mm concrete block
04220	5,280	each	200 mm concrete block
04220	3,700	each	300 mm concrete block

FIGURE 1.1 Quantity Survey

In this method of bidding, the contractors are all bidding on the basis of the same quantities, and the estimator spends time developing the unit prices. For example, the bid may be $123.26 per cubic metre (m³) concrete. Because all the contractors are bidding on the same quantities, they will work on keeping the cost of purchasing and installing the materials as low as possible.

As the project is built, the actual number of units required is checked against the original number of units on which the estimates were made. For example, in Figure 1.1, the original schedule of quantities called for 715 linear metres (m) of concrete curbing. If 722 m were actually installed, then the contractor would be paid for the additional 7 m. If 706 m were used, then the owner would pay only for the 706 m installed and not the 715 m in the original schedule of quantities. This type of adjustment is quite common. When errors do occur and there is a large difference between the original quantity survey and the actual number of units, an adjustment to the unit price is made. Small adjustments are usually made at the same unit rate as the contractor bid. Large errors may require that the unit price be renegotiated.

If the contractor is aware of potential discrepancies between the estimated quantities and those that will be required, he or she may price the bid to take advantage of this situation. With a belief that the estimated quantities are low, the contractor may reduce his or her unit price to be the low bidder. If the assumption is true, the contractor has the potential to make the same profit by distributing the project overhead over a greater number of units.

1.6 Types of Bids

Basically, there are two bidding procedures by which the contractor gets to build a project for owners:

1. Competitive bidding
2. Negotiated bidding

Competitive bidding involves each contractor submitting a lump-sum bid in competition with other contractors to build the project. In most cases, the lowest lump-sum bidder is awarded the contract to build the project, as long as the bid form and proper procedures have been followed and this bidder is able to obtain the required bonds and insurance. Most commonly, the bids must be delivered to the person or place specified by a time stated in the Instruction to Bidders.

Negotiated bidding can involve virtually any combination of arrangements among owners, architects, and contractors. The contractor may team up with an architect to pursue a specific project. The basic underlying difference between this and competitive bidding is that the parties arrive at a mutually agreed upon price, terms and conditions, and contractual relationship. This arrangement often entails negotiations back and forth on virtually all aspects of the project, such as materials used, sizes, finishes, and other items that affect the price of the project. Owners may negotiate with as many contractors as they wish. This type of bidding is often used when owners know which contractor they would like to build the project, in which case competitive bidding would waste time. The biggest disadvantage of this arrangement is that the contractor may not feel the need to work quite as hard to get the lowest possible prices as when a competitive bidding process is used.

1.7 Contract Documents

The bids submitted for any construction project are based on the contract documents. If an estimator is to prepare a complete and accurate estimate, he or she must become familiar with the contract documents. The documents are listed and briefly described in this section. Further explanations of the portions and how to bid them are contained in later chapters.

The contract documents comprise the *Owner-Contractor Agreement*, the *General Conditions of the Contract*, the *Supplementary General Conditions*, and the *drawings and specifications*, including all *addenda* incorporated in the documents before their execution. The combination of all of these documents forms the *contract*.

Agreement. The agreement is the document that formalizes the construction contract and is the basic contract between the owner and the contractor. It incorporates, by reference, all the other documents and makes them part of the contract; it identifies the owner, contractor, and consultant; defines the work; lists the documents comprising the contract; and states the contract price and the Goods and Services Tax on the contract price.

General Conditions. The General Conditions define the rights, responsibilities, and relations of all parties to the construction contract.

Supplementary General Conditions (special conditions). Because conditions may vary by locality and project, the Supplementary General Conditions are used to amend or supplement portions of the General Conditions.

Working Drawings. The actual drawings (plans, elevations, sections, details) from which the project is to be built are known as the working drawings. They contain the dimensions and locations of building elements and materials required and delineate how they fit together.

Specifications. Specifications are written instructions and descriptions concerning project requirements and describe the quality of materials and their expected/desired level of performance, along with the quality of workmanship expected of the trades.

Addenda. The addenda statement is graphical or written information that modifies the basic contract documents after they have been issued to the bidder but prior to the closing date of bids. They may provide clarifications, corrections, or changes in the contract/bidding documents issued for tender.

1.8 Bidding Information

There are several sources of information pertaining to the projects available for bidding. Public advertising (tender calls) is required for many public contracts. The advertisement is generally placed in newspapers, trade magazines, and journals, and notices are posted in public places and on Internet websites, such as Bidnavigator and MERX. Private owners often advertise in the same manner to attract a large cross-section of bidders (Figure 1.2). Included in the advertisement are a description of the nature, extent, and location of the project; the name of the owner; availability of bidding documents; bond requirements; and the time, manner, and place that the bids will be received.

Construction reports, such as the Daily Commercial News (DCN), provide information about projects that are accepting bids or proposals. Reports are issued for particular, defined localities throughout the country and include reports on projects planned for future construction, projects currently being tendered, the initial list of bidders, bid date changes, and addenda. Reports also include the names and bid prices of low bidders and the name of the successful contractor.

In most regions, local construction associations provide plan rooms, where interested parties may review the drawings and specifications of current projects. While most general contractors will obtain several sets of contract documents for bidding, various subcontractors and material suppliers make extensive use of such plan rooms.

INVITATION TO BID FROM NEWSPAPER

NOTICE TO CONTRACTORS
SEALED TENDERS
properly marked:
S-20-26–Cedar Hill Secondary School, Administration Building Expansion.

WILL BE RECEIVED BY:
Board of School Trustees of School District No. 123
2326 West Mountainview
Mountainville, B.C. M4S 7K6
Attention: Office of the Secretary-Treasurer

Up to 1:00 p.m. (local time), May 23, 20_; at which time tenders will be opened in public at the above address. The Contractors are to deposit their bids in the Bid Box located opposite the Boardroom Reception Counter on the 2nd Floor of the Mountainville School Board Education Centre, 2326 West Mountainview, Mountainville, B.C. M4S 7K6

THE SCOPE OF WORK INCLUDES:
demolition of the existing annex and construction of a new three-storey Administration building and associated exterior works.

Tender documents will be available to General Contractors only at:

C. K. Architects, 723 Paradise Street, Mountainville, B.C. on or after 12.00 p.m., May 02, 20_ upon payment by cash or certified cheque made payable to Mountainville School Board, of a refundable fee of $100 per set (including GST).

Each tender shall be accompanied by a Bid Bond duly executed with the name of the Project and the Owner in the amount of ten percent (10%) of the Total Tender Price, and an Undertaking of Surety Company supplying the Bid Bond to provide a Performance Bond and a Labour and Material Payment Bond each in the amount of fifty percent (50%) of the Contract Price.

Plans and specifications may be viewed at the Plan Rooms of the Mountainville Regional Construction Association at:

2626 West 5th Avenue, Mountainville, B.C., or 303 – 2323

77th Avenue, Playville, B.C.

SITE VISIT: Thursday, May 10, 20_. Site visit will begin at 3.00 p.m. It is strongly recommended that all bidders attend. Consultants will be available to answer questions. The visit will commence at the main entrance to the school at 7723 Nissa Street.

Technical inquiries shall be directed to the Consultant:
C.K. Architects
723 Paradise Street
Mountainville, B.C
Tel: 604-662-1180;
Fax: 604-415-4358
Attention: Scott Walker, MAIBC

The lowest or any price will not necessarily be accepted and the Board reserves the right to reject any tender.

CONTACT:
Helen Edwards, Manager,
Purchasing Phone: (604) 471-7212
Fax: (604) 662-1140
E-Mail: Helen.Edwards@ mountainville.ca

FIGURE 1.2 Tender Call

1.9 Availability of Contract Documents

General contractors normally receive two (2) sets of drawings from the owner/architect, and this limitation is generally found in the Invitation to Bid or Instructions to Bidders. Subcontractors, material suppliers, and manufacturers' representatives can usually obtain prints of individual drawings and specification sheets for a fee from the designer, but it should be noted that this fee is rarely refundable. Some construction associations have set up electronic plan rooms, an online document distribution system, allowing users access to project information and documents 24 hours a day, seven days a week. It allows for the storage and distribution of projects on a secure website. Users can view, download, or place orders for prints (or plans on CD) with a click of the mouse.

The owner/architect will require a deposit for each set of contract documents obtained by the prime contractors. The deposit, which may act as a guarantee for the safe return of the contract documents, ranges from $50 to over $200 per set and is usually refundable. It should be realized that the shorter the bidding period, the greater is the number of sets that would be required. Also, a large complex job requires extra sets of contract documents to make an accurate bid.

To obtain the most competitive prices on a project, a substantial number of subcontractors and material suppliers must bid the job. To obtain the most thorough coverage, there should be no undue restrictions on the number of sets of contract documents available. If this situation occurs, it is best to call the owner/architect and discuss the problem.

During the bidding period, the lead estimator needs to be certain that the contract documents are kept together. Never lend out portions of the documents. This practice will eliminate subcontractors' and material suppliers' claims that they did not submit a complete proposal because they lacked part of the information required for a complete bid. The best method available to general contractors to avoid this problem is to set aside space in their offices where the subcontractors' and material suppliers' estimators may work. This arrangement will also encourage the most competitive prices. Although it is true that the lowest prices are desired, it must be remembered at all times that it is in everyone's best interests that all parties concerned make a profit. In this manner, the contract documents never leave the contractor's office and are available to serve a large number of bidders efficiently.

1.10 Sources of Estimating Information

For matters relevant to estimating and costs, the best source of information is your historical data. These figures allow for the pricing of the project to match how the company actually performs its construction. This information takes into account the talent and training of the trade personnel and the management abilities of the field staff. In addition, it integrates the construction companies' practices and methodologies. This is why a careful, accurate accounting system, combined with accuracy in field reports, is so important. If all the information relating to the job is tracked and analyzed, it will be available for future reference. Computerized cost accounting systems are very helpful in gathering this information and making it readily available for future reference.

Several "guides to construction cost" manuals are available; however, a word of caution is offered regarding the use of these manuals. They are only *guides;* the figures should *rarely* be used to prepare an actual estimate. The manuals may be used as a guide in checking current prices and should enable the estimator to follow a more uniform system and, thus, save valuable time. The actual pricing in the manuals is most appropriately used in helping designers check approximate current prices and facilitate their preliminary estimate. In addition to these printed guides, many companies provide electronic databases that can be utilized by estimating software packages. However, the same caution needs to be observed as with the printed version. These databases represent an average or the methodologies of a few contractors. There is no simple way to convert this generalized information to match the specifics of the construction companies' methodologies.

Review Questions

1. What information is contained on the working drawings?
2. What information is contained in the specifications?
3. What is the relationship between the working drawings and the specifications?
4. How does the work involved in being an estimator for a general contractor differ from that of an estimator who works for a subcontractor?
5. What is the difference between doing a quantity takeoff and doing a full, detailed estimate?
6. Why are good math skills important to a person considering a career in construction estimating?
7. What additional skills must the estimator have to be able to take a quantity survey and turn it into a detailed estimate?
8. What is the difference between competitive bidding and negotiated bidding?
9. What are the preliminary methods of estimating? How may they be of help to the general construction estimator?
10. What are the contract documents, and why are they so important?
11. Why is it important to bid only from a full set of contract documents?

CHAPTER **2**

Contracts, Bonds, and Insurance

2.1 The Contract System

Contracts may be awarded either by a single contract for the entire project or by separate contracts for the various phases required for completion of the project. The single contract comprises all work required for the completion of a project and is the responsibility of a single prime contractor. This centralization of responsibility provides that one of the distinctive functions of the prime contractor is to plan, direct, and coordinate all parties involved in completing the project. The subcontractors (mechanical, electrical, and so on) and the material suppliers involved in the project are responsible directly to the prime contractor, who, in turn, is responsible directly to the owner. The prime contractor must ensure that all work is completed in accordance with the contract documents, that the work is completed on time, and that all subcontractors and vendors have been paid. Under the system of separate contracts, the owner signs separate agreements for the construction of various portions of a project. The separate awards are often broken into the following scopes of work:

1. General construction
2. Plumbing
3. Heating, ventilating, and air conditioning
4. Electrical
5. Sewage disposal (if applicable)
6. Elevators (if applicable)
7. Specialties
8. Other

In this manner, the owner retains the opportunity to select the contractors for the various important phases of the project. Also, the responsibility for the installation and operation of these phases is directly between the owner and contractors, rather than through the general contractor. In this contracting scheme, the owner or the owner's agents provide the coordination between the contractors. There is disagreement as to which system provides the owner with the best and the most cost-effective project. Most general contractor trade organizations favour single contracts, but in contrast, most large specialty contract groups favour separate contracts. Owners, however, must critically evaluate their needs and talents and decide which method will provide them with the best product.

Under the single contract, the prime contractor will include a markup on the subcontracted items as compensation for the coordination effort and associated risk. If one of the subcontractors is unable to perform, the prime contractor absorbs the added cost of finding a replacement and any associated delays. It is this markup that encourages the owner to use separate contracts. If no general contractor assumes the responsibility for the management and coordination of the project, then the owner must shoulder this responsibility and its associated risk. If the owner does not have the talents or personnel to accomplish these tasks, he or she must hire them. Typically, the consultant, for an added fee, may provide this service, or a construction management firm that specializes in project coordination may be hired.

2.2 Types of Agreements

The *Owner-Contractor Agreement* formalizes the construction contract. It incorporates, by reference, all other contract documents. The owner selects the type of agreement that will be signed: It may be a standard form of agreement, such as those promulgated by the Canadian Construction Documents Committee (CCDC) and the Canadian Construction Association (CCA).

The agreement generally identifies the parties to the contract and lists the documents comprising the contract and the contract price. Other articles pertaining to amendments, contract documents, payment, and any other items that should be amplified are included. No contract should ever be signed until the solicitors for all parties have had a chance to review the document. Each party's lawyer will normally give attention only to matters that pertain to his or her client's welfare. All contractors should employ the services of a lawyer who understands the nuances of the construction industry and property law.

The construction industry uses several types of construction contracts:

1. Lump-sum agreement (stipulated sum, fixed price)
2. Unit-price agreement
3. Cost-plus-fee agreements

Lump-Sum Agreement (Stipulated Sum, Fixed Price)

In the lump-sum agreements, Stipulated Price Contract - CCDC 2, the contractor agrees to construct the project, in accordance with the contract documents, for a set price arrived at through competitive bidding or negotiation. The contractor agrees that the work will be satisfactorily completed, regardless of the difficulties encountered. This type of

> ... agrees to perform the work in accordance with the contract documents herein described for the lump sum of $275,375.00 ....

FIGURE 2.1 Lump-Sum Agreement

agreement (Figure 2.1) provides the owner with advance knowledge of construction costs and requires the contractor to accept the bulk of the risk associated with the project. The accounting process is simple, and it creates centralization of responsibility in single contract projects. It is also flexible with regard to alternatives and changes required on the project. However, the cost of these changes may be high. When the owner issues a change order, the contractor is entitled to additional monies for the actual work, for additional overhead, as well as additional time. If the original work is already in place, then the cost of the change order includes not only the cost of the new work but also the cost of removing the work that has already been completed. The later in the project that change orders are issued, the greater is their cost. Therefore, changes need to be identified as early as possible in order to minimize their impact on the construction cost and completion date. In addition, the contractor should not begin work on any change orders prior to receiving written authorization from the owner.

The major disadvantages to the lump-sum agreements are risks placed upon the general contractor and the need to guarantee a price even though all the costs are estimated.

Because of the very nature and risks associated with the lump-sum price, it is important that the contractor be able to accurately understand the scope of the project work required at the time of bidding.

Unit-Price Agreement

In a *unit-price agreement*, Unit Price Contract - CCDC 4, the contractor bases the bid on estimated quantities of work and on completion of the work in accordance with the contract documents. The owner of the contracting agency typically provides the quantity takeoff. This type of contracting is most prevalent in road construction. Due to the many variables associated with earthwork, which is the main component of road projects, it is virtually impossible to develop exact quantities. The owner, therefore, provides the estimated quantities, and the contractors are in competition over their ability to complete the work, rather than their estimating ability. Figure 2.2 is an example of unit-price quantities.

Bidders will base their bids on the quantities provided or will use their estimate of the quantities to determine their unit-price bids. If contractors have knowledge of the quantities, they can use that information to their competitive advantage. The contractor's overhead is either directly or indirectly applied to each of the unit-price items. If the contractor believes that the stated quantities are low, the overhead can be spread over

No.	Quantity	Unit	Item
025-254-0300	1,000	m	Curb, straight
025-254-0400	75	m	Curb, radius
022-304-0100	600	m³	Compacted crushed stone base
025-104-0851	290	tonne	40-mm-thick asphalt

FIGURE 2.2 Typical Schedule of Quantities

Unit-Price Bid Tabulation									
				Contractor 1		Contractor 2		Contractor 3	
No.	Quantity	Unit	Item	Bid Unit Price	Estimated Item Cost	Bid Unit Price	Estimated Item Cost	Bid Unit Price	Estimated Item Cost
025-254-0300	305	m	Curb, straight	$28.10	$8,570.50	$28.35	$8,646.75	$29.55	$9,012.75
025-254-0400	23	m	Curb, radius	$48.55	$1,116.65	$48.75	$1,121.25	$46.55	$1,070.65
022-304-0100	456	m³	Compacted base	$65.90	$30,050.40	$33.10	$15,093.60	$77.10	$35,157.60
025-104-0851	263	tonne	40-mm-thick asphalt	$74.95	$19,711.85	$76.95	$20,237.85	$79.00	$20,777.00
Total					$59,449.40		$45,099.45		$66,018.00

FIGURE 2.3 Unit-Price Bid Tabulation

greater quantity, rather than the quantity provided by the owner. This allows the contractor to submit a lower bid while making the same or more profit. In public projects, the low bidder will be determined on the basis of the owner-provided quantities. In Figure 2.3, the unit-price bid tabulation for a portion of the project is shown.

Payments are made on the basis of the price that the contractor bids for each unit of work and actual site measurements of work completed. A site crew that represents the owner must make the verification of the in-place units, meaning that neither the owner nor the contractor will know the exact cost of the project until its completion. The biggest advantages of the unit-price agreement are that:

1. It allows the contractors to spend most of their time working on pricing the labour and materials required for the project while checking for the most economical approach to handle the construction process.
2. Under lump-sum contracts, each contractor does a quantity takeoff, which considerably increases the chances for quantity errors and adds overhead to all the contractors.

Cost-Plus-Fee Agreements

In *cost-plus-fee agreements,* the contractor is reimbursed for the construction costs as defined in the agreement, Cost-Plus-Fee Contract—CCDC 3. However, the contractor is not reimbursed for all items, and a complete understanding of reimbursable and non-reimbursable items is required. This arrangement is often used when speed, uniqueness of the project, and quality take precedence. This contract arrangement allows for construction to begin before all the drawings and specifications are completed, thus reducing the time required to complete the project. The contract should detail accounting requirements, record keeping, and purchasing procedures. There are many types of fee arrangements, any of which may be best in a given situation. The important point is that whatever the arrangement, all parties must clearly understand not only the amount of the fee but also how and when it will be paid to the contractor.

Cost-plus-fee type contracts include a project budget developed by the members of the project team. Although the owner typically is responsible for any expenditure over the project budget, all team members have an intrinsic motivation to maintain the project budget. The members of the project team put their professional reputation at risk. It is unlikely that an owner would repeatedly hire a contractor who does not complete projects within budget. The same holds true for architects if they design projects that are typically over budget; they most likely will not get repeat business. When dealing with owner-developers, there is little elasticity in the project budget. Their financing, equity partners, and rental rates are based on a construction budget, and few sources for additional funds are available.

Percentage Fee. The percentage fee allows the owner the opportunity to profit if prices go down and changes in the work may be readily made. The major disadvantage is that the fee increases with construction costs, so there is little incentive on the contractor's part to keep costs low. The primary incentive for a contractor to keep costs under control is the maintenance of their reputation.

Fixed Fee. The advantages of the fixed fee include the owner's ability to reduce construction time by beginning construction before the drawings and specifications are completed, thus removing the temptation for the contractor to increase costs or cut quality while maintaining a professional status. Also, changes in the work are readily made. The disadvantages are that the exact cost of the project is not known in advance, extensive accounting is required, and keeping costs low depends on the character and integrity of the contractor.

Fixed Fee with Guaranteed Maximum Cost. Advantages of this fixed fee are that a guaranteed maximum cost is assured to the owner; it generally provides an incentive to contractors to keep the costs down, since they share in any savings. Again, the contractor assumes a professional status. Disadvantages include the fact that drawings and specifications must be complete enough to allow the contractor to set a realistic maximum cost. Extensive accounting is required as in all cost-plus-fee agreements.

Sliding Scale Fee. The sliding scale fee provides an answer to the disadvantages of the percentage fee because as the cost of the project increases, the percentage fee of construction decreases. The contractor is motivated to provide strong leadership so that the project will be completed swiftly at a low cost. Disadvantages are that the costs cannot be predetermined, extensive changes may require modifications of the scale, and extensive accounting is required.

Fixed Fee with a Bonus and Penalty. With this type of fixed fee, the contractor is reimbursed the actual cost of construction plus a fee. A target cost estimate is set up, and if the cost is less than the target amount, the contractor receives a bonus of a percentage of the savings. If the cost goes over the target figure, there is a penalty (reduction of percentage).

2.3 Agreement Provisions

Although the exact type and form of agreement may vary, certain provisions are included in all of them. Contractors must check each of those items carefully before signing the agreement.

Scope of the Work. The project, drawings, and specifications are identified; the consultant is identified. The contractor agrees to furnish all material and perform all of the work for the project in accordance with the contract documents.

Time of Completion. Involved in scheduling are the starting and completion times. Starting time should never precede the execution date of the contract. The completion date is expressed either as a number of days or as a specific date. If the number of days is used, it should be expressed in calendar days and not working days, to avoid subsequent disagreements about the completion date.

Contract Sum. Under a lump-sum agreement, the contract sum is the amount of the accepted bid or negotiated amount. The accepted bid amount may be adjusted by the acceptance of alternatives or by minor revisions that were negotiated with the contractor after receiving the bid. In agreements that involve cost-plus-fee conditions, there are generally articles concerning the costs for which the owner reimburses the contractor. Customarily, the owner does not reimburse all costs paid by the contractor; reimbursable and non-reimbursable items should be listed. The contractor should be certain that all costs incurred in the construction are included somewhere. Also, in cost-plus-fee agreements, the exact type of compensation should be stipulated.

Progress Payments. Due to the cost and duration of construction projects, contractors must receive payments as work is completed. These payments are based on completed work and stored materials. However, the owner typically retains a portion of all progress payments as security to ensure project completion and payment of all contractors' financial obligations (holdbacks).

   The due date for payments is any date mutually acceptable to all concerned. In addition, the agreement needs to spell out the maximum time the consultant can hold the contractor's Application for Payment and how soon the owner must pay the contractor after the consultant makes out the Certificate of Payment. There should also be some mention of possible contractor action if these dates are not met. Generally, the contractor has contractural options of stopping work. Some contracts state that if the contractor is not paid when due, the owner must also pay interest at the current bank rates.

Retained Percentage. It is customary for the owner to withhold a certain percentage of the payments, which is referred to as *holdback* and is legal protection for the owner to ensure completion of the contract and payment of the contractor's financial obligations. Pursuant to the Construction Lien Act, the owner and all other "payers" in the construction contract loop are required to hold back a percentage (typically 10 percent, but may vary provincially) calculated on the value of the labour and materials supplied. This obligation to retain funds applies, no matter what the payment terms under the agreement, whether it provides for payment in advance, on progress certificates, or only on final completion. Anyone who advances more than the progress draw/payment before the stipulated release times may have to pay again to settle a claim for lien (a legal instrument that affects an owner's title to property).

Progress Draw Breakdown. The contractor furnishes the consultant with a statement, called a Progress Draw Request with a breakdown that shows bid prices for specific items thus far completed within the project. This statement breaks the project down into quantifiable components on a trade-by-trade basis. Contractors typically overvalue the initial items on the project. This practice is referred to as *front-end loading*. It is, therefore, crucial that the consultant verify the completed work and issue a certificate for payment prior to payment advances by the owner.

Work in Place and Stored Materials. The work in place is usually calculated as a percentage of the work that has been completed. The amounts allowed for each item in the Progress Draw Breakdown are used as the base amounts due on each item. The contractor may also receive payment for materials stored on the site or some other mutually agreed upon location. The contractor may have to present proof of purchase, bills of sale, or other assurances to receive payment for materials stored off the job site.

Acceptance and Final Payment. The acceptance and final payment arrangement sets up a time for final payment to the contractor. When the final inspection, certification of completion, acceptance of the work, and required lien releases are completed, the contractor will receive the final payment, including holdbacks.

2.4 Bonds

Often referred to as surety bonds, bonds are written documents that describe the conditions and obligations relating to the agreement. (In law, a surety is one who guarantees payment of another party's obligations.) The bond is not a financial loan or insurance policy but serves as an endorsement of the contractor. The bond guarantees that the contract documents will be complied with and all costs relative to the project will be paid. If the contractor is in breach of contract, the surety must complete the terms of the contract. Contractors most commonly use a corporate surety that specializes in construction bonds. The owner will reserve the right to approve the surety company and form of bond, as the bond is worth no more than the company's ability to pay.

To eliminate the risk of nonpayment, the contract documents will, on occasion, require that the bonds be obtained from one specified company. To contractors, this may mean doing business with an unfamiliar company, and they may be required to submit financial reports, experience records, projects (in progress and completed), as well as other material that could make for a long delay before the bonds are approved. It is up to the owner to decide whether the surety obtained by the contractor is acceptable or to specify a company. In the latter case, the contractor has the option of complying with the contract documents or not submitting a bid on the project. Standard forms of surety bond are issued by the Surety Association of Canada, endorsed by the Association of Consulting Engineers of Canada (ACEC), the Canadian Construction Association (CCA), the Canadian Council of Professional Engineers (CCPE),and Construction Specifications Canada (CSC) and are applicable to projects where bonds are requested.

Bid Bond

Construction contracts are generally interpreted in accordance with the same principles as all other contracts. The contractual rules governing offer and acceptance form the basis of most contract arrangements in the construction industry. The case of **THE QUEEN v. RON ENGINEERING & CONSTRUCTION (EASTERN) LTD.** represents a major upheaval in Canadian construction law as it applies to bidding. In this case, the contractor, Ron Engineering & Construction (Eastern) Ltd., submitted a bid for a construction project together with a cheque for $150,000 representing a bid deposit as stipulated in the Instruction to Bidders. The tenders were opened, and Ron Engineering was found to be the lowest bidder, and the difference between it and the next lowest bid was well over $600,000. This prompted the contractor to re-examine the submission only to discover that a $750,058 cost had been omitted from the bid. The contractor immediately sent a telex (just over an hour after the opening of the bids) to the owner notifying him of the error and requesting the owner to withdraw the bid without penalty and return the bid deposit. The owner, nevertheless, submitted to the contractor the construction contract. The contractor refused to sign. His position was that since he had notified the owner of the error prior to the acceptance of the bid, the bid could not be accepted as the basis of a contract. The owner, however, decided that this entitled him to retain the bid deposit and accepted the second lowest bid. The contractor sued to recover the bid deposit. The owner counterclaimed for damages caused by having to accept a higher bid.

The Supreme Court of Canada in its decision ruled that the owner was entitled to retain the bid deposit and declared that the process of bidding involves two distinct contracts. When the owner invites bids, he proposes a unilateral contract: He offers to enter into a bidding contract (Contract A) with each and every bidder who follows the rules in the bid documents. Contract A is formed when each bidder submits his bid in the proper form, thus accepting the owner's offer. Another term of the bidding contract is that the bidder will enter into a formal construction contract (Contract B) if the owner awards it to him. If the bidder breaches either of these terms of the bidding contract, he is liable in damages.

The *bid bond* ensures that if a contractor is awarded the bid within the time specified, the contractor will enter into the contract and provide all other specified bonds. If the contractor fails to do so without justification, the bond shall be forfeited to the owner. The amount forfeited shall, in no case, exceed the amount of the bond or the difference between the original bid and the next highest bid that the owner may, in good faith, accept. The contractor's surety usually provides these bonds free of charge or for a small annual service charge. The usual contract requirements for bid bonds specify that they must be 5 to 10 percent of the bid price, but higher percentages are sometimes used. Contractors should inform the surety company once the decision to bid a project is made, especially if it is a larger amount than they usually bid or if they already have a great deal of work. Once a surety writes a bid bond for a contractor, that company is typically obligated to provide the other bonds required for the project. Surety companies, therefore, may do considerable investigation of contractors before they will underwrite a bid bond for them, particularly if it is a contractor with whom they have not done business before or with whom they have never had a bid bond.

Performance Bond

The *performance bond* guarantees the owner that the contractor will perform all work in accordance with the contract documents and that the owner will receive the project built in substantial agreement with the documents. It protects the owner against default on the part of the contractor up to the amount of the bond penalty. Also, the warranty period of one year is usually covered under the bond. The contractor should check the documents to see if this bond is required and in what amount and must also make the surety company aware of all requirements. Most commonly, these bonds must be made out in the amount of 100 percent or 50 percent of the contract price. The rates vary according to the contractor's experience and financial status. The cost to the project is in the region of 1 percent and will vary from contractor to contractor and surety to surety.

Labour and Material Payment Bond

The *labour and material payment bond* guarantees the payment of the contractor's bill for labour and materials used or supplied on the project. It acts as protection for the third parties and the owner, who are exempted from any liabilities in connection with claims against the project. Claims by subcontractors and suppliers must be filed in accordance with the requirements of the bond used. Most often, a limitation is included in the bond, stating that the claimant must give written notice to the general contractor, owner, or surety within 90 days after the last day the claimant performed any work on the project or supplied materials to it.

Subcontractor Bonds. Performance, labour, and payment bonds are those that the subcontractors must supply to the prime contractor. They protect the prime contractor against financial loss and litigation due to default by a subcontractor. Because these bonds vary considerably, prime contractors may require use of their own bond forms or reserve the right to approval of both the surety and form of bond. These types of bonds are often used when the general contractor is required to post a bond for the project. This arrangement protects the general contractors, reduces their risk, and allows them greater bonding capacity.

Lien Bond. The lien bond is provided by the prime contractor and indemnifies the owner against any losses resulting from liens filed against the property.

2.5 Obtaining Bonds

The surety company will thoroughly check out a contractor before it furnishes a bid bond. The surety checks such items as the financial stability, integrity, experience, equipment owned, and professional ability of the firm. The contractor's relations with sources of credit will be reviewed, as will be current and past financial statements. At the end of the surety company's investigations, it will establish a maximum bonding capacity for that particular contractor. The investigation often takes time to complete, so contractors should apply well in advance of the time at which they desire bonding (waits of two months are not uncommon). Each time the contractor requests a bid bond for a particular job, the application must be approved. If the contractor is below the workload limit and there is nothing unusual about the project, the application will be approved quickly. If a contractor's maximum bonding capacity is approached or if the type of construction is new to the particular contractor or is not conventional, a considerably longer time may be required. The surety puts the contractor through investigations before giving a bond for a project to be sure that the contractor is not overextended.

To be successful, the contractor requires equipment, working capital, and organization. None of these should be spread thin. The surety checks the contractor's availability of credit so that if already overextended, the contractor will not take on a project that is too big. The surety will want to know if the contractor has done other work similar to that about to be bid upon. If so, the surety will want to know the size of the project. The surety will encourage contractors to stay with the type of work in which they have the most experience. The surety may also check progress payments and the amount of work to be subcontracted. If the surety refuses the contractor a bond, the contractor must first find out why and then attempt to demonstrate to the surety that the conditions questioned can be resolved.

The contractor must remember that the surety is in business to make money and can only do so if the contractor is successful. The surety is not going to take any unnecessary chances in the decision to bond a project. At the same time, some surety companies are more conservative than others. If contractors believe their surety company is too conservative or not responsive enough to their needs, they should shop around, talk with other sureties, and try to find one that will work with their organization. If contractors are approaching a surety for the first time, they should pay particular attention to what services the company provides. Some companies provide a reporting service that includes projects being bid and low bidders. Also, when contractors are doing public work, the surety company can find out when the particular contractor can expect to get payment and what stage the job is in at a given time. Contractors need to select the company that seems to be the most flexible in its approach and offers the greatest service.

2.6 Insurance

Contractors must carry insurance for the protection of the assets of their business and also because it is often required by the contract documents. The contractor's selection of an insurance broker is of utmost importance because the broker must be familiar with the risks and problems associated with construction projects. While the broker must protect the contractor against the wasteful overlapping of protection, there can be no gaps in the insurance coverage that might cause the contractor serious financial loss. Copies of the insurance requirements in the contract documents should be forwarded immediately to the insurance broker. The broker should be under strict instructions from the contractor that all insurance must be supplied in accordance with the contract documents. The broker will then supply the cost of the required insurance to the contractor for inclusion in the bidding proposal.

Insurance is not the same as a bond. With an insurance policy, the insurance company shoulders the responsibility for specified losses. In contrast, with a bond, the bonding companies will fulfill the obligations of the bond and turn to the contractor to reimburse them for all the money that they expended on their behalf. In addition to the insurance required by the contract documents, the contractor also has insurance requirements. Certain types of insurance are required by statute. For example, the Workplace Safety and Insurance Board requires all employers to obtain workers' compensation. In addition, other insurance, such as Employment Insurance as mandated by Human Resources Development Canada (HRDC), a federal government agency, is also required. This usually includes fire, liability, accident, life, and business interruption insurance. No attempt will be made in this book to describe all the various types of insurance that are available. A few of the most common types are described here.

Workers' Compensation Insurance. A workers' compensation insurance policy provides benefits to employees or their families if they are killed or injured during the course of work. The rates charged for this insurance vary by province, type of work, and the contractor. The contractor's experience rating depends on their own work records with regard to accidents and claims. Contractors with the fewest claims enjoy lower premiums. Workers should be classified correctly to keep rates as low as possible. The rate charged is expressed as a percentage of payroll and will vary considerably. The rates may range from less than 1 percent to over 30 percent, depending on the location of the project and the type of work being performed. The contractor pays the cost of these premiums in full.

Builder's Risk Fire Insurance. Builder's risk fire insurance protects projects under construction against direct loss due to fire and lightning. This insurance also covers temporary structures, sheds, materials, and equipment stored at the site. The cost usually ranges from 0.5 to 1 percent of the contract sum, depending on the project location, type of construction assembly, and the company's past experience with a contractor. If so desired, the policy may be extended to all direct loss causes, including windstorms, hail, explosions, riots, civil commotion, vandalism, and malicious mischief. Also available are endorsements that cover earthquakes and sprinkler leakage. Other policies that fall under the category of project and property insurance are:

1. Fire insurance on the contractor's buildings
2. Equipment insurance
3. Burglary, theft, and robbery insurance
4. Fidelity insurance, which protects the contractor against loss caused by any dishonesty on the part of employees

Review Questions

1. What is a single contract, and what are its principal advantages and disadvantages for the owner?
2. What is a separate contract, and what are the principal advantages and disadvantages for the owner?
3. Describe three options available to the owner, with a separate contract, for managing the contractor's work on the project.
4. List and briefly define the types of agreements for the owner's payment to the contractor which may be used.
5. What is the "time of completion," and why must it be clearly stated in the contract agreement provisions?
6. What are progress payments, and why are they important to the contractor?
7. What is holdback, where is the amount specified, and why is it used?
8. What is a bid bond, and how does it protect the owner?
9. Where would information be found on whether a bid bond was required and, if so, its cost?
10. What are performance bonds? Are they required on all proposals?
11. How are the various surety bonds, which may be required on a specific project, obtained?
12. How does insurance differ from a surety bond?

CHAPTER ■3■

Specifications

After studying this chapter, you will be able to:

1. Describe the system for organizing construction information.

2. Identify the procedures to be followed by bidders for a construction project.

3. Explain how owners request prices for alternative methods or materials of construction.

4. Explain how bidders receive revisions to the bid documents during the bidding period.

5. Describe the process for dealing with discrepancies found in the plans or specifications in the bidding period.

■3.1■ Documents

Specifications are the written construction documents defining the detailed requirements for materials, construction systems, performance and execution of the work, and technical information for the construction of a project.

The contractor submits a bid based on the drawings and specifications. The contractor is responsible for everything contained in the specifications and what is covered on the drawings. The specifications should be read thoroughly and reviewed when necessary. Contractors have a tendency to read only the portions of the specifications that refer to materials and workmanship; however, they are also responsible for anything stated in the Tender Calls, the Instruction to Bidders, the Information to Bidders, the Bid Form, the General Conditions, and the Supplementary General Conditions.

There is a tendency among estimators to simply skim over the specifications. Reading the average set of specifications is time consuming, but many important items are mentioned only in the specifications and not on the drawings. Because the specifications are part of the contract documents, the general contractor is responsible for the work and materials mentioned in them.

The bound volume of specifications contains items ranging from the types of bonds and insurance required, to the type, quality, and colour of materials used on the job. A thorough understanding of the materials contained in the specifications may make the difference between being and not being the lowest bidder.

The practice of skimming the specifications is risky. Either the bids will be too high, because contractors will allow additional money in the price for items that they have missed, or too low, from not including required items.

3.2 ▮ Organizational Format

The bound volume of specifications is generally presented in the *MasterFormat* system— an industry-accepted system of numbers and titles for organizing construction information into a standard order or sequence. This document is used throughout North America and is jointly produced by Construction Specifications Canada (CSC) and the Construction Specifications Institute (CSI) in the United States. The titles and numbers in MasterFormat are grouped under the following headings:

1. Bidding Requirements and Forms
2. Contract Form (Agreement)
3. Conditions of the Contract (General and Supplementary)
4. Specifications Divisions 1–16

Separate contracts and many large projects often have separate bound volumes of specifications for the mechanical and electrical trades.

Construction Specifications Canada and the Construction Specifications Institute have jointly developed a specifications format called *SectionFormat* that divides the major areas involved in building construction into 16 divisions. Each division is subdivided into specific areas. For example, Division 8 covers doors and windows, while the subdivision 08500 deals specifically with metal windows. The CSA/CSI format (shown in Figure 3.1) has found wide acceptance in the construction industry. The CSA/CSI format also ties in easily with computer programs and cost accounting systems. It is not necessary to memorize the major divisions of the format; their constant use will commit them to memory.

3.3 ▮ Invitation to Bid (Tender Calls)

In publicly funded projects, public agencies must conform to regulations that relate to the method they use in advertising for bids. Customarily, a notice of proposed bidding is posted in public places and by advertising for bids on an electronic bulletin board or a commonly read industry publication, such as major daily newspapers (Figure 3.2), trade publications, and construction associations bulletins. Where the tender call is published, how often, and over what period of time it will be published vary considerably according to the jurisdictional regulations. An estimator must not be bashful. If contractors are interested in a certain project, they should never hesitate to call and ask when it will be bid, when and where the agencies will advertise for bids, or for any other information that may be of importance.

Generally, the tender call describes the location, extent, and nature of the work. It will designate the authority under which the project originated. With regard to the bid, it will give the place where bidding documents are available and list the time, manner, and place where bids will be received. It will also list bond requirements and start and completion dates of work.

In private construction, owners often do not advertise for bidders. They may decide to negotiate with a contractor of choice, put the job out to bid on an invitation basis, or put the project out for competitive bidding. If the owner puts a job out for competitive bidding, the designer will call the construction reporting services, which will pass the information on to their members or subscribers.

BIDDING REQUIREMENTS, CONTRACT FORMS, AND CONDITIONS OF THE CONTRACT

00010	PRE-BID INFORMATION
00100	INSTRUCTIONS TO BIDDERS
00200	INFORMATION AVAILABLE TO BIDDERS
00300	BID FORMS
00400	SUPPLEMENTS TO BID FORMS
00500	AGREEMENT FORMS
00600	BONDS AND CERTIFICATES
00700	GENERAL CONDITIONS
00800	SUPPLEMENTARY CONDITIONS
00850	DRAWINGS AND SCHEDULES
00900	ADDENDA AND MODIFICATIONS

Note: Since the items listed above are not specification they are referred to as "Documents" in lieu of "Sections" in the Master List of Section Titles, Numbers and Broadscope Explanations.

SPECIFICATIONS

DIVISION 1 - GENERAL REQUIREMENTS

01010	SUMMARY OF WORK
01020	ALLOWANCES
01025	MEASUREMENT AND PAYMENT
01030	ALTERATIONS/ALTERNATIVES
01040	COORDINATION
01050	FIELD ENGINEERING
01060	REGULATORY REQUIREMENTS
01070	ABBREVIATIONS AND SYMBOLS
01080	IDENTIFICATION SYSTEMS
01090	REFERENCE STANDARDS
01100	SPECIAL PROJECT PROCEDURES
01200	PROJECT MEETINGS
01300	SUBMITTALS
01400	QUALITY CONTROL
01500	CONSTRUCTION FACILITIES AND TEMPORARY CONTROLS
01600	MATERIAL AND EQUIPMENT
01650	STARTING OF SYSTEMS / COMMISSIONING
01700	CONTRACT CLOSEOUT
01800	MAINTENANCE

DIVISION 2 - SITEWORK

02010	SUBSURFACE INVESTIGATION
02050	DEMOLITION
02110	SITE PREPARATION
02140	DEWATERING
02150	SHORING AND UNDERPINNING
02160	EXCAVATION SUPPORT SYSTEMS
02170	COFFERDAMS
02200	EARTHWORK
02300	TUNNELING
02350	PILES AND CAISSONS
02450	RAILROAD WORK
02480	MARINE WORK
02500	PAVING AND SURFACING
02600	PIPED UTILITY MATERIALS
02660	WATER DISTRIBUTION
02680	FUEL DISTRIBUTION
02700	SEWAGE AND DRAINAGE
02760	RESTORATION OF UNDERGROUND PIPELINES
02770	PONDS AND RESERVOIRS
02780	POWER AND COMMUNICATIONS
02800	SITE IMPROVEMENTS
02900	LANDSCAPING

DIVISION 3 - CONCRETE

03100	CONCRETE FORMWORK
03200	CONCRETE REINFORCEMENT
03250	CONCRETE ACCESSORIES
03300	CAST-IN-PLACE CONCRETE
03370	CONCRETE CURING
03400	PRECAST CONCRETE
03500	CEMENTITIOUS DECKS
03600	GROUT
03700	CONCRETE RESTORATION AND CLEANING
03800	MASS CONCRETE

DIVISION 4 - MASONRY

04110	MORTAR
04150	MASONRY ACCESSORIES
04200	UNIT MASONRY
04400	STONE
04500	MASONRY RESTORATION AND CLEANING
04550	REFRACTORIES
04600	CORROSION RESISTANT MASONRY

DIVISION 5 - METALS

05010	METAL MATERIALS
05030	METAL FINISHES
05050	METAL FASTENING
05100	STRUCTURAL METAL FRAMING
05200	METAL JOISTS
05300	METAL DECKING
05400	COLD-FORMED METAL FRAMING
05500	METAL FABRICATIONS
05580	SHEET METAL FABRICATIONS
05700	ORNAMENTAL METAL
05800	EXPANSION CONTROL
05900	HYDRAULIC STRUCTURES

DIVISION 6 - WOOD AND PLASTICS

06050	FASTENERS AND ADHESIVES
06100	ROUGH CARPENTRY
06130	HEAVY TIMBER CONSTRUCTION
06150	WOOD-METAL SYSTEMS
06170	PREFABRICATED STRUCTURAL WOOD
06200	FINISH CARPENTRY
06300	WOOD TREATMENT
06400	ARCHITECTURAL WOODWORK
06500	PREFABRICATED STRUCTURAL PLASTICS
06600	PLASTIC FABRICATIONS

DIVISION 7 - THERMAL AND MOISTURE PROTECTION

07100	WATERPROOFING
07150	DAMPROOFING
07190	VAPOR AND AIR RETARDERS
07200	INSULATION
07250	FIREPROOFING
07300	SHINGLES AND ROOFING TILES
07400	PREFORMED ROOFING CLADDING AND SIDING
07500	MEMBRANE ROOFING
07570	TRAFFIC TOPPING
07600	FLASHING AND SHEET METAL
07700	ROOF SPECIALTIES AND ACCESSORIES
07800	SKYLIGHTS
07900	JOINT SEALERS 1

DIVISION 8 - DOORS AND WINDOWS

08100	METAL DOORS AND FRAMES
08200	WOOD AND PLASTIC DOORS
08250	DOOR OPENING ASSEMBLIES
08300	SPECIAL DOORS
08400	ENTRANCES AND STOREFRONTS
08500	METAL WINDOWS
08600	WOOD AND PLASTIC WINDOWS
08650	SPECIAL WINDOWS
08700	HARDWARE
08800	GLAZING
08900	GLAZED CURTAIN WALLS

DIVISION 9 - FINISHES

09100	METAL SUPPORT SYSTEMS
09200	LATHE AND PLASTER
09230	AGGREGATE COATINGS
09250	GYPSUM BOARD
09300	TILE
09400	TERRAZZO
09500	ACOUSTICAL TREATMENT
09540	SPECIAL SURFACES
09550	WOOD FLOORING
09600	STONE FLOORING
09630	UNIT MASONRY FLOORING
09650	RESILIENT FLOORING
09680	CARPET
09700	SPECIAL FLOORING
09780	FLOOR TREATMENT
09800	SPECIAL COATINGS
09900	PAINTING
09950	WALL COVERINGS

DIVISION 10 - SPECIALTIES

10100	CHALKBOARDS AND TACKBOARDS
10150	COMPARTMENTS AND CUBICLES
10200	LOUVERS AND VENTS
10240	GRILLES AND SCREENS
10250	SERVICE WALL SYSTEMS
10260	WALL AND CORNER GUARDS
10270	ACCESS FLOORING
10280	SPECIALTY MODULES
10290	PEST CONTROL
10300	FIREPLACES AND STOVES
10340	PREFABRICATED EXTERIOR SPECIALTIES
10350	FLAGPOLES
10400	IDENTIFYING DEVICES
10460	PEDESTRIAN CONTROL DEVICES
10500	LOCKERS
10520	FIRE PROTECTION SPECIALTIES
10530	PROTECTIVE COVERS
10550	POSTAL SPECIALTIES
10600	PARTITIONS
10650	OPERABLE PARTITIONS
10670	STORAGE SHELVING
10700	EXTERIOR SUN CONTROL DEVICES
10750	TELEPHONE SPECIALTIES
10800	TOILET AND BATH ACCESSORIES
10880	SCALES
10900	WARDROBE AND CLOSET SPECIALTIES

DIVISION 11 - EQUIPMENT

11010	MAINTENANCE EQUIPMENT
11020	SECURITY AND VAULT EQUIPMENT
11030	TELLER AND SERVICE EQUIPMENT
11040	ECCLESIASTICAL EQUIPMENT
11050	LIBRARY EQUIPMENT
11060	THEATER AND STAGE EQUIPMENT
11070	INSTRUMENTAL EQUIPMENT
11080	REGISTRATION EQUIPMENT
11090	CHECKROOM EQUIPMENT
11100	MERCANTILE EQUIPMENT
11110	COMMERCIAL LAUNDRY AND DRY CLEANING EQUIPMENT
11120	VENDING EQUIPMENT
11130	AUDIO-VISUAL EQUIPMENT
11140	SERVICE STATION EQUIPMENT
11150	PARKING CONTROL EQUIPMENT
11160	LOADING DOCK EQUIPMENT
11170	SOLID WASTE HANDLING EQUIPMENT
11190	DETENTION EQUIPMENT
11200	WATER SUPPLY AND TREATMENT EQUIPMENT
11280	HYDRAULIC GATES AND VALVES
11300	FLUID WASTE TREATMENT AND DISPOSAL EQUIPMENT
11400	FOOD SERVICE EQUIPMENT
11450	RESIDENTIAL EQUIPMENT
11460	UNIT KITCHENS
11470	DARKROOM EQUIPMENT
11480	ATHLETIC, RECREATIONAL AND THERAPEUTIC EQUIPMENT
11500	INDUSTRIAL AND PROCESS EQUIPMENT
11600	LABORATORY EQUIPMENT
11650	PLANETARIUM EQUIPMENT
11660	OBSERVATORY EQUIPMENT
11700	MEDICAL EQUIPMENT
11780	MORTUARY EQUIPMENT
11850	NAVIGATION EQUIPMENT

DIVISION 12 - FURNISHINGS

12050	FABRICS
12100	ARTWORK
12300	MANUFACTURED CASEWORK
12500	WINDOW TREATMENT
12600	FURNITURE AND ACCESSORIES
12670	RUGS AND MATS
12700	MULTIPLE SEATING
12800	INTERIOR PLANTS AND PLANTERS

DIVISION 13 - SPECIAL CONSTRUCTION

13010	AIR SUPPORTED STRUCTURES
13020	INTEGRATED ASSEMBLIES
13030	SPECIAL PURPOSE ROOMS
13080	SOUND, VIBRATION, AND SEISMIC CONTROL
13090	RADIATION PROTECTION
13100	NUCLEAR REACTOR
13120	PRE-ENGINEERED STRUCTURES
13150	POOLS
13160	ICE RINKS
13170	KENNELS AND ANIMAL SHELTERS
13180	SITE CONSTRUCTED INCINERATORS
13200	LIQUID AND GAS STORAGE TANKS
13220	FILTER UNDERDRAINS AND MEDIA
13230	DIGESTION TANK COVERS AND APPURTENANCES
13240	OXYGENATION SYSTEMS
13260	SLUDGE CONDITIONING SYSTEMS
13300	UTILITY CONTROL SYSTEMS
13400	INDUSTRIAL AND PROCESS CONTROL SYSTEMS
13500	RECORDING INSTRUMENTATION
13550	TRANSPORTATION CONTROL INSTRUMENTATION
13600	SOLAR ENERGY SYSTEMS
13700	WIND ENERGY SYSTEMS
13800	BUILDING AUTOMATION SYSTEMS
13900	FIRE SUPPRESSION AND SUPERVISORY SYSTEMS

DIVISION 14 - CONVEYING SYSTEMS

14100	DUMBWAITERS
14200	ELEVATORS
14300	MOVING STAIRS AND WALKS
14400	LIFTS
14500	MATERIAL HANDLING SYSTEMS
14600	HOISTS AND CRANES
14700	TURNTABLES
14800	SCAFFOLDING
14900	TRANSPORTATION SYSTEMS

DIVISION 15 - MECHANICAL

15050	BASIC MECHANICAL MATERIALS AND METHODS
15250	MECHANICAL INSULATION
15300	FIRE PROTECTION
15400	PLUMBING
15500	HEATING, VENTILATING, AND AIR CONDITIONING (HVAC)
15550	HEAT GENERATION
15650	REFRIGERATION
15750	HEAT TRANSFER
15850	AIR HANDLING
15880	AIR DISTRIBUTION
15950	CONTROLS
15990	TESTING, ADJUSTING, AND BALANCING

DIVISION 16 - ELECTRICAL

16050	BASIC ELECTRICAL MATERIALS AND METHODS
16200	POWER GENERATION
16300	HIGH VOLTAGE DISTRIBUTION (Above 600-Volt)
16400	SERVICE AND DISTRIBUTION (600-Volt and Below)
16500	LIGHTING
16600	SPECIAL SYSTEMS
16700	COMMUNICATIONS
16850	ELECTRIC RESISTANCE HEATING
16900	CONTROLS
16950	TESTING

FIGURE 3.1 CSC/CSI Divisions

INVITATION TO BID FROM NEWSPAPER

NOTICE TO CONTRACTORS

SEALED TENDERS
properly marked:
S-20-26–Cedar Hill Secondary
School, Administration Building
Expansion.

WILL BE RECEIVED BY:
Board of School Trustees of
School District No. 123
2326 West Mountainview
Mountainville, B.C. M4S 7K6
Attention: Office of the
Secretary-Treasurer

Up to 1:00 p.m. (local time),
May 23, 20_; at which time tenders will be opened in public at
the above address. The
Contractors are to deposit their
bids in the Bid Box located
opposite the Boardroom
Reception Counter on the 2nd
Floor of the Mountainville
School Board Education
Centre, 2326 West
Mountainview, Mountainville,
B.C. M4S 7K6

The scope of work includes:
demolition of the existing annex
and construction of a new 3-
storey Administration building
and associated exterior works.

Tender documents will be available to General Contractors
only at:

C. K. Architects, 723 Paradise
Street, Mountainville, B.C. on or
after 12.00 p.m., May 02, 20_
upon payment by cash or certified cheque made payable to
Mountainville School Board, of
a refundable fee of $100 per
set (including GST).

Each tender shall be accompanied by a Bid Bond duly executed with the name of the
Project and the Owner in the
amount of ten percent (10%) of
the Total Tender Price, and an
Undertaking of Surety Company
supplying the Bid Bond to provide a Performance Bond and a
Labour and Material Payment
Bond each in the amount of fifty
percent (50%) of the Contract
Price.

Plans and specifications may
be viewed at the Plan Rooms of
the Mountainville Regional
Construction Association at:

2626 West 5th Avenue,
Mountainville, B.C., or 303 –

2323 77th Avenue, Playville,
B.C.

SITE VISIT: Thursday, May 10,
20_. Site visit will begin at 3.00
p.m. It is strongly recommended
that all bidders attend.
Consultants will be available to
answer questions. The visit will
commence at the main entrance
to the school at 7723 Nissa
Street.

**Technical inquiries shall be
directed to the Consultant:**
C.K. Architects
723 Paradise Street
Mountainville, B.C.
Tel: 604-662-1180
Fax: 604-415-4358
Attention: Scott Walker, MAIBC

The lowest or any price will not
necessarily be accepted and
the Board reserves the right to
reject any tender.

CONTACT:
Helen Edwards, Manager,
Purchasing
Phone: (604) 471-7212
Fax: (604) 662-1140

E-Mail: Helen.Edwards@
mountainville.ca

FIGURE 3.2 Invitation to Bid

3.4 Instructions to Bidders

Instructions to Bidders is the document that states the procedures to be followed by all bidders. It states in what manner the bids must be delivered; the time, date, and location of bid opening; and whether it is a public opening. (Bids may either be opened publicly and read aloud or be opened privately.) The *Instructions to Bidders* states where the drawings and specifications are available and the deposit required. It also lists the form of owner-contractor agreement to be used, bonds required, times of starting and completion of project, and any other bidder requirements.

Each set of instructions is different and should be read carefully. In reading the Instructions to Bidders, contractors should note special items. Figure 3.3 is an example of a set of instructions to bidders.

Bids. Be sure to check exactly where the bids are being received. Be sure to check each addendum to see if the time or location has been changed. It is rather embarrassing (as well as unprofitable) to wind up in the wrong place at the right time, or the right place at the wrong time. Typically, bids will be returned unopened if they are submitted late. Figure 3.4 is an example of a bid form.

GENERAL REQUIREMENTS

INSTRUCTIONS TO BIDDERS

These instructions to Bidders form a part of the Specifications and Contract Documents.

TENDER FOR STIPULATED PRICE BID SUBMISSION

Sealed Bids for "Stipulated Price Bid" will be received by **The Board of School Trustees of School District No. 123, 2326 West Mountainview, Mountainville, British Columbia,** for the supply of all labour, materials, equipment and services required for the erection and completion of the **New Administration Building at Cedar Hill Secondary School, 7723 Nissa Street, Mountainville, British Columbia** in accordance with the drawings and specifications prepared by **C. K. Architects,** Mountainville, British Columbia.

Bids shall be submitted in duplicate on Stipulated Price Bid Forms furnished by the Architects, fully filled out in ink or typewritten with signature in longhand. Signatures shall be those of the authorized officers of the corporation with the corporate seal affixed. Completed bid form shall be without interlineations or alteration.

**TENDER
ADMINISTRATION BUILDING EXPANSION
CEDAR HILL SECONDARY SCHOOL
MOUNTAINVILLE, BRITISH COLUMBIA**

For

**THE BOARD OF SCHOOL TRUSTEES OF SCHOOL DISTRICT NO. 123,
MOUNTAINVILLE, BRITISH COLUMBIA**

Prior to the submission of the Stipulated Price Bid, all bidders shall carefully examine the bid form, the contract documents, visit the site of the proposed work and fully inform themselves of the existing conditions and limitations of the work. If there exists doubt in the bidder's mind as to the intent of any information shown or requested on the above listed documents, **he must request clarification from the Architect prior to submission of his Stipulated Price Bid.**

Submitted Stipulated Price Bid shall cover the cost of all items contemplated by the contract and no allowance shall be made subsequently in this connection on behalf of the Contractor for any error or negligence on his part.

No oral, telephonic or electronic proposals will be considered.

SALES TAX
(A) Provincial Sales Tax (PST)

THE STIPULATED PRICE BID SUBMITTED FOR THIS CONTRACT SHALL INCLUDE ALL APPLICABLE (EXISTING AND KNOWN REVISIONS TO) RETAIL SALES TAXES.

(B) Federal Goods & Services Tax (GST)
The Stipulated Price Bid submitted for this contract shall include the Federal Goods and Services Tax (GST). For purposes of calculating costs of extra work performed, any GST. paid by the Contractor to Suppliers or Sub-Contractors shall be deducted prior to any mark-up, profit or overhead by the Contractor. The Contractor will not be permitted to add any mark-up for overhead or profit to the GST amount or to claim for any time involved in processing or collecting the GST and for its remittance to Revenue Canada.

ACCEPTANCE OR REJECTION

Stipulated Price Bids shall remain open to acceptance for a period of sixty (60) calendar days commencing on and including the date set for receipt of Stipulated Price Bids, and the Owner may at any time within this period accept any of the Stipulated Price Bids received.

The right to reject any or all Stipulated Price Bids in whole or in part, or to accept the Stipulated Price Bid or parts thereof judged most satisfactory is expressly reserved by The Board of School Trustees of School District No. 123, Mountainville, British Columbia without liability on the part of the Board and the Architect.

The Owner and the Architect shall not be responsible for any liabilities, costs, expenses, loss or damage incurred, sustained or suffered by any Bidder prior or subsequent to or by reason of the acceptance or the non-acceptance by the Owner of any Bid or by reason of any delay in the acceptance of a Bid. Bids are subject to a formal Contract being prepared and executed.

The Owner reserves the right to reject any or all tenders and to waive formalities as the interests of the Owner may require without stating reasons therefore and the lowest or any Bid will not necessarily be accepted.

BONDS

As per Article GC 11.2, Sub-sections 11.2.1 and 11.2.2 of the General Conditions and 11.2.3 of the Amended General Conditions.

(a)   Furnish with Stipulated Price Bid a Bid Bond payable to The Board of School Trustees of School District No. 123, Mountainville, British Columbia in the amount of 10 % of the total tender price. Bid Bond shall remain in force the

FIGURE 3.3 Instructions to Bidders

complete tender acceptance period noted above, and shall be forfeited to the Board if the Bidder refuses to enter into a formal contract for the performance of the work if so requested by the Board during this period. Include with the Bid Bond an Agreement to provide the required Performance Bond from the Bonding Company.

(b) The successful General Contractor shall:

1. Furnish a 50% Bond covering the faithful performance of the contract in strict accordance with the Architect's plans and specifications including the correction after completion provided for in Section GC 11.2 of the General Conditions covering all obligations arising under the Contract.

2. Furnish a 50% Labour and Materials Payment Bond.

The bonds must be provided by the successful bidder before commencing work.

3. The successful Bidder shall execute a Contract in writing with the Owner within ten (10) days after being notified in writing by the Owner or his agent of the acceptance of his Bid. In the event that such Contract is not executed within the said period, the Bid Bond of the Bidder whose Bid has been accepted and who has failed to execute a Contract shall be forfeited to the Owner and thereafter the Contract between such Bidder and the Owner shall be forthwith terminated, forfeited and ended.

PLANS, SPECIFICATIONS & TENDER FORMS

Plans, Specifications and Stipulated Price Bid Forms may be obtained from C.K. Architects upon deposit of a certified cheque in the amount of **One Hundred Dollars ($100.00)** payable to C.K. Architects. This cheque will be refunded **only if:**

(a) A Stipulated Price Bid form properly completed, signed, complying in every respect with the conditions outlined in these Instructions to Bidders, or all drawings and specifications are returned to the Architect a minimum of 72 hours prior to time of closing of tenders.

AND

(b) Drawings and Specifications are returned to the Architects **office WITHOUT MARKS, NOTES OR OTHER MUTILATIONS** thereon, within **SEVEN** (7) **days** after bid call date. The successful General Contractor will be given twenty (20) complete sets of contract documents. If additional sets are required, they shall be paid for by the General Contractor.

BID FORM

The bid form will be the current edition of the Stipulated Price Bid Form, Canadian Standard Construction Document; CCDC Document 10, revised as issued by C.K. Architects.

TIME OF COMPLETION

Time is the essence of this Contract. The General Contractor shall state in the place provided on the Stipulated Price Bid Form the time required to substantially complete the work. The time stated to substantially complete the work will be considered in selecting the successful bidder.

EXPLANATION TO BIDDERS & ADDENDA

Bidders may be advised during the tender period of omissions, additions or alterations in the drawings or specifications. Such advice will be contained in addenda issued by the Architect and the cost of all such revisions shall be included in the Stipulated Price Bid without exception. Receipt of such addenda, if issued, shall be acknowledged in the proper place in the bid form.

The requirements of the Instructions to Bidders, General Conditions as amended and the Supplementary General Conditions govern all phases of the work and the Stipulated Price Bid shall include all costs that might arise from compliance with such regulations.

Bidders are responsible for acquainting all subcontractors and supply bidders with the requirements of the Instructions to Bidders, General Conditions as amended, Supplementary General Conditions, Contract Forms and Documentation and General Requirements.

No allowance will be made after award of contract for errors or omissions due to subcontractors or suppliers not being familiar with such requirements.

SUPERVISORY PROJECT PERSONNEL

Upon completion of tendering, bidders upon request shall be prepared to furnish the Architect and Owner the experience and qualifications and all other pertinent information deemed necessary regarding supervisory personnel. If in the opinion of the Architect, such personnel are not competent to carry out their work, the Contractor shall replace these personnel immediately upon written request of the Architect.

FIGURE 3.3 Instructions to Bidders *continued*

EXAMINATION OF SITE

The General Contractor shall visit and examine the site and become familiar with all features, characteristics, conditions and suitability of the work affecting the work of the contract. No allowance will be made by the Owner for any errors, misjudgments and/or difficulties encountered by the General Contractor due to any feature or peculiarity of the site or surrounding property that exists at the time the General Contractor's Stipulated Price Bid is submitted.

BASIS OF TENDER PRICE

Tender price shall include the complete cost of all work outlined in the Specification Divisions as listed in the Index and/or shown on the Drawings dated May 0I, 20_ for Project No. S-20-26.

END OF SECTION

FIGURE 3.3 Instructions to Bidders *continued*

Commencement and Completion. Work on the project will have to commence within a specified period after execution of the contract. The contractor has to determine the number of calendar days to complete the project. The completion date must be realistic, as most contractors have a tendency to be overly optimistic with work schedules. At the same time, they should not be overly conservative, since it may cause the owner concern over their ability to expedite the work.

Responsibility of Bidders. Contractors should read the responsibilities to bidders section thoroughly, as it points to the importance of checking for all the drawings and specifications. The contractor clearly has the responsibility to visit the site.

Award or Rejection of Bid. The owner reserves the right to:

1. Reject any or all bids.
2. Accept a bid that is not the lowest.
3. Reject any bid not prepared and submitted in accordance with the contract documents.

These are common stipulations; in effect, owners may contract to any bidder they select. Needless to say, it causes some hard feelings, disputes, and possible litigation when an owner does not accept the lowest bid.

3.5 Bid Forms

A prepared *bid form* (see Figure 3.4) is included in the bound set of specifications, and the contractor must use this form to present a bid. By using a prepared bid form, the owner can evaluate all bids on the same basis.

The bid form stipulates the price for which the contractor agrees to perform all the work described in the contract documents. It also ensures that if the owner accepts the bid within a certain time, the contractor must enter into an agreement or the owner may call in the bid bond.

The bid must be submitted according to the requirements of the Instruction to Bidders. Any deviation from these requirements for the submission of the bid may result in the bid being rejected.

Contractors must fill in all blanks of the bid form, acknowledge receipt of all addenda, submit the required number of copies of the bid, supply proper bid

---

STIPULATED PRICE BID

Project Number: S-20-26

Project: Cedar Hill Secondary School, Administration Building Expansion.
Located: 7723 Nissa Street, Mountainville, British Columbia.

Submitted To: Board of School Trustees of School District No. 123, 2326 West Mountainview, Mountainville, British Columbia M4S 7K6

Bidder _____

Legal Name _____

Address _____

City _____ **Province** _____ **Postal Code** _____

Bid Price
Having examined the Bid Documents as listed In Appendix "A" to this Stipulated Price Bid, and Addenda No. _____ to No. _____ inclusive, all as issued by C.K. Architects

Consultant
and having visited the Place of the Work; we hereby offer to enter into a Contract to perform the Work required by the Bid Documents for the stipulated price of

Dollars ($_____) in Canadian funds, which price includes Value Added Taxes.

Interest
Should either party fail to make payments as they become due under the terms of the Contract or in an award by arbitration or court, interest at _____ percent (%) per annum above the bank rate on such unpaid amounts shall also become due and payable until payment. Such interest shall be compounded on a monthly basis. The bank rate shall be the rate established by the Bank of Canada as the minimum rate at which the Bank of Canada makes short-term advances to the chartered banks.

Declarations
We hereby declare that:
(a) we agree to perform the work in compliance with the required completion schedule stated in the Bid documents.
(b) no person, firm, or corporation other than the undersigned has any interest in this bid or in the proposed contract for which this Bid is made.
(c) this Bid is open to acceptance for a period of 60 days from the date of bid closing.

Signatures
SIGNED AND SUBMITTED for and on behalf of:

_____
Name of Bidder

_____
Signature

_____
Name and title of person signing

_____
Signature

_____
Name and title of person signing

_____
Date:

_____

_____
Witness

_____
Signature

_____
Name and title of person signing

FIGURE 3.4 Bid Form

security, and be at the right place at the correct time. Countless bids have never been opened because they were delivered a few minutes late; others have been rejected because a blank space was not filled in.

3.6 Form of Owner-Contractor Agreement

The owner-contractor agreement form spells out exactly the type or form of agreement between the owner and the contractor. The agreement may be a standard form published by the CCDC, in which case reference may be made to it and a copy may or may not be actually included. Government agencies controlling the work usually have their own forms of agreement; the same is true of many corporations. In these cases, a copy of the agreement is included in the specifications. If the contract agreement is unfamiliar, the contractor must have his or her lawyer review it before submitting the bid. If the form of agreement is unacceptable, the contractor may prefer not to bid that particular project. Types of agreements are discussed in Chapter 2.

3.7 General Conditions

The *CCDC Standard Construction Document* is the most commonly used standard form. It has found wide acceptance throughout the industry. Many branches of government as well as large corporations have also assembled their own versions of general conditions. The contractor must carefully read each article and make appropriate notes. In some cases, it is best for the contractor to give a copy of the general conditions to a lawyer for review and comment. In the event the contractor decides it is not in his or her best interests to work under the proposed set of general conditions, the consultant should be informed of the reasons and asked if he or she would consider altering the conditions. If not, the contractor may decide it is best not to bid the project. Typically, all general conditions will include, in one form or another, the 14 topics included in the CCDC General Conditions. The General Conditions clearly spell out the rights and responsibilities of all the parties. Obviously, the most stringent demands are placed on the contractors because they are entrusted with the responsibility of actually building the project.

3.8 Supplementary General Conditions

The *Supplementary Conditions* to the General Conditions of the CCDC Standard Construction Document are the amendments and/or supplements portions of the General Conditions. It is through these supplemental conditions that the General Conditions are geared to all the special requirements of geography, local requirements, and individual project needs. Part of the supplemental conditions cancel or amend the articles in the General Conditions, while the remaining portion adds extra articles.

Contractors must carefully check the supplementary conditions, as each set is different. Items that may be covered in this section include insurance, bonds, and safety requirements. Also included may be comments concerning:

1. Pumping and shoring
2. Dust control
3. Temporary offices
4. Temporary enclosures
5. Temporary utilities
6. Temporary water
7. Material substitution
8. Soil conditions

9. Signs

10. Cleaning

11. Shop drawings—drawings that illustrate how specific portions of the work shall be fabricated and/or installed

12. Surveys

As the contractor reviews the Supplementary General Conditions, he must make notes of the many requirements included in them, as they may be costly, and must include an amount to cover these items in the estimate. When actually figuring the estimate, the contractor should go through the entire Supplementary General Conditions carefully, noting all items that must be covered in the bid and deciding on how much to allow for each item.

3.9 Technical Trade Sections

The *technical trade sections* generally follow the CSC/CSI format. These specifications include the type of materials required, their required performance, and the method that must be used in order to obtain the specified result. When a particular method is specified, the contractor should base the bid on that methodology. Although deviations may be allowed once the contract has been signed, those items can be handled through the use of change orders. If an alternative method is assumed in the estimate and later denied by the consultant, the contractor would have to shoulder any losses.

The materials portion of the specifications usually mentions the physical properties, performance requirements, handling, and storage requirements. Often, specific brands or types of material are listed as the standard of quality required. Sometimes, two or three acceptable brands are specified, and the contractor can choose which to supply. If the contractor wishes to substitute another manufacturer's materials, it must be done in accordance with the contract documents.

On occasion, a trade section may include the required performance of a product or an assembly, and it may be up to the contractor to decide how that will be achieved. This is known as a "performance specification." Results that may be specified include such items as the texture of the material, appearance, noise reduction factors, allowable tolerances, heat loss factors, and colours.

3.10 Alternatives

In many projects, the owner requests prices for alternative methods or materials of construction (Figure 3.5). These alternatives are generally spelled out on a separate listing in the specifications, and they are listed on the proposal form. The alternates may be either an *add price* or a *deduct price,* which means that contractors either add the price to the base bid or deduct it from the base bid. The price for any alternatives must be complete and include all taxes, overhead, and profit. When an owner has a limited budget, the system of alternatives allows a choice on how to best spend the available money.

Since lump-sum contracts are awarded on the basis of the total base bid, plus or minus any alternatives accepted, there is always concern that the owner will select alternatives in a way that will help a particular contractor become the low bidder. This concern has become so great that some contractors will not bid projects with a large number of alternatives. To relieve the contractor of this concern, many architects include in the contract documents the order of acceptance of the alternatives. Alternatives

ALTERNATE PRICE

The following are our prices for the alternative work listed hereunder. Such alternative work and amounts are NOT included in our Bid Price. These prices for the alternative work include value added taxes.

Description of Alternative Work	Effect on Stipulated Price ($)	
	Addition	Deduction
Alternate Price No. 1 - Supply and install 100 mm face brick with 200 mm block exterior walls (as detailed in section A-1) in lieu of the painted 300 mm concrete block exterior walls		

FIGURE 3.5 Alternatives

deserve the same estimating care and consideration as the rest of the project; so contractors should not rush through them or leave them until the last minute.

3.11 Addenda

The period after the basic contract documents have been issued to the bidders and before the bids are due is known as the *bidding period*. Any amendments, modifications, revisions, corrections, and explanations issued by the consultant during the bidding period are effected by issuing the *addenda* (Figure 3.6). The statements and any drawings included serve to revise the basic contract documents. They notify the bidder of any corrections in the documents, interpretations required, and any additional requirements, as well as other similar matters. The addenda are issued in writing.

Because the addenda become part of the contract documents, it is important that all prime contractors promptly receive copies of them. Many design offices send copies to all parties who have received the contract documents (including the plans rooms). The addenda are also of concern to the subcontractors, material suppliers, and manufacturers' representatives who are preparing proposals for the project as the revisions may affect their bids.

All bidders and plans rooms receive the addenda not less than four days prior to bid closing. It is suggested that the contractor call the consultant's office once in the several days before bids are to be received and again the day before to check that all the addenda have been received. Most bid forms have a space provided in which the contractor must list the addenda received. Failure to complete this space may result in the bid being disqualified.

3.12 Taxes and Duties

Included in the bid form and specifications are clear and precise instructions that taxes and duties (excluding value-added taxes) already in effect prior to the bid closing date are to be included in the bid price. Value-added taxes, such as the federal Goods and Services Tax (GST) and Quebec Sales Tax (QST), should be quoted as an extra to the contract all the way through the process as a line item, since there is a process for submitting these taxes that may result in a credit to the collector. The GST, collected by the contractor on behalf of the Canada Customs and Revenue Agency (formerly Revenue Canada), is not considered part of the contract price.

May 10, 20_ _

**BOARD OF SCHOOL TRUSTEES OF
SCHOOL DISTRICT NO. 123, MOUNTAINVILLE**

ADDENDUM #2 TO CONTRACT NO.: S-20-26

FOR: Cedar Hill Secondary School, Administration Building Expansion.

LOCATION: 7723 Nissa Street, Mountainville.

Bidders are to make the following changes to the above noted Contract No. S-20-26 as indicated below:

Revised Closing: Tuesday May 25, 20_ **1:00** p.m. (Local Time**)**

Revised specification:

IT-11.1.2 Article "ALTERNATE NO. 2" - Delete in full as written and substitute the following:

The Contractor shall state in the Bid Form the amount to be added or deducted from the base bid to furnish and install all materials as shown on drawing A L/1 and in accordance with all other bidding documents.

SP-21.3.3 Article "MATERIALS" - Amend by adding the following paragraph as follows:

All ends of the double tees that rest on steel beams shall have a bearing plate 3 mm x 69 mm x 125 mm attached to the reinforcing bar and field welded to the steel beam.

This addendum shall remain attached to and form part of the contract documents.

Yours truly,

Scott Walker MAICB
C.K. Architects

File: Contract # -S-20-26

FIGURE 3.6 Addendum

3.13 Errors in the Specifications

Ideally, the final draft of the technical trade specifications sections should be written concurrently with the preparation of the working drawings. The specification writer and production staff should keep each other posted on all items so that the written and graphic portions of the documents complement and supplement each other.

Many designers have been highly successful in achieving this difficult balance. Unfortunately, there are still offices that view specification writing as dull and dreary. For this reason, they sometimes assign a person not sufficiently skilled to this extremely important task. Also, many times the specifications are put off until the last minute and then rushed so that they are published before they have been proofread. Some designers brush off the errors that arise with the phrase, "We'll pick it up in an addendum."

Another practice that results in errors is the designer's use of the "cut and paste" specification, which involves cutting portions of an old specification out and inserting them to cover the new job. Usually, items are inadvertently left out in such cases.

The real question is what to do when such errors are found. Some specifications include clauses regarding conformance to local building codes and require that contractors notify the designer of discrepancies found in the plans or specifications. If an error is discovered early in the bidding period and no immediate answer is needed, the error is noted on a specific list of all errors and omissions. Contractors are strongly urged to keep a list solely for the purpose of keeping a record of all the errors and omissions so that it can be accessed whenever needed. Most specifications require that all requests for interpretations must be made in writing, stating how many days before the date set for opening bids they must be received. The contractor must check for this (often in the Instruction to Bidders) and note it on the errors and omissions sheet. It is further stipulated in the specifications that all interpretations shall be made in writing in the form of addenda and sent to all bidders. In actual practice, it is often accepted that estimators will telephone the designer's office and request clarifications (interpretations, really). If the interpretation will materially affect the bid, the contractor must ensure receiving it in writing to avoid later problems. If there are contradictions between the drawings and the specifications, he should also attempt to get them resolved as early in the bidding period as possible.

In keeping a list of all discrepancies (errors) and any items not thoroughly understood, contractors should make notes about where on the drawings and specifications the problems occur. In this manner, when clarifications are required, it is relatively easy to explain exactly what is needed. The designers should not be contacted about each problem separately, but about a few at a time. Often, the contractors will answer the questions themselves as they become more familiar with the drawings and specifications. When calling designers, contractors need to be courteous, with the understanding that everyone makes mistakes. Being courteous will help keep the designers on the contractors' side. Besides, the information may have been included, but simply overlooked. Contractors should not wait until bids are due to call with questions, since verbal interpretations often will not be given, and even if they are, the person who knows the answers may not be available at the time. Regardless of the project or type of estimate, the keys to success are cooperation and organization.

Review Questions

1. What types of information are found in the specifications?
2. Why is it important for the estimator to review carefully the entire set of specifications?
3. Describe the CSC/CSI format and how it is used.
4. What types of information are in the Invitation to Bid?
5. Why is it important that the bids be delivered at the proper time and place?
6. How do the Supplementary ~~General~~ Conditions differ from the General Conditions?
7. What information is contained in the technical trade sections of the specifications?

8. Explain what an alternative is and how it is handled during the bidding process.

9. Why is it important to prepare the alternative amounts carefully and thoroughly?

10. Explain what an addendum is and when it is used.

11. Why is it important that the estimator be certain that all addenda have been received before submitting a bid?

12. How should the estimator handle any errors or omissions that may be found in the contract documents?

CHAPTER 4

The Estimate

After studying this chapter, you will be able to:

1. Identify a plan for completing a construction estimate in an organized manner.

2. Determine the factors considered by a contractor in deciding on which projects to submit a bid.

3. Describe the steps taken by an estimator to compile a detailed estimate.

4. Provide a checklist to be used by a contractor when attending a site visit.

5. Describe the bid depository procedures for receiving trade bids.

4.1 Organization

The estimator must maintain a high degree of organization throughout the estimate development. A well-organized estimate improves the probability of getting the work, facilitating the actual work in the field, and completing the work within budget. The organization required includes a plan for completing the estimate and maintaining complete and up-to-date files. It must include a complete breakdown of costs for each project, both of work done by one's own forces and of work done by subcontractors. Using appropriate software can be an effective way to keep organized. The estimate information should include quantities, material prices, labour conditions, costs, weather conditions, job conditions, delays, equipment costs, overhead costs, and salaries of foremen and superintendents. All data generated during the development of the estimate must be filed in an orderly manner. Some firms put the information into a computer for quick recall, while still other firms manually file their data in a convenient, orderly method for quick reference.

The estimate of the project being bid must be systematically done and be neat, clear, and easy to follow. The estimator's work must be kept organized to the extent that in an unforeseen circumstance (such as illness or accident), someone else might step in, complete the estimate, and submit a bid on the project. If the estimator has no system, if the work done cannot be read and understood, then there is no possible way that anyone can pick up where the original estimator left off. The easiest way for you to judge the organization of a particular estimate is to ask yourself if someone else could pick it up, review it and the contract documents, and be able to complete the estimate. Ask yourself: Are the numbers labelled? Are calculations referenced? Where did the numbers come from? What materials are being estimated?

4.2 Planning the Estimate

The need for organization during the estimating process is critical. There are many decisions that need to be made concerning the logistics of who will do which portion and when. Figure 4.1 is a diagrammatic representation of the steps that are required to complete an estimate. Another helpful tool when preparing an estimate is a bar chart schedule that details when the activities concerning the estimate will be completed. In addition, the persons who are responsible for those activities should be listed on the schedule. Figure 4.2 is a sample bar chart schedule for completing an estimate. The bars and milestones will be darkened in as the activities are completed.

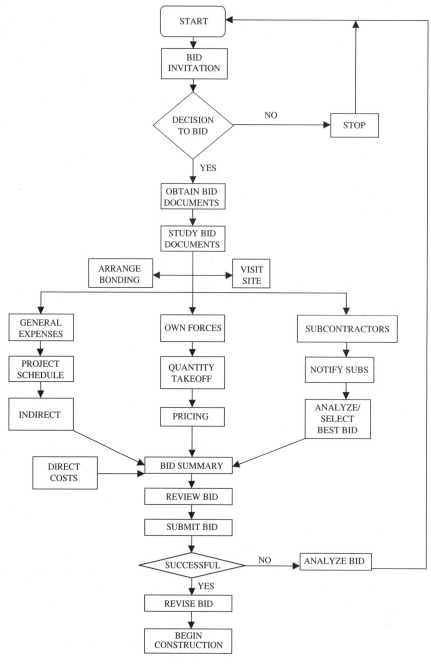

FIGURE 4.1 The Estimate Process

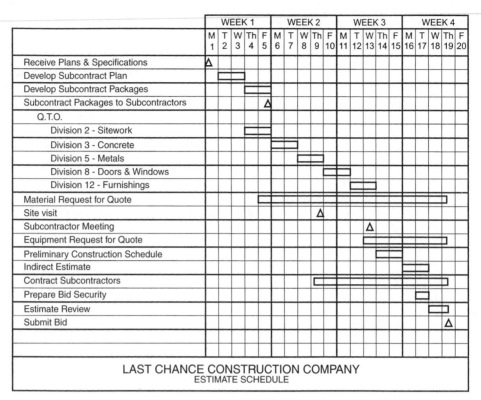

	WEEK 1					WEEK 2					WEEK 3					WEEK 4				
	M 1	T 2	W 3	Th 4	F 5	M 6	T 7	W 8	Th 9	F 10	M 11	T 12	W 13	Th 14	F 15	M 16	T 17	W 18	Th 19	F 20
Receive Plans & Specifications	△																			
Develop Subcontract Plan		▭																		
Develop Subcontract Packages				▭																
Subcontract Packages to Subcontractors					△															
Q.T.O.																				
Division 2 - Sitework				▭																
Division 3 - Concrete						▭														
Division 5 - Metals								▭												
Division 8 - Doors & Windows										▭										
Division 12 - Furnishings										▭										
Material Request for Quote							▭▭▭▭▭▭▭▭▭▭▭▭												
Site visit								△												
Subcontractor Meeting												△								
Equipment Request for Quote												▭▭▭▭▭▭▭								
Preliminary Construction Schedule												▭								
Indirect Estimate																▭				
Contract Subcontractors										▭▭▭▭▭▭▭▭▭▭										
Prepare Bid Security																▭				
Estimate Review																	▭			
Submit Bid																			△	

LAST CHANCE CONSTRUCTION COMPANY
ESTIMATE SCHEDULE

FIGURE 4.2   Sample Estimate Schedule

Since the preparation of an estimate is a collaborative effort, it is essential that all persons have input into when certain items are required and that they understand the inter-relationships among the responsible parties. Therefore, one of the first things that needs to be done when preparing the estimate is to bring together all the estimate team members to develop the overall estimate schedule.

4.3 ▮ Notebook

A *notebook* should be kept for each estimate prepared. The notebook should be broken down into several areas: the workup sheets, summary sheets, errors and omissions sheets, quotations received from subcontractors, quotations received from material suppliers and manufacturers' representatives, and notes pertaining to the project. Also, a listing of all calls made to the designer should be kept together specifying who called, who was contacted at the designer's office, the date of the call, and what was discussed. The notebook should be neat and easy to read and understand.

Every page of the estimate should be numbered and initialled by the person who prepared that portion of the estimate. In addition, every page of the estimate should be checked and verified, and that person's initials should be placed on the page. This rather cumbersome procedure is required to help answer questions that may arise at a later date. When construction begins and the estimate is used to purchase materials and there are questions concerning a specific item, the estimator can be found and asked to clarify any questions.

4.4 ■ To Bid or Not to Bid

It is impossible for a contractor to submit a bid for every project that is tendered. Through personal contact and the reporting services, the contractor finds out what projects are out for bid and then must decide on which projects to submit a proposal. Many factors must be considered: the type of construction involved compared with the type of construction the contractor is usually involved in, the location of the project, the size of the project in terms of total cost and in relation to bonding capacity, the designer/consultant, the amount of work currently under construction, the equipment available, and whether qualified personnel are available to run the project.

There are also certain projects for which a contractor is not allowed to submit a proposal. The owners may accept proposals only from contractors who are invited to bid; other projects may have certain conditions pertaining to work experience or years in business that must be met.

4.5 ■ The Estimate

Once the contractor has decided to bid on a particular project, he needs to make arrangements to pick up/receive the contract documents. Then, he should proceed with the estimate in a manner that will achieve the greatest accuracy and completeness possible. The accuracy required must be in the range of 98 to 99 percent for all major items on the estimate.

Listed below are the steps in working up a detailed estimate. These steps should form a basis for estimating, so it is important to read and understand them.

1. Carefully check the drawings and the specifications to be sure that you have everything, including all addenda. Not all architectural and engineering offices number their drawings in the same manner, so sometimes there can be confusion as to whether you have all the drawings. Architectural drawings are usually prefixed with the letter A. Structural drawings may be prefixed with the letter S, or they may be included with the architectural drawings. Mechanical drawings may be prefixed with M, P, or HVAC. Electrical drawings typically use the designation E. Some jobs have no prefixes before the numbers, but in these circumstances, the pages are typically labelled "sheet 1 of 25." Usually, the front of the specifications contains a list of all the drawings included in the set. Check all sources to ensure that you have received all the drawings. If there are any discrepancies, check with the designer and complete your set. Follow the same procedure with the specifications. Check the list in the front of the specifications against what was received.

2. Scan the drawings to get a feel for the project. How large is it? What shape is it in? What are the principal materials? Pay particular attention to the elevations. At this step, it is important that the estimator understand the project. Make a mental note of exterior finish materials, the amount of glass required, and any unusual features.

3. Review the floor plans, again getting the "feel" of the project. The estimator should begin to note all unusual plan features of the building. Look it over; follow through the rooms, starting at the main entrance. Again, make a mental note of what types of walls are used. Note whether enlarged floor plans show extra dimensions or whether special room layouts are required.

Calculate the gross floor area of the building so that a job unit price (cost per square metre/square foot) can be established and compared with current "market" rates or the cost of similar buildings constructed by the company.

4. Begin to examine the wall sections for a general consideration of materials, assemblies, and makeup of the building. Take special note of any unfamiliar details and assemblies; circle them lightly with a red pencil so that you can refer to them readily.

5. Review the structural drawings. Note what types of structural systems are being used and what types of construction equipment will be required. Once again, if the structural system is unusual, make a mental note to spend extra time on this area.

6. Review the mechanical drawings paying particular attention to how they will affect the general construction, underground work requirements, outlet requirements, chases in walls, and other cutting and patching items. Even under separate contracts, the mechanical portions must be checked.

7. The submitted bid is based on the drawings and specifications. You are responsible for everything contained in the specifications, as well as what is covered on the drawings. Read and study the specifications thoroughly and review them, when necessary. Take notes on all unusual items contained in the specifications.

8. Visit the site after making a preliminary examination of the drawings and specifications. The visit should be made by the estimator or other experienced persons, including members of the proposed project execution team. By including these persons on the site visit, expertise and estimate ownership will be enhanced. The information that is obtained from the site visit will influence the bidding of the project.

9. Even though estimators must rely on their own experience in construction, it is imperative that they create and maintain a close liaison with the other office personnel and field superintendents. After the estimator has become familiar with the drawings and specifications, he should call a meeting with the people who would most likely hold the key supervisory positions if the bid has been successful. Be sure to allow these people time before the meeting to become familiar with the project. During this meeting, the project should be discussed in terms of construction methods that could be followed, the most desirable equipment to use, the time schedules to be followed, and personnel needed on the project.

10. Check carefully through the General Conditions and Supplementary General Conditions making a list of all the items contained in the specifications that will affect the cost of the project.

11. Send a copy of all insurance requirements for the project to your insurance company and all bonding requirements for the project to your bonding company.

12. Prepare a checklist of pertinent trades and select which subtrades will be used.

13. The estimator may now begin to do a takeoff of the quantities required. Each item must be accounted for, and the estimate itself must be as thorough and complete as possible. The items should be listed in the same

manner and with the same units of measure in which the work will be constructed on the job. Whenever possible, the estimate should follow the general setup of specifications. This work is done on a workup sheet. As each item is estimated, the type of equipment to be used for each phase should be listed. The list will vary depending on the equipment owned and what is available for rent. Prices on equipment to be purchased or rented must be included.

14. At the time the estimator is preparing the quantity takeoff on workup sheets, the following tasks can also be ongoing:

    a. Notify subcontractors, material suppliers, and manufacturers' representatives that the company is preparing a bid for the project and ask them if they intend to submit bids on the project.
    b. Begin to make a list of all items of overhead that must be included in the project. This will speed up the future pricing of these items.

15. The information on the workup sheet is carried over to the summary sheet. Work carefully; double-check all figures. If possible, have someone go over the figures with you. The most common error is the misplaced decimal point. Other common errors include:

    a. Errors in addition, subtraction, multiplication, and division.
    b. Omission of such items as materials, labour, equipment, or overhead.
    c. Omission of transportation costs and storage costs.
    d. Errors in estimating the length of time required to complete the project.
    e. Errors in estimating construction waste.
    f. Errors in estimating quantities of materials.
    g. Errors in transferring numbers from one sheet to another.
    h. Failure to allow for weather conditions or equipment breakdowns.
    i. Inadequate provisions for escalation in labour costs and material prices.

16. Having priced everything, make one last call to the designer's office to check the number of addenda issued to be sure you have received them all. Double-check the time, date, and place that bids are being received. Double-check that all the requirements for the submission of the proposal have been followed; be sure the bid form is complete.

4.6 Site Investigation

The contract documents often stipulate that the contractor attend a mandatory site visit. The importance of the visit and the items to be checked vary depending on the type of project and its location. As a contractor expands to relatively new and unfamiliar areas, the importance of the preliminary site investigation increases, as does the list of items that must be checked. Examples of the type of information that should be collected are as follows:

1. Site access
2. Availability of utilities (electric, water, telephone)
3. Site drainage
4. Transportation facilities
5. Any required protection or underpinning of adjacent property

6. A rough layout of the site locating proposed storage trailer and equipment locations

7. Soil conditions, including any possible contamination

8. Local ordinances and regulations, and any special requirements (permits, licences, hoardings), by-laws regulating working hours, noise, and so on.

9. The local labour situation and local union rules

10. Availability of construction equipment rentals, the type and conditions of what is available, as well as the cost

11. The conditions of the roads leading to the project, low bridges, and load limits on roads or bridges

4.7 Subcontractors

A *subcontractor* is a separate contractor hired by the prime contractor to perform certain portions of the work. The amount of work that the prime contractor will subcontract out often exceeds 80 percent. Contractors today can construct entire projects without having any forces/personnel of their own. The use of subcontractors has gained popularity as a means to reduce risk and overhead; however, the contractor gives up a substantial amount of control when subcontracting the entire project.

The contractor must be certain to notify potential subcontractors and material suppliers early in the bidding process so that they have time to prepare a complete, accurate quotation. If rushed, subcontractors tend to bid high just for protection against what might have been missed. Some contractors will review the scope of work thoroughly with the subcontractors well before the job closing and only require the price at the closing.

An estimate of the work should be prepared by the prime contractor for any subcontract sections of the work. If the subcontractor does not have that section included, then the prime contractor can insert the amount required. The bid tabulation sheet should have the scope of work broken down into sections. All subcontractors' bids (Figure 4.3) are compared with the estimator's price; it is important that a subcontractor's price is neither too high nor too low. If either situation exists, and if time permits, the estimator should call the subcontractor and discuss the quotation with him.

The subcontractor's bid is often phoned or faxed in to the general contractor's office at the last minute because of the subcontractor's fear that the contractor will tell other subcontractors the bid price and encourage lower bids. This practice is commonly referred to as bid peddling, or bid shopping, and is highly unethical and should be discouraged. To prevent bid shopping, subcontractors submit their final price only minutes before the bids close, which leads to confusion and makes it difficult for the estimator to analyze all bids carefully. Subcontractors who submit unsolicited bids compound this confusion. These bids come from subcontractors who were not contacted or invited to submit a bid; but they find out which contractors are bidding the project and submit a bid. Since these companies are not prequalified, there is an element of risk associated with accepting one of these bids.

In checking subcontractor bids, note especially what is included and what is left out. Each subsequent bid may add or delete items. Often, the bids set up certain conditions, such as use of water, heat, or hoisting facilities. The estimator must compare all the proposals and select the one that is the most economical.

```
┌─────────────────────────────────────────────────────────────┐
│                        McBill Precast                          │
│                       1215 Miriam Rd.                          │
│                 Mountainville, British Columbia                │
│  May 23, 20___                                                 │
│                                                                │
│  Ace Construction                                              │
│  501 Hightower St.                                             │
│  Mountainville, British Columbia                               │
│                                                                │
│  RE:  Cedar Hill Secondary School, Administration Building Expansion │
│                                                                │
│  Bids Due:  May 23, 20___                                      │
│                                                                │
│  Gentlemen:                                                    │
│                                                                │
│  We propose to furnish and install all precast double tees for the above-mentioned project for │
│  the lump sum of:  $13,250.00                                  │
│                                                                │
│  Furnish double tee fillers (for ends of tees, literature enclosed):    $3.75 each │
│                                                                │
│  All materials are bid in accordance with the contract documents, including Addendum #1. │
│                                                                │
│  Sincerely,                                                    │
│                                                                │
│                                                                │
│  Charles McBill                                                │
└─────────────────────────────────────────────────────────────┘
```

FIGURE 4.3 Subcontractor's Proposal

All costs must be included somewhere. If the subcontractor does not include an item in the bid, it must be considered elsewhere. A tricky task for the prime contractor is the comparison of the individual subcontractor price quotes. Throughout the estimate process, the prime contractor should be communicating with the specific subcontractors concerning the fact that they will submit a price quote and what scope of work is be to included within that quote. However, subcontractors will include items that they were not asked to bid and will exclude items that they were asked to bid. A bid tabulation or "bid tab" is used to equalize the scope between subcontractors so that the most advantageous subcontractor's bid can be included in the prime contractors bid. Figure 4.4 is an example of a bid tabulation form.

4.8 Bid Depositories

Regional Associations of the Canadian Construction Association (CCA) have established offices of certain local construction associations for the administration of the *bid depository*. This is a system of assembling sealed bids from trade contractors and distributing them to prime contractors, thus enabling them to obtain firm quotations in writing and in adequate time to compile their tenders. This is a fair and equitable process and is in the best interests of the owners, bid-calling authorities, and general and trade contractors.

It is a requirement of the bid depository system that general contractors enter into a contract with the selected bidders whose bids have been properly deposited and accepted for subcontract purposes. Following the submission of general contractor bids, substitution or replacement of the selected bidders is subject to specific approval of the tender-calling authority and/or the owner. The advantage in using the bid depositories is that it helps to combat the unethical practice of "bid shopping." The disadvantage is that prime contractors may not necessarily wish to work with the lowest subtrade bidder.

BID TABULATION

Project: _____			Estimate No. _____
Location: _____			Sheet No. _____
Architect: _____			Date: _____
Subcontract Package: _____			By: _____ Checked: _____

Scope of Work	Subcontractor 1	Subcontractor 2	Subcontractor 3	Subcontractor 4	Subcontractor 5
Base Bid					
Adjustments					
1					
2					
3					
4					
5					
6					
7					
8					
9					
10					
11					
Adjusted Bids					

Comments

FIGURE 4.4 Subcontract Bid Tabulation

A tender-calling authority using this system must include the following format and language for the bid depository section of the Instructions to Bidders.

Sample Clause

Bids for the following sections or divisions shall be submitted in accordance with the current Construction Association Rules of Procedure.

Division 15 - Mechanical
Division 16 - Electrical

Bid depositories require trade contractors to submit their bids at a specified time, usually two days before the prime bid closing. Prior to finalizing the closing dates, it is recommended that the bid-calling authority contact the local construction association office to ensure that there are no other projects closing on the same date and that it is not a designated industry holiday or the day after a designated holiday. This will help ensure that competitive bids are received from as many contractors as possible.

General contractors must name the selected subtrade on the Bid Form in compliance with the following instruction to bidders:

Prime contractors shall list on the Bid Form the names of the proposed trade contractors for all the specified bid depository sections or divisions. Only bids that list *one* trade contractor for each section or division submitted in accordance with the bid depository rules of procedure for those sections or divisions specified shall be subject to a recommendation of acceptance from the bid-calling authority to the owner. Any other bids shall be rejected. No change of trade contractors will be allowed following the opening of bids without the written permission of the owner.

Prime contractors intending to perform any of the work of the specified bid depository sections or divisions must have complied with the requirements of the bid depository rules of procedure, and shall list their own forces for the appropriate sections or divisions on the Bid Form. General contractors naming their own forces shall be required to demonstrate their ability to perform the work to the owner's satisfaction.

The standard practice stipulates that Trade Bids must be submitted on the Official Bid Form. Bids are submitted in pink and green envelopes as follows:

1. One envelope containing one of the bid forms is forwarded to the prime contractor.

2. One envelope containing one of the bid forms is retained by the bid depository.

3. All the above envelopes are sealed in a white envelope and delivered to the bid depository at the stipulated time.

The bid depository rules may vary from association to association, and the estimator must be acquainted with the rules applicable to the specific location.

4.9 Materials

For each project being bid, the contractor will request quotations from materials suppliers and manufacturers' representatives for all materials required. Although on occasion, a manufacturer's price list may be used, it is more desirable to obtain written quotations that spell out the exact terms of the freight, taxes, time required for delivery, materials included in the price, and the terms of payment (Figure 4.5). The written proposals should be checked against the specifications to ensure that the specified material was bid.

United Block Company
713 Charles Blvd.
Mountainville, British Columbia

May 23, 20____

Ace Construction
501 Hightower St.
Mountainville, British Columbia

Re: Cedar Hill Secondary School, Administration
    Building Expansion

Gentlemen:

We are pleased to quote on the materials required for the above referenced project.
All of the materials listed below meet the requirements as specified in the drawings and specifications.

200 × 200 × 400	Concrete Block	$1.18 ea.
200 × 100 × 400	Concrete Block	$.82 ea.
200 × 150 × 400	Concrete Block	$.98 ea.

All prices quoted exclude PST and GST.
Terms: 2% - 30 Days
FOB Jobsite

FIGURE 4.5 Materials Price Quote

All material costs entered on the workup and summary sheets (see Sections 4.10 and 4.11) must be based on delivery to the job site, not including tax. The total will include all necessary freight, storage, transportation insurance, and inspection costs. When dealing with material quotes, contractors must consider the Provincial Sales Tax (PST) and the Goods and Services Tax (GST).

4.10 Workup Sheets

The estimator uses two basic types of manual takeoff sheets: the workup sheet and the summary sheet. The *workup* sheet can be a variety of forms contingent upon what is being quantified. Figure 4.6 is an example of a workup sheet that could be used to quantify reinforcing steel.

ESTIMATE WORK SHEET **REINFORCING STEEL**												

Project Little Office Building
Location Mountainville, BC
Architect C.K. Architects
Items Foundation Concrete

Estimate No. 1234
Sheet No. 1 of 1
Date 23/12/20xx
By DCK   **Checked** SRS

Cost Code	Description	Dimensions				Count	Bar Size	Linear Metre	kg/m		Quantity	Unit
		L	W	Space								
		mm					M					
	Continuous Footings											
	Perimeter - Long Bars	102.00	4	2		8	20	816.00	2.355		1922	kg
	Perimeter - Short Bars	0.86	337	2		674	20	579.64	2.355		1365	kg
	Interior - Long Bars	23.00	3	2		6	20	138.00	2.355		325	kg
	Interior - Short Bars	0.81	76	2		152	20	123.12	2.355		290	kg
	Dowels											
	Perimeter	1.20	337	1		337	20	404.40	2.355		952	kg
	Interior	1.20	77	1		77	20	92.40	2.355		218	kg
	Foundation Walls											
	Perimeter - Long Bars	102.00	4	2		8	20	816.00	2.355		1922	kg
	Perimeter - Short Bars	1.12	336	2		672	20	752.64	2.355		1772	kg
	Interior - Long Bars	23.00	8	2		16	20	368.00	2.355		867	kg
	Interior - Short Bars	2.44	76	2		152	20	370.88	2.355		873	kg
	Spread Footings											
	Dowels	1.20	4	3		12	20	14.40	2.355		34	kg
	Column Piers - Vertical Bars	1.12	4	3		12	20	13.44	2.355		32	kg
	Column Piers - Stirrups	1.01	5	3		15	10	15.15	0.785		12	kg
	Drilled Piers											
	Vertical Bars	102.00	6	3		18	20	1836.00	2.355		4324	kg
	Horizontal Bars	6.00	31	3		93	10	558.00	0.785		438	kg
	Grade Beams											
	Grade Beams - Long Bars	21.00	6	1		6	20	126.00	2.355		297	kg
	Grade Beams - Stirrups	1.12	69	1		69	10	77.28	0.785		61	kg

FIGURE 4.6 Estimate Workup Sheet: Reinforcing Steel

The workup sheet is used to make calculations and sketches and to generally "work up" the cost of each item. Material and labour costs should always be estimated separately. Labour costs vary more than material costs, and the labour costs will vary in different stages of the project. For example, a concrete block will cost less for its first 1.20 metres than for the balance of its height; and the labour cost goes up as the scaffold goes up, and yet material costs remain the same.

When beginning the estimate on workup sheets, the estimator must be certain to list the project name and location, the date that the sheet was worked on, and the estimator's name. All sheets must be numbered consecutively, and when completed, the total number of sheets is noted on each sheet (e.g., if the total number was 56, sheets would be marked "1 of 56" through "56 of 56"). The estimator must account for every sheet because if one is lost, chances are that the costs on that sheet will never be included in the bid price. Few people can write legibly enough for everyone to easily understand what they have written; it is, therefore, suggested that the work be printed. If a computer spreadsheet program is being used, contractors must be very careful to verify that all formulas are correct and that the page totals are correct. Errors in spreadsheet programs can be costly. Sophisticated estimating software with versatile spreadsheets and reporting capabilities offer tremendous assistance to the estimator. Never alter or destroy calculations; if they need to be changed, simply draw a line through them and rewrite. Numbers that are written down must be clear beyond a shadow of a doubt. Too often a "4" can be confused with a "9" or "2" with a "7" and so on. All work done in compiling the estimate must be totally clear and self-explanatory. It should be clear enough to allow another person to come in and follow up on all work completed and all computations made each step of the way.

When taking off the quantities, contractors must make a point to break each item down into different sizes, types, and materials, which involves checking the specifications for each item they are listing. For example, in listing concrete blocks, they must consider the different sizes required, the bond pattern, the colour of the unit, and the colour of the mortar joint. If any of these items varies, it should be listed separately. It is important that the takeoff be complete in all details; for example, do not simply write "wire mesh," but write "welded wire mesh 152 × 152 MW9.1 × MW9.1," as the size and type are very important. If the mesh is galvanized, it will increase your material cost by about 20 percent, so this should be noted on the sheet. Following the CSA/CSI format helps organize the estimate and provides a checklist.

4.11 Summary Sheet

All costs contained on the workup sheets are condensed, totalled, and included on the summary sheet. All items of labour, equipment, material, plant, overhead, and profit must likewise be included. The workup sheets are often summarized into summary sheets that cover a particular portion of the project. Figure 4.7 is an example of a summary sheet used to summarize the cost of concrete to be used in the project. Every item on this page is supported by workup sheets.

ESTIMATE SUMMARY SHEET

Project	Little Office Building		Estimate No.	1234
Location	Mountainville, BC		Sheet No.	1 of 1
Architect	C.K. Architects		Date	23/12/20xx
Items	Foundation Concrete		By	DCK Checked SRS

Cost Code	Description	Q.T.O.	Waste Factor %	Purch. Quan.	Unit	Crew	Prod Rate	Wage Rate	Work Hours	Unit Cost Labour	Unit Cost Material	Unit Cost Equipment	Labour	Material	Equipment	TOTAL
	Continuous Footings 20M Bars															
	(1922 +1365 +325+ 90)/1000	3.70	10	4.07	Tonne		15	25.50	61.05	382.50	950.00		1,556.78	3,866.50		5,423.28
	Dowels from Ftg to Wall															
	(952+218)/1000	1.17	10	1.29	Tonne		15	25.50	19.31	382.50	950.00		492.28	1,222.65		1,714.93
	Foundation Walls															
	(1922 +1772 +867+ 873)/1000	5.43	10	5.973	Tonne		11	25.50	65.70	280.50	950.00		1,675.43	5,674.35		7,349.78
	Spread Footing w/Dowels															
	(34)/1000	0.03	10	0.03	Tonne		15	25.50	0.50	382.50	950.00		12.62	31.35		43.97
	Column Piers															
	10 M Bars	0.01	10	0.01	Tonne		24	25.50	0.26	612.00	950.00		6.73	10.45		17.18
	20 M Bars	0.03	10	0.03	Tonne		24	25.50	0.79	612.00	950.00		20.20	31.35		51.55
	Drilled Piers															
	10 M Bars	0.44	10	0.48	Tonne		24	25.50	11.62	612.00	950.00		296.21	459.80		756.01
	20 M Bars	4.32	10	4.75	Tonne		24	25.50	114.05	612.00	950.00		2,908.22	4,514.40		7,422.62
	Grade Beams															
	10 M Bars	0.06	10	0.07	Tonne		22	25.50	1.45	561.00	950.00		37.03	62.70		99.73
	20 M Bars	0.30	10	0.33	Tonne		22	25.50	7.26	561.00	950.00		185.13	313.50		498.63
	Slab on Grade	2.00	10	2.20	Tonne		13	25.50	28.60	331.50	950.00		729.30	2,090.00		2,819.30
	Wire Mesh	1	0	1	Roll		1	25.50	1.00	25.50			25.50	0.00		25.50
	TOTALS								311.59				7,945.42	18,277.05	0.00	26,222.47

FIGURE 4.7 Estimate Summary Sheet (Portion)

In addition to summarizing portions of the project, it is helpful to summarize the entire estimate onto a single page. Figure 4.8 is an example of a summary sheet that could be used to summarize all costs for the entire project.

The summary sheet should list all the information required but none of the calculations and sketches that were used on the workup sheets. It should list only the essentials, and still provide information complete enough for the person pricing the job not to have to continually look up required sizes, thicknesses, strengths, and similar types of information.

4.12 Errors and Omissions

No matter how much care is taken in the preparation of the contract documents, it is inevitable that certain errors will occur. Errors in the specifications were discussed in Section 3.13, and the same note-taking procedure is used for all other discrepancies, errors, and omissions. The Instruction to Bidders or Supplementary General Conditions ordinarily state that if there are discrepancies, the specifications take precedence over drawings and dimensioned figures, and detailed drawings take precedence over scaled measurements from drawings. All important discrepancies (those that affect the estimate) should be checked with the designer's office.

ESTIMATE SUMMARY					

Project Little Office Buildin g **Estimate No.** 1234
Location Mountainville, BC **Date** 23/12/20xx
Architect C.K. Architects **By** DCK **Checked** SRS

DIV	DESCRIPTION	LABOUR $$$	MATERIAL	EQUIPMENT	SUBCONTRACT $	TOTAL $
	TRADE COSTS					
2	SITEWORK					
3	CONCRETE					
4	MASONRY					
5	METALS					
6	WOOD & PLASTICS					
7	THERMAL AND MOISTURE PROTECTION					
8	DOORS AND WINDOWS					
9	FINISHES					
10	SPECIALTIES					
11	EQUIPMENT					
12	FURNISHINGS					
13	SPECIAL CONSTRUCTION					
14	CONVEYING SYSTEMS					
15	MECHANICAL					
16	ELECTRICAL					
	TOTAL TRADE COSTS	000			0	0
	GENERAL EXPENSES					
	SITE STAFF					
	TEMPORARY OFFICES					
	TEMPORARY FACILITIES					
	TEMPORARY UTILITIES					
	PROTECTION					
	CLEANING					
	MISC. EQUIPMENT					
	PERMITS					
	BONDS					
	INSURANCES					
	LABOUR BURDENS					
	PST					
	TOTAL GENERAL EXPENSES	000			0	0
	HEAD OFFICE COSTS					
	LAST-MINUTE CHANGES					
	TOTAL LAST-MINUTE CHANGES	000			0	0
	PROFIT					
	TOTAL PROJECT COSTS	000			0	0

COMMENTS

FIGURE 4.8 Overall Estimate Summary Sheet

Web Resources

www.construction.com
www.e-builder.net
www.rsmeans.com
www.dcd.com
www.costbook.com
www.frankrwalker.com
www.bidnavigator.com
www.corecon.com

Review Questions

1. Why should a notebook of the estimate be kept? What items should be kept in it?

2. Why might a contractor decide not to bid a particular project?

3. Why should estimators check carefully to be certain that they have all the contract documents before bidding a project?

4. One of the requirements of most contract documents is that the contractor visits the site. Why is the site investigation important?

5. Explain what a subcontractor is and the subcontractor's relationship to the owner and general contractor.

6. Why should material quotes be in writing? What items must be checked in these proposals?

7. What is the difference between workup sheets and summary sheets?

8. Go to a local construction association plans room, stop at the desk, and introduce yourself. Explain that you are taking a class in estimating and ask if you could be given a tour of their facilities.

CHAPTER 5

Overhead and Contingencies

1. Identify the types of overhead costs on a construction project.

2. Determine the difference between general and site overhead costs.

3. Describe the steps taken in scheduling a construction project.

4. Explain how estimators use contingency provisions in an estimate.

5. Provide a checklist for estimating job overhead costs.

5.1 Overhead

Overhead costs are generally divided into *head office overhead costs* and *site overhead costs*. The head office overhead costs include items that cannot be readily charged to any one project but represent the cost of operating the construction company. The site overhead costs include all costs not chargeable to a specific item of work on a project. The major difference between the two is that the head office overhead costs are incurred regardless of any specific project.

Site overhead costs constitute a large percentage of the total cost of a construction project. The site overhead costs can range from 15 to 40 percent of the total project cost. Because these costs are such a large portion of the total project cost, they must be estimated with the same diligence and precision as the direct costs. Simply applying a percentage for project overhead degrades the overall accuracy of the estimate. If the direct portion of the estimate is quantified and priced out with diligence and precision but the project overhead is only guessed at, then the overall accuracy of the estimate is only as good as the guess of the project overhead. Estimators must consider overhead costs carefully, make a complete list of all required items, and then estimate the cost for each of these items.

5.2 Head Office Overhead

Head office overhead costs are costs that are not readily chargeable to one particular project. These costs are fixed expenses that must be paid by the contractor and are the costs of operating the business. These expenses must be shared proportionally among the projects undertaken; usually the head office cost items are estimated on the basis of a fiscal year budget and reduced to a percentage of the anticipated annual revenue. The following items should be included in a head office overhead budget:

Office. Rent (if the building is owned, the cost plus return on investment), electricity, heat, water, office supplies, postage, insurance (fire, theft, liability), taxes (property), voice and data communications, office machines, and furnishings.

Salaries. Office employees, such as executives, accountants, estimators, and secretarial and clerical staff.

Miscellaneous. Advertising, literature (magazines, books for library), legal fees (not applicable to one particular project), professional services (architects, engineers, quantity surveyors), donations, travel (including company vehicles), and club and association dues.

Depreciation. Expenditures on office equipment, calculators, computers, and any other equipment. A certain percentage of the cost is written off as depreciation each year and is part of the general overhead expense of running a business. A separate account should be kept for these expenses.

Figure 5.1 is a breakdown of typical head office costs for a general contractor for one year. Obviously, for smaller contractors, the list would contain considerably fewer items, and for large contractors, it could fill pages, but the idea is the same. From this,

Non-Reimbursable Salaries		
Office Employees		
President	$100,000	
Vice President for Operations	$95,000	
Comptroller	$60,000	
Chief Estimator	$60,000	
Estimator	$29,000	
Director of Human Resources	$22,000	
Secretaries (2)	$40,000	
Payroll Clerk	$15,000	
Accounts Payable Clerk	$15,000	
Total Office Labour Costs	**$436,000**	$436,000
Benefits @ 38%		$165,680
Office Rent (Gross Lease) 200 sq. m @ $120.00		$24,000
Telephone		$3,600
Office Supplies		$1,200
Office Equipment		$11,500
Advertising		$5,000
Trade Journals		$200
Donations		$15,000
Legal Services		$2,000
Accounting Services		$3,600
Insurance on Office Equipment		$800
Club & Assoc. Dues		$1,000
Travel & Entertainment		$12,000
Cars (2) w/Insurance		$9,000
Anticipated Office Expense for Year		$690,580

FIGURE 5.1 Estimated Head Office Costs for One Year

it should be obvious that the more work that can be handled by site operations, the smaller is the amount that must be charged for general overhead and the better are the chances of being the low bidder. As a contractor's operation grows, he should pay attention to how much and when additional staff should be added. The current staff may be able to handle the extra work if the additional workload is laid out carefully and through the use of selective spot overtime. Another consideration to adding new staff is the cost of supporting those persons with computers, communications equipment, office space, and furniture.

Once the head office overhead has been estimated, it becomes necessary to estimate the dollar volume of sales for the year. If it is planned that the amount should rise over the coming year, then the plan must state how to make that happen and the associated costs included in the budget. Will this growth come about by bidding for additional jobs, and will that require additional estimators? Will the growth come by expanding into new markets, and if so, what are the costs of becoming known in these new markets? These are very important strategic issues that need to be addressed by the key people in the construction company. Once the sales for the year are estimated, a percentage can be developed. This percentage can then be applied to all work that is pursued. If this is not included in the estimate, then from an accounting viewpoint, that project is not making a contribution to the company overheads. Example 5.1 takes the head office annual costs and shows how it would be allocated to specific projects.

E X A M P L E   5.1   Head Office Cost Allocation

Anticipated sales volume for fiscal year = $9.5 million.

$$\text{Head office cost allocation} = \frac{\underline{\text{Annual estimated head office costs}}}{\text{estimated annual revenue}}$$

$$= \frac{\$690,580}{\$9,500,000} = 7.3\%$$

Some contractors do not allow for the category "General Overhead Expense" separately in their estimates; instead, they figure a larger percentage for profit, or they group overhead and profit together. This, in effect, "buries" part of the expenses. From a job costing viewpoint, it is desirable that all expenses be listed separately so that they may be analyzed periodically. In this manner, the amount allowed for profit is actually figured as profit—the amount left after all expenses are figured.

5.3 Site Overhead (General Expenses, Direct Overhead)

Also referred to as *general expenses* or *direct overheads,* site overhead comprises all costs that can be readily charged to a specific project but cannot be charged to a specific item of work on that project. The list of site overhead items is placed as the first of the general estimate sheets under the heading of General Expenses or Direct Overheads. Most of these items are a function of the project duration; therefore, having a good estimate of the project duration is critical in developing a good site overhead estimate.

Itemizing each cost gives the estimator a basis for determining the amount of that expense and also provides for comparisons between projects. A percentage should not be added to the cost simply to cover overhead. It is important that each portion of the estimate be analyzed for accuracy to determine whether the estimator, in the future, should bid an item higher or lower.

Salaries. Salaries include those paid to the site superintendent, assistant site superintendent, site clerk, all foremen and site labourers required, and security, if needed. The cost must include all fringe benefits and labour burdens.

The salaries of the various workers required are estimated per week or per month, and that amount is multiplied by the estimated time it is expected that each will be required on the project. The estimator must be neither overly optimistic nor pessimistic with regard to the time each person will be required to spend on the job. Figure 5.2 is a good example of how a bar chart schedule can be used to estimate the labour costs and then used during construction to control these costs.

Temporary Office. The cost of providing a temporary site office for use by the contractor and consultants during the construction of the project should also include office expenses, such as electricity, gas, heat, water, voice and data communications, and office equipment. Check the specification for special requirements pertaining to the office. A particular size may be required; the consultant may require a temporary office, or other requirements may be included.

If the contractor owns the temporary office, a charge is still made against the project for depreciation and return on investment. If the temporary office is rented, the rental cost is charged to the project. Because the rental charges are generally based on a monthly fee, careful estimate of the number of months is required. At $250 per month, three extra months amount to $750 from the profit of the project. Check whether the monthly fee includes setup and return of the office. If not, these costs must also be included.

Temporary Buildings, Hoardings, Enclosures. The cost of temporary buildings includes all tool sheds, workshops, and material storage spaces. The cost of building and maintaining hoardings and providing signal lights in conjunction with the hoardings must also be included. Necessary enclosures include fences, temporary doors and windows, ramps, platforms, and protection over equipment.

Temporary Utilities. The costs of temporary water, electricity, and heat must also be included. For each of these items, the specifications must be read carefully to determine arrangements for the installation of the temporary utilities and who will pay for the actual amounts of each item used (power, fuel, water).

PROJECT STAFF PLAN AND ESTIMATE																
Title	Monthly Gross Pay	Months on Project	Cost	Month												
				1	2	3	4	5	6	7	8	9	10	11	12
Project Manager	$6,500	10	$65,000												
Assistant Project Manager	$5,500	10	$55,000												
Site Superintendent	$5,000	10	$50,000												
Assistant Site Superintendent	$4,000	5	$20,000												
Purchasing Agent	$2,500	4	$10,000												
Clerical	$1,500	8	$12,000												
Payroll	$1,400	12	$16,800												
Total Project Staff Costs			$228,800												

FIGURE 5.2 Site Staff Plan and Estimate

Water may have to be supplied to all subcontractors by the general contractor. This information is included in the specifications and must be checked for each project. The contractor may be able to connect into existing water mains. In this case, a plumber will have to be hired to make the connection. Sometimes, the people who own the adjoining property will allow use of their water supply, usually for a fee. The water source is one of the items the estimator should investigate at the site. Many local authorities charge for the water used and the estimator will have to estimate the volume of water that will be required and price it accordingly.

Electrical requirements may include lighting and power for the project. These items are often covered in the electrical specifications, and the estimator must review it to be certain that all requirements of the project will be met. On small projects, it may be sufficient to tap existing power lines, have them run to a meter, and string out extension lines through the project, from which lights will be added and power taken off. On large projects, it may be necessary to install pole lines, transformers, and extensive wiring so that all power equipment deployed to construct the project will be supplied. If the estimator finds that the temporary electricity provisions stated in the specifications would not meet the project requirements, the first call should be to the consultant to discuss the situation. The consultant may decide to issue an addendum revising the specifications. If not, the estimator will have to include these costs in the general expenses portion of the estimate. Typically, the temporary wiring and transformers are included in the scope of the electrical subcontractor's work. All power consumed must be paid for, and how this is handled should be in the specifications; often the cost is split on a percentage basis among the contractors, and it is necessary to factor this into the estimated allowance.

Heat is required if the project will be under construction in the winter. Much of the construction process requires maintenance of temperatures at a certain point. Portable heaters using oil, gas, or propane for small projects may supply the required heat. The total cost includes the costs of equipment, fuel, and required labour attending the equipment. On large projects, the heating system is sometimes put into use before the rest of the project is complete; however, costs to run and maintain the system must be included in the estimate. Filters may have to be replaced on heating systems used by the contractor in the construction phase prior to handing over the building to the owner on substantial completion of the project.

Sanitary Facilities. All construction sites must provide and maintain, in compliance with applicable codes and by-laws, temporary toilet facilities for the workers. The most common type in use is the portable toilet, which is most often rented. Large projects will require several portable toilets throughout the project. This item is also one of the first things that must be on the job site. Another important aspect with regard to portable toilets is waste hauling—how often the waste will be removed and if weekend service is available. If the contractor or one of the subcontractors works on weekends, the portable toilets must be cleaned at that time, which often will incur an additional charge. This information needs to be included in the estimate in order to ensure that there will be adequate funds in the project budget.

Drinking Water. The cost of providing drinking water in the temporary office and throughout the project must be included. It is customary to provide cold or ice water throughout the summer. Keep in mind that the estimated cost must include the containers, cups, and someone to maintain them. If construction is taking place in a hot and dry area, special electrolyte liquids may be required to maintain the health of the labour force.

Photographs. Many project specifications require photographs at various stages of construction, and even if they are not required, it is strongly suggested that a still camera and a video camera be kept at each job site. The superintendent should make use of them at all important phases of the project to record progress. In addition to the cost of the above items, the cost of developing and any required enlarging of the film should be considered. The use of webcams is becoming increasingly popular. These devices come with a wide variety of features and options. These cameras allow persons to view the project via the Internet. If a webcam is desired or required, there are operating costs that are to be added to the cost of the hardware.

Surveys. If a specification requests a survey of the project location on the property, the estimator must include the cost for the survey in the estimate. Check the specifications to see if a licensed registered land surveyor is required, and then ask several local surveyors to submit a proposal.

Cleanup. Throughout the construction's progress, rubbish will have to be removed from the project site. The estimator needs to estimate how many trips will be required and a cost per trip. In addition, a plan needs to be devised concerning where the rubbish can be dumped. As landfills become sparse, it is becoming more difficult and expensive to dump construction waste. Because some landfills will not accept construction waste, the debris may have to be hauled for long distances. In addition, recycling of construction waste and "green construction" are areas that are receiving lots of attention. As the cost and difficulty of disposing of construction waste increases, so does the feasibility of some alternative methods of disposal.

Before acceptance of the project by the owner, the contractor will have to clean all floors, walls, windows, doors, ceilings, and visible finished surfaces. This is typically performed by a cleaning service and estimated by the square metre/square foot.

Winter Construction. When construction is scheduled through the winter months, several items of extra cost must be considered, including the cost of temporary enclosures, heating the enclosure, heating concrete and materials, and protecting equipment from the cold-weather elements.

Protection of Adjacent Property. Miscellaneous items that should be contemplated include the possibility of damaging adjacent buildings, such as breaking windows, the possibility of undermining foundations, or damages caused by workmen or equipment. Protection of the adjacent property is critical. It is recommended that the contractor's insurance carrier be contacted prior to commencing construction, as the insurance carrier can survey the adjacent structures for existing damages. This service is normally provided at no charge. If the insurance carrier does not provide this service, then a consultant should be hired to take a photographic inventory. All sidewalks and paved areas that are torn up or damaged during construction must be repaired. Many items of new construction require protection to avoid their damage during construction, including wood floors, carpeting, finished hardware, and wall finishes in heavily travelled areas. New work that is damaged will have to be repaired or replaced. Often, no one will admit to damaging an item, so the contractor must absorb this cost.

The supplementary general conditions should be checked carefully for other requirements that will add to the general expenses. Examples of these are project signs, billboards, building permits, testing of soil and concrete, and written progress reports. Section 5.6 is an indirect estimate checklist that is helpful in estimating this portion of the project.

5.4 Scheduling

A major underlying determinant of indirect costs associated with a project is how long it will take to complete the project. The time for completion may be an owner-stipulated requirement, but the estimator must prepare a schedule to determine a realistic project duration. This length of time impacts overhead items, wages paid to supervisory and head office personnel, rental on trailers and toilet facilities, and costs for any guards, flagmen, and hoarding required. It also affects the estimate in terms of how long equipment will be required on the project.

Traditionally, the estimator assumes approximate project duration that is the basis for the estimated project overhead costs. There are many computer programs, such as Microsoft Project, SureTrak, and Primavera Project Planner, that are useful tools in sequencing the construction process and assist in developing the project duration. In the competitive bid process, this work is time consuming and expensive. The estimator and the appropriate management personnel must make a choice as to how much time and money should be expended in the estimating phase of the project. The site overhead costs very often impact whether a particular contractor is the low bidder. The speed at which the project is constructed has a heavy impact on the ultimate project cost. At minimum, a bar chart schedule should be developed to estimate a reasonable project duration. One approach is to bring together proposed members of the project team and have them work together to develop this schedule. The advantage of this approach is that the persons who are actually going to direct the work in the field will have ownership and input in the construction schedule. This cooperative effort will avoid any conflict between the head office and the field and will add credibility to the schedule.

However, if the work is being negotiated and the decision focuses not upon which contractor to select but the economic feasibility of the project, then detailed network schedules may be appropriate. In addition to forcing everyone to think about the construction process, it shows the owner that the contractor is serious about planning the project and then executing it according to the plan.

The basics of scheduling the project can be broken down into four steps:

1. List all activities required for the completion of the project.
2. Assign a duration to each of the activities listed in step 1. It is most important that *all* the times be reasonably accurate. If the work is to be subcontracted, contact the subcontractor for his or her input.
3. Write each activity and its duration on a "post-it" note and have the construction team develop a network diagram that shows the sequence in which the activities will be performed. The most popular type of network diagram used today is the precedence diagram.
4. Perform a forward pass calculation (this is a schedule calculation) to determine the estimated project duration. If one of the commercially available scheduling programs is available, the "post-it" network will become the guide for entering the data.

It is beyond the scope of this text to show the complete workings of computerized scheduling. By using a small example, however, the basics can be explained. The small office shown in Figure 5.3 will be used. Using those simple drawings the activity list in Figure 5.4 was developed.

Once the activity list has been developed, it can be organized into a precedence diagram network schedule. A discussion of this methodology is clearly outside the

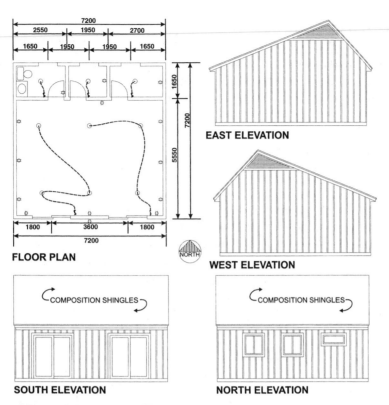

FIGURE 5.3 Sample Office Building

Activity Id.	Description	Duration (Days)
10	Clear site	2
20	Scrape topsoil	2
30	Gravel fill	3
40	Plumbing rough-in	2
50	Form, concrete slab	2
60	Pour and finish concrete	2
70	Rough carpentry	10
80	Electrical, rough-in	2
90	Insulation	3
100	Roofing	3
110	Plumbing, top out	1
120	Drywall	4
130	Interior trim	2
140	Exterior trim	5
150	Telephone, rough-in	1
160	Plumbing, finish	1
170	HVAC, rough-in	3
180	HVAC, finish	3
190	Painting	4
200	Stain, exterior	3
210	Carpet	1
220	Windows	1
230	Glass doors	1
240	Wood doors	1
250	Final grade	1
260	Seed	1
270	Electrical finish	1

FIGURE 5.4 Activity List

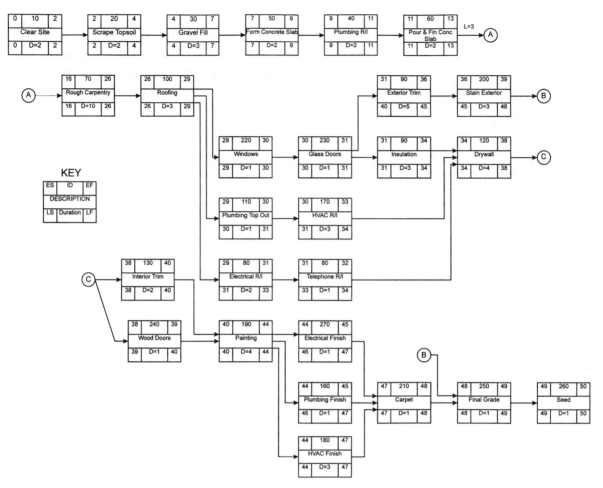

FIGURE 5.5 Sample Schedule

scope of this book, and Figure 5.5 is included for reference only. There are many excellent scheduling books on the market that cover scheduling in great detail.

Once the project duration has been determined, it needs to be converted into calendar days. In the example in Figure 5.5, the project duration is 50 days. If that is a five-day workweek, then that makes the duration 10 weeks or 70 calendar days. If there were nonworking periods in those intervening 70 days, the calendar duration would be extended by the number of nonworking periods. Typical nonworking periods are holidays, such as New Year's Day or Christmas.

5.5 Contingencies

On virtually every construction project, some items are left out or not foreseen when the estimates are prepared. In some cases, the items left out could not have been anticipated at the time of estimating. Should a contingency amount be included? That is, should a sum of money (or percentage) be added to the bid for items overlooked or left out? This money would provide a fund from which the items could be purchased.

If the money is allowed, then it is not necessary to be quite so careful in the preparation of an estimate. But if an accurate estimate is not made, an estimator will never know how much to allow for these forgotten items.

Contingencies are often an excuse for using poor estimating practices. When estimators use them, they are not truly estimating a project. Instead of adding this amount, the proper approach is to be as careful as possible in listing all items from the plans and specifications. This listing should include everything the contractor is required to furnish, it should estimate labour costs carefully and accurately, and it should include the job overhead expense. To these items, you can then add the profit you honestly hope to get from the job. The most rational use for contingencies is an allowance for price escalation.

5.6 Checklist

Undistributed Labour

- Project manager
- Site superintendent
- Assistant site superintendent
- Engineers
    Field engineers (layout surveyors)
- Buyers
- Timekeepers
- Material/Site clerks
- Watchmen
- Project meetings
- Submittal coordination
- Administrative staff

To Estimate Cost

Labour

$$(\text{\# of weeks on the project}) \times (\text{Weekly rate})$$

Expenses

If the company policy is for a per-diem rate to cover travel expenses and so on, the cost would be,

$$(\text{Duration of each staff person on the project}) \times (\text{Rate per diem})$$

If it is a remote project, the cost of relocation should be included:

$$(\text{Persons to be relocated}) \times (\text{Cost per person})$$
$$(\text{Get estimated cost from moving company})$$

If staff does not relocate but commutes on a regular basis, a policy on the frequency of trips home needs to be established and that policy integrated into the estimate.

Cost of travel for executives for job site visits must be included.

Temporary Buildings, Enclosures, and Facilities

- Temporary fences
- Temporary sheds
- Storerooms
- Storage and handling
- Temporary enclosures
- Ladders and stairs, used prior to permanent ones being installed

- Temporary partitions, used to separate new construction from existing facilities.
- Temporary closures for doors and windows
- First aid
- Construction elevators, hoists, and cranes
- Noise control
- Dust control
- Water control
- Pest/rodent control

To Estimate Cost

Labour

$$(\text{Quantity required}) \times (\text{Productivity rate})$$

Material

$$(\text{Quantity required}) \times (\text{Unit cost})$$

Temporary Office

- Temporary site office
- Temporary office for consultant
- Telephone
- Heat
- Lights
- Computers
- Stationery
- Project sign and associated signage

To Estimate Cost

Labour

Typically none, except for office setup.

Materials/Equipment

$$(\text{Duration required at job site}) \times (\text{Monthly cost})$$

Barricades and Signal Lights

- Building and maintaining barricades
- Cost of maintaining signal lights

To Estimate Cost

Labour

$$(\text{Quantity of materials}) \times (\text{Productivity rate})$$

Material

$$(\text{Quantity of materials}) \times (\text{Unit price})$$

Many contractors either rent hoardings or contract for supply and maintenance; if so:

(Quantity) × (Monthly rate)

Temporary Utilities

- Temporary toilets: Local codes, union rules, or accepted company ratios usually govern the quantity and types of toilets that must be supplied. Chemical, portable toilets will usually suffice. In the case of a renovation project, the existing toilet facilities might be available for use; however, permission is typically needed.

To Estimate Cost

(Number of toilets) × (Number of months required) × (Monthly rate)

- Temporary water: On most projects, the general contractor must provide water for all trades. This item can become costly if working in remote or extremely hot climates. If temporary water comes from a municipality, there may be some sewer requirements.

To Estimate Cost

- Function of project location and environment
- Temporary light and power: This is a function of power requirements and may vary from project to project. If high power electrical equipment is being used in the construction, then a transformer may be required to supply power to operate the office and power tools. The best information comes from previous projects that had similar power requirements. In addition, there is the need to include temporary lights, which are needed after the building is closed in, and any associated power cords and bulbs.

The estimator needs to also be aware of the anticipated startup of the mechanical equipment. If this equipment is started early, in order to improve the work environment, a substantial utility cost will be associated with keeping this equipment running during the remainder of the construction process.

To Estimate Cost

Determine the power requirements and get cost information from the utility supplier. If familiar with working in the area, a historical monthly allowance would be acceptable.

- Temporary heat: If the project runs through the winter, it may be necessary to rent heaters and provide needed fuel.

To Estimate Cost

(Number of heaters) × (Number of months required) × (Monthly rental rate)

Repairs and Protection

- Repairs to streets and pavement: This covers the cost of repairing all streets damaged during construction.

- Damage to adjoining structures: There is always the chance of damaging adjacent structures, such as windows, foundations, and walls improperly shored. These items should be considered when estimating the project.
- Protecting new work from damage during construction: During construction, it becomes necessary to protect certain classes of work, such as cut stone, marble, terrazzo, granite, and all types of floors and wood products.
- Repairing new work damaged during construction: Patching damaged plaster, replacing broken glass, and so forth are the responsibility of the contractor. General contractors should keep a close watch on these items, as they can be back-charged to a subcontractor if they broke or damaged an item.

Relocating Utilities

- Water lines
- Electric lines

To Estimate Cost

Identify items and solicit prices or get the cost from the works department responsible for the lines.

Cleaning

- Removal of rubbish (typically weekly)
- Cleaning for final acceptance

To Estimate Cost

Estimate the number of loads or number of dumpsters required. Get a price quote on a per-haul basis.

Cleaning for final acceptance: Get a per square metre/square foot cost from a cleaning service.

Permits

To Estimate Cost

Contact the municipality or city buildings department to determine what is required and the cost. This should be done when going to or coming from the pre-bid site visit.

Professional Services

- Surveys
- Photographs
- Testing

To Estimate Cost

The amount of surveying required is a function of the layout, which will be performed by the site engineer and by a professional engineer/surveyor. It is best to contact a local firm and find out what their hourly billing rates are for these services.

The photographic requirements are part of the specifications and company policy. With the advent of low-priced video cameras, this item is becoming less costly.

The testing requirements for materials are typically found in the specifications. This provides for the quantity of tests required. The cost of these tests can be determined by contacting a testing lab to determine what its charges for specific types of tests are. In addition, the estimator must know what the lab charges to collect samples. The cost of collecting the samples can exceed the costs of the test.

Labour Burdens and Taxes

- Workplace Safety and Insurance Board (WSIB) rates
- Canada Pension Plan (CPP) employer contributions
- Employment Insurance (EI)
- Employer Health Tax (EHT)
- Benefits
- Provincial Sales Tax (PST) and Goods and Services Tax (GST)

To Estimate Cost

Labour burdens are a function of labour costs. Statutory-related payroll burden rates are rather dynamic and typically change whenever there is any federal or provincial legislation. These taxes are paid partly by the employee and the employer. The best source for the most accurate information is the Canada Customs and Revenue Agency. This agency has offices in most regions. Through the use of their website (www.ccra-adrc.gc.ca), payroll tax information can be found. Provincial Retail Sales Tax and Employer Health Tax are functions of the provinces. Benefits vary from company to company. The benefits are specified in union agreements for union contractors, whereas in nonunion organizations, they are set by policy. The benefits typically include an allowance for health insurance, retirement, and any other perks given to employees.

Bonds and Insurance

- Bid bond
- Performance bonds
- Labour and material payment bonds
- Warranty bonds
- Contractor's general liability insurance
- All-risk property insurance
- Boiler and machinery insurance

To Estimate Cost

Bonds and insurance are functions of project requirements and experience rating of the contractor. Contacting the contractor's insurance or bonding agent can determine the exact cost of these items.

Miscellaneous Equipment and Services

- Pickup truck(s)
- Flatbed truck(s)
- Pumps
- Fire protection

To Estimate Cost

Determine the quantity of service vehicles and the approximate duration of their use on the job site. This can then be converted into a cost item by charging the cost of the vehicles to the project on a per-month basis. If the trucks are to be rented, the cost would be the number of months per vehicle times the monthly rate. If a long-term lease agreement is used, there may be substantial penalties for returning the vehicle early. These costs could be charged to the project if the vehicle is not needed on another project.

Determine the quantity of fire extinguishers required on the job site. Allow for maintenance and testing to the approval of the authorities having jurisdiction.

Web Resources

www.ccra-adrc.gc.ca — Canada Customs and Revenue Agency
www.rev.gov.on.ca — Government of Ontario
www.gov.mb.ca — Government of Manitoba
www.gov.nf.ca — Government of New Foundland
www.rev.gov.bc.ca — Government of British Columbia

Review Questions

1. What is overhead? Why must it be included in the cost of the project?
2. What is the difference between general and site overhead costs?
3. How are the items that the estimator will include in each type of overhead determined?
4. Why is the length of time it will take to complete the project so important in determining the overhead costs?
5. What competitive advantage is there for a company that has kept its general overhead low?
6. How may the weather, climate, and season during which the project is to be constructed affect the overhead costs?
7. Where would the cost of temporary utilities be put in the estimate?
8. How would the estimator determine who is responsible for what cost portion of supplying the temporary utilities?
9. How can a preliminary work schedule of the project help the estimator?
10. Define contingency amounts and how some estimators use them.
11. If a full set of contract documents is available, make a list of overhead expenses required for the project. If a set is not available, make another trip to the plan room and review a set of documents on file there.
12. Using the plans and outline specifications in Appendix D, prepare a bar chart schedule for the construction of the project.
13. Using the plans for the proposed two-storey building in Appendix B, prepare a bar chart schedule for the construction of the project.

CHAPTER 6
Labour

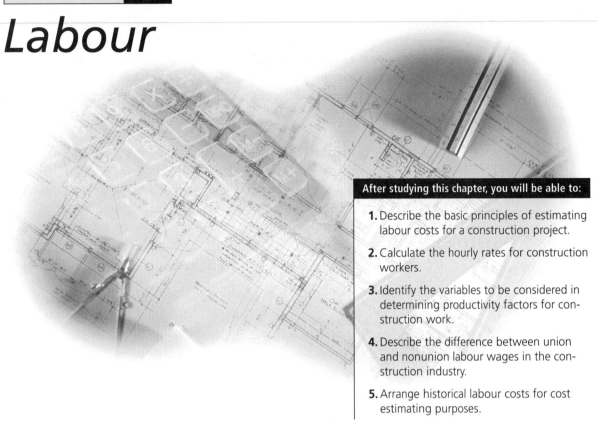

After studying this chapter, you will be able to:

1. Describe the basic principles of estimating labour costs for a construction project.

2. Calculate the hourly rates for construction workers.

3. Identify the variables to be considered in determining productivity factors for construction work.

4. Describe the difference between union and nonunion labour wages in the construction industry.

5. Arrange historical labour costs for cost estimating purposes.

6.1 General

The basic principles of estimating labour costs are discussed in this chapter; they form a basis for the labour costs, which will be illustrated in each chapter that covers quantity takeoff. Estimating labour requires determining the number of work hours to do a specific task and then applying a wage rate. Determining the work hours requires knowing the quantity of work to be placed and the productivity rate for the specific crew that will perform the work. The crew is an aggregation of construction trades working on a specific task. The productivity rate is often expressed as a number of work hours per unit of work. The productivity rates can come from a number of sources, but the most reliable are historical data. The advantage of historical information is that it reflects how a particular company's personnel perform the tasks. Formula 6.1 is used to determine the number of work hours.

Formula 6.1 Work Hours

Work hours = Quantity takeoff × Productivity rate

EXAMPLE 6.1 Work Hours

Type of work – Placing 200 × 200 × 400 mm concrete masonry units (CMUs)

Productivity rate – 1.18 work hours / square metre
Quantity takeoff (QTO) – 100 m²
Crew – 3 masons & 2 helpers

Work hours = 100 m^2 $\times$ 1.18 Work hours per m^2 = 118 Work hours
60 percent of the hours will be worked by masons and 40 percent by their helpers
Mason work hours = 0.60 $\times$ 118 = 71 Mason work hours
Mason helper work hours = 0.40 $\times$ 118 = 47 Mason helper work hours

The productivity rate that is used if derived from historical data is for the average or standard project. On many occasions, the project that is being bid deviates from these standard conditions. Therefore, the work hours need to be modified to take into consideration how the project that is being bid deviates from the standard condition. Using a productivity factor does this. Formula 6.2 is the mathematical means by which the productivity factor is applied.

Formula 6.2 Adjusted Work Hours

Adjusted work hours = Work hours $\times$ Productivity factor

The productivity factor is a combination of several variables. This is perhaps one of the most complicated determinations. There are no hard and fast rules concerning productivity factors. Experience and instinct are perhaps the best qualities required of estimators when determining productivity factors.

Availability and Productivity of Workers. When there is plenty of work available and workers are scarce, less-trained trade persons are accepted. These less-trained persons will require more time or work hours to complete the required task. Conversely, when construction projects are scarce, workers may become motivated, and the contractor can be selective and hire only the most qualified. This will result in producing more work per hour.

Climatic Conditions. Cold, heat, wind, rain, snow, and combinations of these all affect the amount of work that can be produced in an hour. Typically, any weather extremes will slow down the work pace and may require additional precautions that add indirect work hours to the project. The estimator must try to factor in each of these to determine the most cost-effective approach. Can the project be scheduled such that the concrete can be poured before the winter cold sets in? If not, extra time and materials will be necessary to ensure that the concrete does not freeze after it is poured. On the other hand, if it is too hot, precautions need to be taken to prevent rapid evaporation of water from the concrete surface.

Working Conditions. The jobsite working conditions can have a great effect on the rate of work. A project being built in the city with little working space, limited storage space, and difficult delivery situations typically has less work accomplished per work hour, just due to the difficulty of managing the resources available. The same may be true of high-rise construction, where workers may have to wait for the crane to deliver material to them, have difficulty in moving from floor to floor, and take extra time just to get from where they punch the time clock to where they will be working.

Projects that are far removed from the supply of workers and materials often have similar situations that the estimator must consider. How can material deliveries be made in a timely fashion? Where will the material be stored until needed? Will extra equipment and workers be required to transport the material from the storage area to where it will be installed? Will storage sheds be necessary for material that cannot be left out in the weather? If so, who will be responsible for receiving inventory and moving it to where it will be installed? Worker availability for remote jobs must also be considered. Whether workers are available, at what costs, and any special incentives required must all be considered.

Other Considerations. Workers seldom work a full 60 minutes during the hour. Studies of the actual amount of time worked per hour averaged 30 to 50 minutes. Keep in mind that the time it takes to "start up" in the morning, coffee breaks, trips to the bathroom, breaks for a drink of water, conversations about the big game or date last night, lunches that start a little early and may end a little late, and cleanup time all tend to shorten the work day. This list of variables is long, but these items must be considered. Of most importance to the estimator are those items that can be done to make it "convenient" to work. This will involve sufficient equipment to run the project; prompt awarding of material contracts; sufficient restroom facilities and water containers; and anything else that may help. Regardless of the number of units of work to be accomplished, the startup time and the cleanup time tend to remain the same. Larger volumes of work can be cheaper on a per-unit basis, since these costs can be spread out over a greater number of units.

6.2 Unions—Wages and Rules

The local labour situation must be surveyed carefully in advance of making the estimate. Local unions and their work rules should be given particular attention, since they may affect the contractor in a given locality. The estimator will have to determine whether the local union is cooperative or not and whether the union mechanics tend to be militant in their approach to strike or would prefer to talk first and strike as a last resort. The estimator will also have to determine if the unions can supply the skilled workers required for the construction. These items must be considered in determining how much work will be accomplished on any particular job in one hour.

While surveying the unions, the estimator will have to get information on the prevailing hourly wages, fringe benefits, and holidays, as well as the date that raises have been negotiated for, when the present contract expires, and what the results and attitudes during past negotiations have been. If the contract will run through the expiration date, the estimator will have to include enough in the prices to cover all work done after the expiration of the contract. Often, this takes a bit of research into price trends throughout the country; at best, it is risky, and many contractors refuse to bid just before the expiration of union contracts.

6.3 Nonunion

The construction unions have experienced a decline in membership over the past decade. The nonunion contractor or subcontractor does not have to deal with restrictive union work rules, which gives the contractor greater flexibility and allows trade persons with multiple skills to stay on the project longer and perform a greater portion of the work. The downside of this arrangement is that the quality of the trade person is not known when he is hired. The union carpenter has gone through a structured apprentice program to become a union carpenter. On the other hand, nonunion contractors cannot go to the local union looking for trade persons. Rather, they must directly hire all their trade personnel, which adds to higher turnover and training costs if trade labour is in short supply.

6.4 Pricing Labour

To price labour, first the estimator must estimate the work hours required to do a unit of work. These work hours can then be multiplied by an hourly wage rate to develop the labour cost for that unit of work. However, when the crew is made up of different trades being paid different wage rates, a weighted average wage rate must be determined. This is done by determining the total cost for the crew for an hour and then dividing that amount by the number of persons on the crew.

EXAMPLE 6.2 Weighted Average Wage Rate

Quantity	Trade	Hourly Wage Rate	Cost/Hour
3	Mason	$28.50	$ 85.50
2	Labourer	$24.29	$ 48.58
5	**Total**		**$134.08**
Average Wage Rate			$ 26.82

Once the average crew wage rate has been found, it can be multiplied by the number of work hours to determine the labour cost. Formula 6.3 is used to determine the basic labour costs.

Formula 6.3 Labour Cost

Labour cost = Adjusted work hours × Weighted average wage rate

EXAMPLE 6.3 Determining Labour Cost

From Example 6.1, there were 118 work hours (assume the productivity factor is 1). The labour cost would be:

Labour cost = 118 Work hours × $26.82 / hour = $3,164.76

The wage rate needs to be increased by the burden and benefit rate to determine the total labour rate. Chapter 5 included an in-depth discussion of trade benefits. Typically, these benefits and burdens run between 18 to 30 percent for trade personnel.

EXAMPLE 6.4 Pricing the "All-In" Construction Labour Hourly Wage Rate

In calculating the "all-in" construction labour rate, the components of trade, statutory and union fringe benefits are factored into the calculations. Obtain the latest wage information publication to determine the hours of work, basic hourly wages, and vacation pay. The publication *The Construction Wage Rates*, published by the contractors' associations, identifies the union fringe benefits, such as welfare, pension, and training funds, payable by the employer as per the collective bargaining agreement. The statutory component of the contractor's payroll burden includes the federal government provisions for Canada Pension Plan (CPP) and Employment Insurance (EI). The cost to the contractor for workers' compensation is dependent on the Workplace Safety and Insurance Board (WSIB) rates. Other statutory requirements may vary from province to province. In Ontario, employers are required to pay an Employer Health Tax (EHT). The tax rate for a payroll of over $400,000 is 1.95 percent. Figure 6.1 illustrates a breakdown of these three major components of a typical "all-in" construction wage rate.

Group "A" Labourer	As at April 30, 20xx	
Trade		**$**
A. Basic Hourly Wage Rate	$24.29	
B. Basic Weekly Wage Rate	A × 40 hours	971.60
C. Vacation Pay	B × 10%	97.16
		$1,068.76
Statutory		
D. WSIB Rate	8.19% of max. $59,200.00/52	93.24
E. Canada Pension Plan	3.20% of max. $33,400.00/52	20.55
F. Employment Insurance	1.4 × 2.7% of Max. $39,000.00/52	28.35
G. Health Premium (Total package A + C + H + I + J) × 40 Hours × 1.95%	31.50 × 40 hours × 1.95%	24.57
		$1,235.47
	All-In Hourly Rate/40	**$30.89**
Union Fringe Benefits		
H. Welfare		1.65
I. Pension		2.62
J. Other		0.51
	Total Hourly Rate	**$35.67**

FIGURE 6.1 Calculating the "All-In" Construction Labour Rate

6.5 Historical Labour Costs

The challenge of any construction cost control system is in obtaining factual distribution of labour. The coding of labour items is generally the weak link in a cost control system. Cost Control reports cannot be effective unless the data presented are reliable and accurate. Without such accuracy, no amount of management or sophistication will improve it. If there are no appropriate resources on the construction site to track, record, and allocate labour quantities and costs completed, accurate distribution of costs will not likely happen.

Proper coding of labour begins when the labour information is entered on time cards (sometimes called time sheets). On smaller projects, the construction site superintendent may be responsible to ensure that all labour is properly coded, whereas on larger projects, an assistant site superintendent, foreman, or site clerk may be employed to oversee this task. The recording of the cost allocation code may be a joint effort as the foreman may track who did what in his own journal and give a copy to the site clerk to record all information on the time sheet. Whoever is responsible, care must be taken to ensure that the appropriate labour cost codes are applied to the correct items breakdown.

Despite the style or setup of the daily time cards, labour distribution should be done on a daily basis to ensure maximum accuracy. Recording the time and the activity on a daily basis by the designated site personnel will likely result in consistent accurate information. If the time card system is computerized, individual time cards may be summarized to produce a weekly summary report system.

Construction personnel tend to think of time sheets as a system that is required by the accounting department to generate the payroll for the site employees. The time card system is also a very important step in the cost control process. In the construction stage, cost report data may be conveyed to the estimator on a regular basis during the construction phase. This will help the estimator with current and future pricing and bidding. Data are compared with averaged historical data and may by used to change historical figures for productivity. In the postconstruction stage, cost reports as well as work item files are used for productivity analysis for future estimates. Productivity rates, not unit prices, are kept. With productivity rates used for pricing, a unit price can be built up using localized labour rates, crew mixes, and location factors using formula 6.4.

Formula 6.4 Unit Price

Unit price = productivity $\times$ Labour rate

Review Questions

1. What unit of time is used to measure labour? What does it represent?
2. How do climatic conditions influence the amount of work actually completed in an hour?
3. What effect can upcoming labour union negotiations have on a bid?
4. Why do many contractors hesitate to bid just before the expiration of a union contract?
5. What effect could an extreme shortage of skilled workers have on the cost of a project?
6. How can working conditions on the jobsite affect worker productivity?
7. How may union work rules affect the pricing of labour?
8. How can crews be used in the estimating of labour? How does this compare with using individual workers?
9. What are time cards, and who makes them out?

CHAPTER 7

Equipment

7.1 General

One problem an estimator faces is the selection of equipment suitable to use on a given project. The equipment must pay for itself. Unless a piece of equipment will earn money for the contractor, it should not be used.

Because it is impossible for contractors to own all types and sizes of equipment, the selection of equipment will be primarily from what they own already. However, new equipment can be purchased if the cost can be justified. If the cost of the equipment can be charged off to one project or written off in combination with other proposed uses of the equipment, the equipment will pay for itself and should be purchased. For example, if a piece of equipment costing $15,000 will save $20,000 on a project, it should be purchased, regardless of whether it will be used on future projects or whether it can be sold at the end of the current one.

Figuring the cost of equipment required for a project presents the same problems to estimators as figuring the cost of labour. It is necessary for the estimator to decide what equipment is required for each phase of the work and for what length of time it will have to be used.

If the equipment is to be used for a time and then will not be needed again for a few weeks, the estimator should ask: What will be done with it? Will it be returned to the main plant yard? Is there room to store it on the project site? If rented, will it be returned so that the rental charge will be saved?

Equipment that is required throughout the project is included under equipment expenses because it cannot be charged to any particular item of work. This equipment often includes hoist towers, as well as material-handling equipment, such as forklift trucks. As the estimator does a takeoff of each item, all equipment required should be listed so that the cost can be totalled in the appropriate column.

Equipment required for one project only or equipment that might be used only infrequently is often purchased for the one project and sold when it is no longer needed. In addition to the operating costs, the difference between the purchase price and the selling price would then be charged to the project.

7.2 Operating Costs

The costs of operating the construction equipment should be calculated on the basis of the working hours, since the ownership or rental cost is also a cost per hour. Included are such items as fuel, grease, oil, electricity, miscellaneous supplies, and repairs. Operators' wages and mobilization costs are not included in equipment operation costs.

Costs for power equipment are usually based on the horsepower of the equipment. Generally, a gasoline engine will use between 0.23 to 0.26 litres of gasoline per horsepower per hour when operating at full capacity. However, the equipment will probably operate at 55 to 80 percent of full capacity per working hour and will not operate for the full hour, but only for a portion of it.

E X A M P L E 7.1   Fuel Cost

What is the estimated fuel cost of a 120-horsepower (90 kw) payloader? A job-condition analysis indicates that the unit will operate about 45 minutes per hour (75 percent) at about 70 percent of its rated horsepower.

Assumptions:
  Fuel cost − $0.30 per litre
  Consumption rate − 0.23 litres per hp per hour
  Power utilization − 70%
  Use factor − 75%

Fuel cost per machine hour = hp rating × Power utilization × Use factor × Consumption rate × Fuel cost

$4.35 = 120 hp × 0.70 × 0.75 × 0.23 litres per hp per hr × $0.30

A diesel engine requires about 0.15 to 0.23 litres of fuel per horsepower per hour when operating at full capacity. Because the equipment is usually operated at 55 to 80 percent of capacity and will not operate continuously each hour, the amount of fuel actually used will be less than the full per-hour requirement. The full capacity at which the equipment works, the portion of each hour it will be operated, the horsepower, and the cost of fuel must all be determined from a job condition analysis.

Lubrication. The amount of oil and grease required by any given piece of equipment varies with the type of equipment and job conditions. A piece of equipment usually has its oil changed and is greased every 100 to 150 hours. Under severe conditions, the equipment may need much more frequent servicing. Any oil consumed between oil changes must also be included in the cost.

EXAMPLE **7.2** Equipment Lubrication

A piece of equipment has its oil changed and is greased every 120 hours. It requires 6 litres of oil for the change. The time required for the oil change and greasing is estimated at 2.5 hours.

Assumptions:
Oil cost − $1.30 per litre
Oiler labour rate − $17.50 per hour

Lubrication cost = 6 litres of oil × $1.30 per litre = $7.80
Labour cost = 2.5 hours × $17.50 per work hour = $43.75
Total cost for oil change = $7.80 + $43.75 = $51.55
Cost per machine work hour = $51.55 / 120 hours = $0.43 per hour

Tires. The cost of tires can be quite high on an hourly basis. Because the cost of tires is part of the original cost, it is left in when figuring the cost of interest, but taken out for the cost of repairs and salvage values. The cost of tires, replacement, repair, and depreciation should be figured separately. The cost of the tires is depreciated over the useful life of the tires and the cost of repairs taken as a percentage of the depreciation, on the basis of past experience.

EXAMPLE **7.3** Tires

Four tires for a piece of equipment cost $5,000 and have a useful life of about 3,500 hours; the average cost for repairs to the tires is 15 percent of depreciation. What is the average cost of the tires per hour?

Tire depreciation = $5,000 / 3,500 hours = $1.43 per hour
Tire repair = 15% × $1.43 per hour = $.21 per hour
Tire cost = $1.43 + $.21 = $1.64 per machine work hour

7.3 Depreciation

As soon as a piece of equipment is purchased, it begins to decrease *(depreciate)* in value. As the equipment is used on the projects, it begins to wear out, and in a given amount of time, it will have become completely worn out or obsolete. If an allowance for depreciation is not included in the estimate, then when the equipment is worn out, there will be no money set aside to purchase new equipment. This is not profit, and the money for equipment should not be taken from profit.

On a yearly basis, for tax purposes, depreciation can be figured in a number of ways. But for practical purposes, the total depreciation for any piece of equipment will be 100 percent of the capital investment minus the scrap or salvage value, divided by the number of years it will be used. For estimating depreciation costs, assign the equipment a useful life expressed in years, hours, or units of production, whichever is the most appropriate for a given piece of equipment.

E X A M P L E **7.4**   **Depreciation**

If a piece of equipment had an original cost of $67,500 and an anticipated salvage value of $10,000 and an estimated life of five years, what would be the annual depreciation cost?

Depreciable cost = Original cost − Salvage value
Depreciable cost = $67,500 − 10,000 = $57,500
Depreciable cost per year = Depreciable cost / Useful life
Depreciable cost per year = $57,500 / 5 = $11,500 per year

If the piece of equipment should last 10,000 machine work hours, the hourly cost would be found as follows:

Depreciable cost per hour = Depreciable cost / Useful life (hours)
Depreciable cost per hour = $57,500 / 10,000
Depreciable cost per hour= $5.75 per machine work hour

7.4 Interest

The estimator must check interest rates. The interest should be charged against the entire cost of the equipment, even though the contractor paid part of the cost in cash. Contractors should figure that the least they should get for the use of their money is the current rate of interest. Interest is paid on the unpaid balance. On this basis, the balance due begins at the cost price and decreases to virtually nothing when the last payment is made. Since the balance on which interest is being charged ranges from 100 to 0 percent, the average amount that the interest is paid on is 50 percent of the cost.

$$\text{Approximate interest cost} = \frac{C \times I \times L}{2}$$

when

C = Amount of loan
I = Interest rate
L = Life of loan

Because the estimator will want to have the interest costs in terms of cost per hour, the projected useful life in terms of working hours must be assumed. The formula used to figure the interest cost per hour would be

$$\text{Approximate interest cost} = \frac{C \times I \times L}{2 \times H}$$

when

H = Useful life of equipment (working hours)

The last formula is the one used to determine interest costs toward the total fixed cost per hour for a piece of equipment, as shown in the discussion of ownership costs (Section 7.5). Remember that 8 percent, when written as a decimal, is 0.08, since the percent is divided by 100.

Other fixed costs are figured in a similar manner to that used for figuring interest. The costs to be considered include insurance, taxes, storage, and repairs. Depreciation (Section 7.3) must also be considered. These items are taken as percentages of the cost of equipment minus the cost of the tires and are expressed as decimals in the formula.

When the expenses are expressed in terms of percent per year, they must be multiplied by the number of years of useful life to determine accurate costs.

7.5 Ownership Costs

To estimate the cost of using a piece of equipment owned by the contractor, the estimator must consider depreciation, major repairs, and overhaul, as well as interest, insurance, taxes, and storage. These items are most often taken as a percentage of the initial cost to the owner. Also to be added later is the cost for fuel, oil, and tires. The cost to the owner should include all freight costs, sales taxes, and preparation charges.

EXAMPLE 7.5   Cost of Ownership

Estimate the cost of owning and operating a piece of equipment on a project with the costs following:

Assumptions:
Actual cost (delivered) – $47,600
Horsepower rating – 150 hp (110 kw)
Cost of tires – $4,500
Salvage or scrap value – 3%
Useful life – 7 years or 14,000 hours
Total interest – 8% per year
Length of loan – 7 years
Total insurance, taxes, and storage – 6% per year
Fuel cost – $0.30 per litre
Consumption rate – 0.23 litres per hp per hour
Power utilization – 62%
Use factor – 70%
Lubrication – 4 litres oil at $1.25
Oiler labour – 2 hours labour at $16.50
Lubrication schedule – every 150 hours
Life of tires – 4,000 hours
Repair to tires – 12% of depreciation
Repairs to equipment – 65% over useful life

Fixed Cost (per hour):

$$\text{Approximate interest cost} = \frac{\$47,600 \times .08 \times 7}{(2 \times 14,000)}$$

Approximate interest cost = $0.95 per equipment work hour
Salvage value = $47,600 × 3% = $1,428.00
Depreciable cost = $47,600 − $1,428 = $46,172
Depreciable cost per work hour = $46,172 / 14,000
Depreciable cost per work hour = $3.30 per equip. work hour
Repairs = (47,600 x.65) / 14,000 = $2.21 per equip. work hour
Insurance taxes & storage = (6% × 7 × $47,600) / 14,000
Insurance taxes & storage = $ 1.43 per equip. work hour

(See Figure 7.1 for total fixed cost.)

continued

Recap of Fixed Costs	
Item	$/Work Hour
Approximate Interest Cost	0.95
Depreciable Cost	3.30
Repairs	2.21
Insurance Taxes and Storage	1.43
Total Fixed Cost	7.89

FIGURE 7.1 Fixed Costs

Operating Cost (per hour)

Tire depreciation = ($4,500 / 4,000 Hrs.) = $1.13 per equip. work hour

Tire repair = 12% × $1.13 = $0.14 per equip. work hour

Fuel cost = 150 × 0.23 × 0.62 × 0.7 × 0.30 = $4.49 per equip. work hour

Lubrication cost = 4 litres of oil × $1.50 per litre = $6

Lubrication labour = 2 work hours × $16.50 per work hour = $33

Lubrication cost per hour = ($6.00 + $33.00) / 150

Lubrication cost per hour = $.26 per equip. work hour

(See Figure 7.2 below for total operating cost.)

Recap of Operating Costs	
Item	$/Work Hour
Tire Depreciation	1.13
Tire Repair	0.14
Fuel Cost	4.49
Lubrication Cost	0.26
Total Operating Cost	6.02

FIGURE 7.2 Operating Costs

Total ownership cost = Fixed cost + Operating cost

Total ownership cost = $7.89 + $6.02

= $13.91 per equipment work hour

7.6 Rental Costs

If a project is a long distance from the contractor's home base or if the construction involves the use of equipment that the contractor does not own and will not likely use after the completion of this one project, the estimator should seriously consider the cost of renting equipment. In considering the renting of equipment, the estimator must investigate the available rental agencies for the type and condition of equipment available, the rates, and the services the rental firm provides. The estimator must be certain that all terms of rental are understood, especially those concerning repair of the equipment.

Contractors tend to buy equipment even when it is more reasonable to rent. Many rental firms have newer equipment than a contractor might purchase. They also may have a better maintenance program. Estimators should check the rental firms carefully, especially when doing work in a given locale for the first time. The price of the rental is important, but the emphasis should be on the equipment's condition and service. If no reputable rental agency is available, the contractor may be forced to purchase the required equipment.

Equipment is generally rented for a short time, and lease agreements are arranged when that time extends to one year or more. Rental rates are usually quoted by the month, week, or day. These costs must be broken down into costs per hour or per unit of work so that they may be accurately included in the estimate and checked during construction. The rental charge will be based on a day of eight hours (or less). If the equipment is to be used more than eight hours per day, a proportional charge will be added. To this must be added the other costs of operating the equipment; usually these include mobilization, repairs (except ordinary wear and tear), and day-to-day maintenance, as well as the costs of fuel, insurance, taxes, and cleaning.

7.7 Miscellaneous Tools

Examples of miscellaneous tools are wheelbarrows, shovels, picks, crowbars, hammers, hoses, buckets, and ropes. The mechanics who work on the projects have their own small tools, but the contractor will still need a supply of miscellaneous tools and equipment. The estimator should list the equipment required and estimate its cost. The life of this type of equipment and tools is generally taken as an average of one year. Loss of miscellaneous tools and equipment due to theft is common, and all attempts must be made to avoid this loss.

7.8 Cost Accounting

The costs for equipment cannot come from thin air; the estimator must rely heavily on equipment expense data for future bids. Especially in heavy construction, cost accounting is important, since the contractor has a great deal of money invested and the equipment costs are a large percentage of the costs of the project. It is important that equipment costs be constantly analyzed and kept under control.

Small, miscellaneous equipment and tools are not subject to this cost control analysis and are generally charged to each project on a flat-rate basis. The procedure for determining equipment expenses varies from contractor to contractor, but the important point is that the expenses must be determined. Generally, the equipment expense is broken down into a charge per hour or a charge per unit of work. Site reports of equipment time must include only the time during which the equipment is in use. When excessive idle time occurs, estimators must check to see whether it can be attributed to bad weather, poor working conditions, or management problems on the project. Management problems sometimes include poor site supervision, poor equipment maintenance, poor equipment selection, and an excessive amount of equipment on the project.

A report on quantities of work performed is required if a cost per unit of work is desired. Generally, the work is measured on a weekly basis; sometimes, the work completed is estimated as a percentage of the total work to be performed. This type of report must be stated in work units that are compatible with the estimate.

7.9 Mobilization

The estimate must also include the cost of transporting all equipment required for the project to the jobsite and then back again when the work is completed. Obviously, this cost will vary with the distance, type and amount of equipment, method of transportation used, and the amount of dismantling required for the various equipment. Mobilization costs must be considered for rental equipment also, since it must be brought to the jobsite. The cost of erecting some types of equipment, such as hoists, scaffolding, or cranes, must also be included, as well as the costs of loading the equipment at the contractor's plant yard and unloading it at the jobsite.

7.10 Checklist

Equipment listings are given in each chapter and are considered in relation to the work required on the project. Equipment that may be required throughout the project includes:

Lifting cranes	Heaters
Hoisting engines	Lift trucks
Hoisting towers	Scaffolding

Review Questions

1. What are the advantages, to a small contractor, of renting equipment instead of owning?

2. What is depreciation on equipment?

3. What operating costs must be considered?

4. Why should interest be included in the equipment costs if the contractor paid cash for the equipment?

5. What is the total hourly ownership cost if the cost of equipment is $75,000, the interest rate is 9.5 percent, the life of the loan is six years, the tire cost is $7,500, the fuel cost is $0.40 per litre, and assuming all other data as in Example 7.5?

6. Why must mobilization be included in the cost of equipment?

7. Why is it important that reports from the site pertaining to equipment be kept?

8. If there is excessive idle time for equipment on the job, what factors may this be attributed to?

CHAPTER 8

Excavation

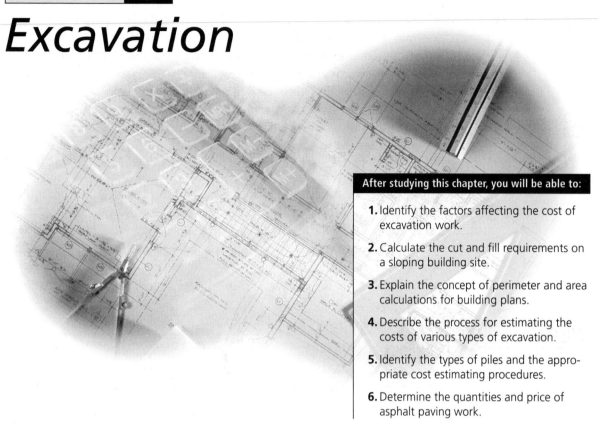

After studying this chapter, you will be able to:

1. Identify the factors affecting the cost of excavation work.

2. Calculate the cut and fill requirements on a sloping building site.

3. Explain the concept of perimeter and area calculations for building plans.

4. Describe the process for estimating the costs of various types of excavation.

5. Identify the types of piles and the appropriate cost estimating procedures.

6. Determine the quantities and price of asphalt paving work.

8.1 General

Figuring the quantities of earthwork that must be excavated is considered to be one of the most difficult portions of the estimator's task. Figuring the "Excavation" section of the specifications often involves a great deal of work. The number of cubic metres/yards to excavate is sometimes easy enough to compute, but figuring the cost for this portion of the work is difficult because of the various hidden items that may affect the cost. These include such variables as the type of soil, the required slope of the bank in the excavated area, whether shoring will be required, and whether water will be encountered and dewatering required.

8.2 Specifications

The estimator must carefully check the specifications to see exactly what is included under the title "Excavation." Several questions demand answers: What is the extent of work covered? What happens to the excess excavated material? Can it be left on the site, or must it be removed? If the excess must be removed, how far must it be hauled? All these questions and more must be considered. Who does the clearing and grubbing? Who removes trees? Must the topsoil be stockpiled for future use? If so, where? Who is responsible for any trenching required for the electrical and mechanical trades?

If the owner is procuring the work under a construction management contract, then it is important that the estimator understand exactly what work each contractor is performing. On the other hand, if the work is being performed under a stipulated lump-sum contract, the general contractor is responsible for addressing all the coordination issues.

8.3 ▌ Soil

When preparing estimates on any structure involving foundations or earthwork, the estimator must ascertain the type of soil which will be encountered at various depths below the ground surface and determine whether or not water will be found and at what depth. This information is shown on the Geotechnical Report included in the Specification or made available at the consultant's office in order to assist the excavation contractors in preparing their bids. Geotechnical investigation reports usually caution contractors bidding the works to decide on their own investigations, as well as their own interpretations of the factual borehole results so that they can draw their own conclusions as to how the subsurface conditions may affect their execution of the work. It is a common practice for the estimator to investigate the soil conditions when visiting the site. Bringing a long-handled shovel or a post-hole digger will allow the estimator to personally check the soil and then record all observations in the project notebook.

8.4 ▌ Calculating Excavation

Excavation is measured by the cubic metre (m³) for the quantity takeoff. Before excavation, when the soil is in an undisturbed condition, it weighs about 1,600 kilograms per m³; rock weighs about 2,400 kilograms per m³.

The site plan is the key drawing for determining earthwork requirements, and dimensions are typically shown in millimetres (mm) or metres (m). Remember that when estimating quantities, the computations need not be worked out to an exact answer.

EXAMPLE ▌ 8.1 ▌ Required Accuracy

Given the following dimensions, determine the quantity to be excavated:

Length − 16.10 m
Width − 23.09 m
Depth − 1.93 m
Volume (m³) = L × W × D
16.10 m × 23.09 m × 1.93 m = 717.48 m³

Use 717 m³

Swell and Compaction. Once excavation begins, the soil is disturbed and begins to swell. This expansion causes the soil to assume a larger volume; this increase represents the amount of *swell* and is generally expressed as a percentage gained above the original volume. When loose material is placed and compacted (as fill) on a project, it will be compressed into a smaller volume than when it was loose; this reduction in volume is referred to as *shrinkage* or *compaction*. Figure 8.1 is a table of common swell and shrinkage factors for various types of soils.

Percentage of Swell & Shrinkage		
Material	Swell	Shrinkage
Sand and Gravel	10 to 18%	95 to 100%
Loam	15 to 25%	90 to 100%
Dense Clay	20 to 35%	90 to 100%
Solid Rock	40 to 60%	85%

FIGURE 8.1 Swell and Shrinkage Factors

EXAMPLE **8.2** **Determining Swell and Haul**

If 1,000 bank cubic metres (in place at natural density) of dense clay (30 percent swell) need to be hauled away, how many cubic metres would have to be hauled away by truck?

$$m^3 \text{ of haul} = \text{In-place quantity} \times \text{Swell percentage}$$
$$1,000 \text{ m}^3 \times 1.3 = 1,300 \text{ m}^3$$

If 8 m³ dump trucks will be used to haul this material away, how many loads would be required?

$$\text{Loads} = m^3 \text{ of haul} / m^3 \text{ per load}$$
$$1,300 \text{ m}^3 / 8 \text{ m}^3 \text{ per load} = 163 \text{ loads}$$

EXAMPLE **8.3** **Determining Shrinkage and Haul**

If 500 m³ in-place of sand / gravel is required, how many loads would be required?

$$\text{Required } m^3 = \text{Required in-place } m^3 / \text{Shrinkage percentage}$$
$$\text{Required } m^3 = 500 \text{ in-place } m^3 / 0.95 = 526 \text{ m}^3$$

If the same 8 m³ dump truck were used, the following would be the number of loads required:

$$526 \text{ m}^3 / 8 \text{ m}^3 \text{ per load} = 66 \text{ loads}$$

8.5 Equipment

Selecting and using suitable equipment is of prime importance. The methods available vary considerably, depending on the size of the project and the equipment owned by the contractor. Hand digging should be kept to a bare minimum, but almost every job requires some handwork. Equipment used includes trenching machines, bulldozers, power shovels, scrapers, front-end loaders, backhoes, and clamshells. Each piece of equipment has its use, and as the estimator does the takeoff, the appropriate equipment for each phase of the excavation must be selected. If material must be hauled some distance, either as excavated material hauled out or fill material hauled in, such equipment as trucks or tractor-pulled wagons may be required.

The front-end loader is frequently used for excavating basements and can load directly into the trucks to haul excavated material away. A bulldozer and front-end loader are often used in shallow excavations, provided the soil excavated is spread out near the excavation area. If the equipment must travel over 30 metres in one direction, it will probably be more economical to select other types of equipment.

The backhoe is used for digging trenches for strip footings and utilities and for excavating individual pier footings, inspection chambers, catch basins, and septic tanks. The excavated material is placed alongside the excavation. For large projects, a trenching machine may be economically used for footing and utility trenches.

A power shovel is used in large excavations as an economical method of excavating and loading the trucks quickly and efficiently. On large-sized grading projects, tractor-hauled and self-propelled scrapers are used for the cutting and filling requirements.

8.6 Earthwork—New Site Grades and Rough Grading

Virtually every project requires a certain amount of earthwork. It generally requires cutting and filling to reshape the grade. *Cutting* consists of bringing the ground to a lower level by removing earth. *Filling* is bringing soil in to build the site to a higher elevation.

The estimator with little or no surveying or engineering knowledge can still handle smaller, uncomplicated projects but should obtain help from someone more experienced if it is a complex project.

Cut and fill sheets similar to the one found in Figure 8.2 are helpful in performing the quantification of rough grading. Regardless of the type of form used, it is essential that the cut and fill quantities be kept separate to allow the estimator to see if the available cut material can be used for fill. In addition, the estimator needs these quantities to estimate the amount of effort required to convert the cut material into fill material. For example, if the cut material is to be used for fill material, it must be hauled to where it will be used and then compacted. This requires decisions concerning what type of hauling equipment will be used and the need for compaction equipment and their associated operators.

The primary drawing for site excavation is the site plan. This drawing typically shows contour lines and spot elevations, and locates all site improvements. Contour lines connect points of equal elevation. Contour lines intervals are usually based on multiples of 100 mm, with contours of small lots normally shown in 500 mm increments. If the sites are large or very hilly, contour lines may be at 1 m intervals. If the sites are relatively level, intervals may be given at 300 mm or less. A typical site plan will indicate the north point and the benchmark taken from a local city datum point. Most commonly, the existing *natural grade* elevations are shown with solid lines, while the proposed new elevations are denoted with dashed lines. The finished grade elevations are shown in small rectangles. In Figure 8.3, the drainage currently slopes from a ridge at elevation 105.000 to the top right-hand corner and the bottom left-hand corner of the site. The revised contour lines change this slope and create a level area at elevation 104.000.

Spot elevations detail an exact elevation of a point or object on the site. Because the contour lines are typically shown in multiples of 100 mm increments, the elevation of any point in between those lines must be interpolated. Through the use of spot elevations, the designer increases the accuracy of the site drawings. For example, the top of grate elevation on a catch basin may be denoted on the drawing as elevation 104.300. Because that elevation is critical for the drainage of the parking lot, it is specified as an exact dimension.

ESTIMATE WORK SHEET

Project	Little Office Building		Estimate No.	1234
Location	Mountainville, BC		Sheet No.	1 of 1
Architect	C.K. Architects	**CUT & FILL WORK SHEET**	Date	1/1/20xx
Items	Rough Grading		By DCK	**Checked** SRS

Grid	Fill									Cut								
	Fill at Intersections						Average	Area	Total	Cut at Intersections						Average	Area	Total
	1	2	3	4	5	Points				1	2	3	4	5	Points			
1																		
2																		
3																		
4																		
5																		
6																		
7																		
8																		
9																		
10																		
11																		
12																		
13																		
14																		
15																		
16																		
17																		
18																		
19																		
20																		
21																		
22																		
23																		
24																		
25																		
26																		
27																		
28																		
29																		
30																		
31																		
32																		
33																		
34																		
35																		
36																		
37																		
38																		
39																		
40																		
41																		
42																		

TOTAL FILL - Cubic metres		TOTAL CUT - Cubic metres	
Shrinkage Factor		Swell Factor	
Required Cubic Metres of Fill		Cubic Metres of Cut to Haul	
Net Cubic Metres to Import			

FIGURE 8.2 Cut and Fill Workup Sheet

Cross-Section Method

The cross-section method entails dividing the site into a grid and then determining the cut or fill for each of the grids. The size of the grid should be a function of the site, the required changes and the required level of accuracy. If the changes in elevation are substantial, the grid should be small. The smaller the grid, the more accurate will be the quantity takeoff. In Figure 8.4, the site was divided into a 10 m grid in both directions. Each line on the grid should be given a number or letter designation. If the horizontal lines were numeric, then the vertical lines would be alphabetic. The opposite is also true. By using this type of labelling convention, points on the site plan can be easily found and referenced. In addition to this numbering system, it is also helpful to number each resulting grid square.

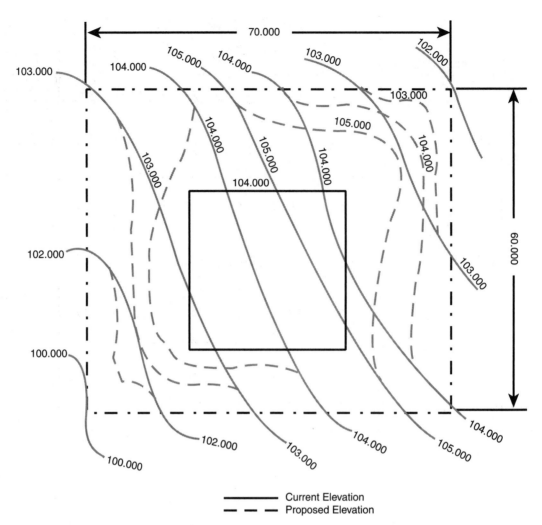

FIGURE 8.3 Sample Site Plan

The next step is to determine the approximate current and proposed elevation for each grid line intersection. Once these are noted, the cut and fill elevation changes can also be noted. Figure 8.5 shows the labelling convention that should be used for this process.

Because contour lines rarely cross the grid intersections, it is necessary to interpolate the current and proposed elevations at each of the grid intersection points. If the proposed elevation is greater than the current elevation, then fill will be required. Conversely, if the proposed elevation is less than the current elevation, then cutting will be needed. Figure 8.6 shows the previous site plan with all the elevations and the cut and fill requirements. Once the entire grid has been laid out with existing and proposed elevations, cut or fill, examine it to see which grids contain both cut and fill. This is done by checking the corners of the individual grid boxes. In Figure 8.6, these are grids 3, 4, 10, 11, 12, 17, 18, 19, 25, 27, 32, 34, 39, and 41. These grids require special consideration. In these grids, some of the materials will be converted from cut to fill or vice versa. The quantity of cut and fill needs to be separated out within these grids so that the cost of converting materials from cut to fill or fill to cut can be estimated for that specific grid.

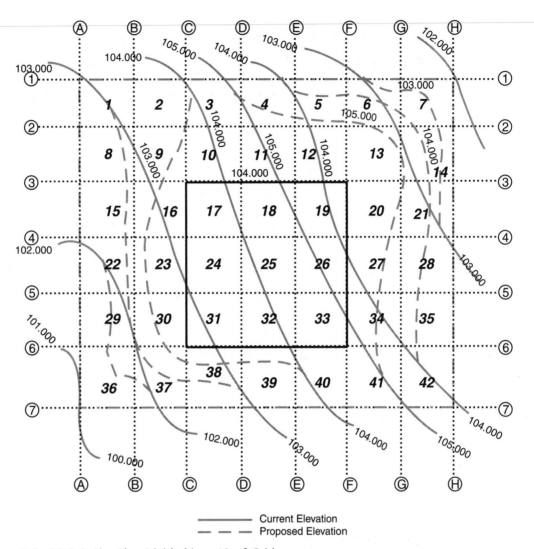

FIGURE 8.4 Site Plan Divided into 10 m² Grid

For the squares that contain only cut or fill, the changes in elevation are averaged and then multiplied by the grid area to determine the required volume of cut or fill. Those quantities are then entered in the appropriate columns on the cut and fill work sheets.

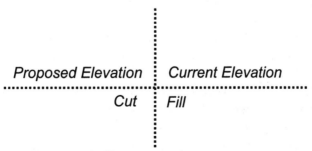

FIGURE 8.5 Labelling Convention

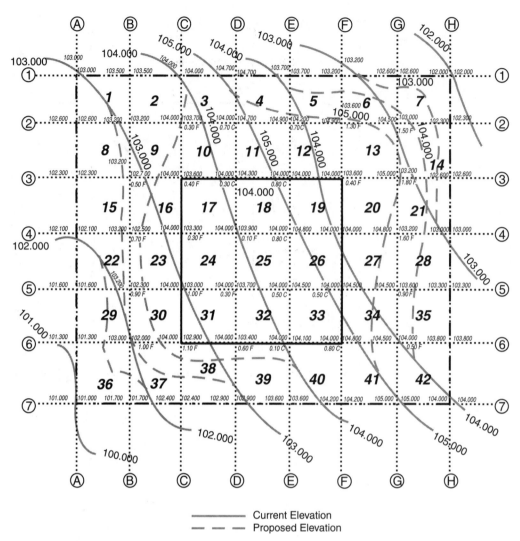

FIGURE 8.6 Grid with Elevations

EXAMPLE **8.4** **Fill Volume**

Using grid 13 (Figure 8.7) from Figure 8.6 as an example, determine the fill quantity. From Figure 8.7, the following information is known about grid 13:

Point	Proposed Elevation	Existing Elevation	Fill (m)
F2	104.900	103.600	1.300
G2	104.500	103.000	1.500
F3	104.000	103.600	0.400
G3	105.000	103.200	1.800

continued

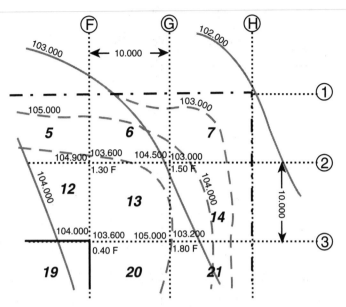

FIGURE 8.7 Excerpt of Grid 13

Volume of fill = Sum of fill at Intersections / Number of intersections × Area

$$\frac{(1.300 + 1.500 + 0.400 + 1.800)}{4} \times 100 \text{ m}^3 = 125.00 \text{ m}^3 \text{ of fill}$$

That amount of fill is then entered in the fill column of the cut and fill work sheet.

E X A M P L E   **8.5**   **Cut Volume**

The volume of cut is determined in exactly the same fashion for cut as fill. Using grid 40 as an example (Figure 8.8), the following information is known:

Point	Proposed Elevation	Existing Elevation	Cut (m)
E6	104.000	104.100	0.100
F6	104.000	104.800	0.800
E7	103.600	103.600	0.000
F7	104.200	104.200	0.000

continued

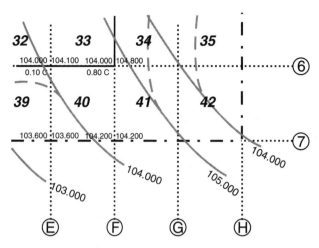

FIGURE 8.8 Excerpt of Grid 40

0.100 + 0.800 + 0.000 + 0.000 / 4 × 100 m³ = 22.50 m³ of cut

That amount of cut is then entered in the cut column on the cut and fill work sheet.

When a specific grid contains both cut and fill, that grid needs to be divided into grids that contain only cut, only fill, or no change. These dividing lines occur along theoretical lines that have neither cut nor fill. These lines of no change in elevation are found by locating the grid sides that contain both cut and fill. Theoretically, as one moves down the side of the grid, there is a transition point where there is neither cut nor fill. These transition points, when connected, develop a line that traverses the grid dividing it into cut and fill areas and in some instances areas of no change.

E X A M P L E   **8.6**   **Cut and Fill in the Same Grid**

Grid 10 (Figure 8.9) from Figure 8.6 is an example of a square that contains both cut and fill.

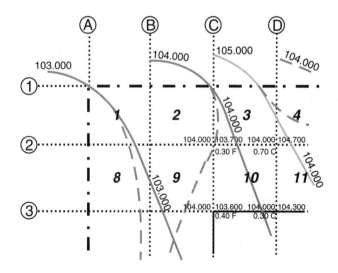

FIGURE 8.9 Grid 10

continued

Along line 2, somewhere between lines C and D, there is a point where there is no change in elevation. First, this point is found by determining the total change in elevation and dividing that amount by the distance between the points. Second, determine the change in elevation per metre of run.

Total change in elevation (C–D) = 0.300 + 0.700 = 1.000 Change in elevation
Change in elevation per metre of run (C–D) = 1.000 / 10 = 0.10 per metre of run

Because the elevation change is 0.10 metres per metre of run, the estimator can now determine how many metres must be moved along that line until there has been a 0.300 metres change in elevation.

0.300 / 0.10 per metre of run = 3 m

This means that as one moves from point C2 toward point D2, at 3 metres past point C2, there is the theoretical point of no change in elevation or the transition point. Because the same thing occurs along line 3 between points C3 and D3 the same calculations are required.

Total change in elevation (C–D) = 0.400 + 0.300 = 0.700 Change in elevation
Change in elevation per metre of run (C–D) = 0.700 / 10 = 0.07 per metre of run

From this calculation, the distance from point C3 to the point of no change in elevation can be found.

0.400 / 0.07 per metre of run = 5.71 m

Given this information, grid 10 can be divided into two distinct grids, one for cut and one for fill. Figure 8.10 details how the grid would be divided.

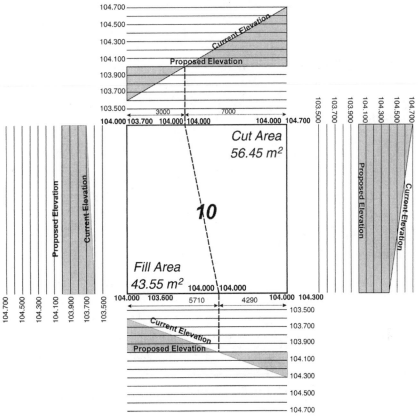

FIGURE 8.10 Grid 10 Layout

continued

The next step is to determine the area of the cut and fill portions. A number of methods are available. Perhaps the most simple is to divide the areas into rectangles and/or triangles.

$$\text{Fill area (average width)} = (3.000 + 5.710) / 2 = 4.355\text{m}$$
$$\text{Fill area (length)} = 10.00 \text{ m}$$
$$\text{Fill area} = 4.355 \text{ m} \times 10.00 \text{ m} = 43.55 \text{ m}^2$$

$$\frac{(0.300 + 0.400 + 0.000 + 0.000)}{} \times 43.55 \text{ m}^2 = 7.62 \text{ m}^3 \text{ of fill}$$

$$\text{Cut area} = 100.00 \text{ m}^2 - 43.55 \text{ m}^2 = 56.45 \text{ m}^2$$

$$\frac{(0.300 + 0.700 + 0.000 + 0.000)}{} \times 56.45 \text{ m}^2 = 14.11 \text{ m}^3 \text{ of cut}$$

EXAMPLE 8.7 Cut and Fill

Occasionally, when the grid is divided, a portion of the grid will be neither cut nor fill. Grid 3 is an example of such an occurrence. Figure 8.11 is an excerpt from the site plan. In that grid, the change from fill to cut occurs on line 2 between C and D.

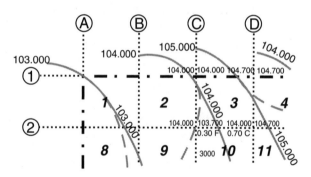

FIGURE 8.11 Grid 3

$$\text{Total change in elevation (C–D)} = 0.300 \text{ m} + 0.700 \text{ m} =$$
$$1.000 \text{ m change in elevation}$$
$$\text{Change in elevation per metre of run (C–D)} = 1.000 \text{ m} / 10 \text{ m} =$$
$$0.10 \text{ m per metre of run}$$

From this calculation, the distance from point C2 to the point of no change in elevation can be found.

$$0.300 / 0.10 \text{ per metre of run} = 3 \text{ m}$$
$$\text{Fill area} = 0.5 \times 10 \text{ m} \times 3 \text{ m} = 15.00 \text{ m}^2$$
$$\text{Cut area} = 0.5 \times 10 \text{ m} \times 7 \text{ m} = 35.00 \text{ m}^2$$

$$\frac{0.300 + 0.000 + 0.000}{} \times 15.00 \text{ m}^2 = 1.50 \text{ m}^3 \text{ of fill}$$

$$\frac{0.700 + 0.000 + 0.000}{} \times 35.00 \text{ m}^2 = 8.17 \text{ m}^3 \text{ of cut}$$

continued

Figure 8.12 shows the dimensions and proportions among cut, fill, and the unchanged area of grid 3. The remaining 50 m² theoretically has no cut or fill. Figure 8.13 is the entire site plan with the areas of no cut and fill shown. Figure 8.14 is the completed cut and fill work sheet for the entire plot.

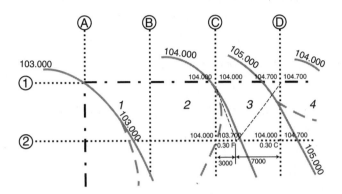

FIGURE 8.12 Cut and Fill Area

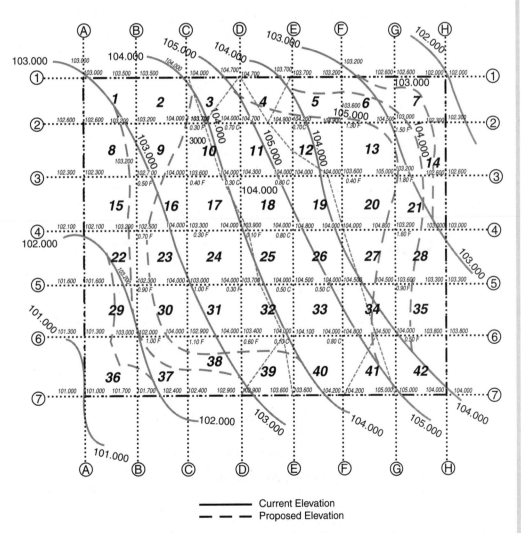

FIGURE 8.13 Complete Site Plan with Areas of No Change Noted

continued

ESTIMATE WORK SHEET

Project	Little Office Building	Estimate No.	1234
Location	Mountainville, BC	Sheet No.	1 of 1
Architect	C.K. Architects — **CUT & FILL WORK SHEET**	Date	1/1/20xx
Items	Rough Grading	By DCK	Checked SRS

Grid	Fill at Intersections 1	2	3	4	5	Points	Average	Area	Total	Cut at Intersections 1	2	3	4	5	Points	Average	Area	Total
1						0	0.000		0.00						0	0.000		0
2	0.00	0.00	0.30	0.00		4	0.075	100.00	7.50						0	0.000	0.00	0.00
3	0.00	0.00	0.30			3	0.100	15.00	1.50	0.00	0.70	0.00			3	0.233	35.00	8.17
4	0.00	0.00	0.70			3	0.233	25.00	5.83	0.00	0.70	0.00			3	0.233	25.00	5.83
5	0.00	0.00	1.30	0.70		4	0.500	100.00	50.00						0	0.000		0.00
6	0.00	0.00	1.50	1.30		4	0.700	100.00	70.00						0	0.000		0.00
7	0.00	0.00	0.00	1.50		4	0.375	100.00	37.50						0	0.000		0.00
8	0.00	0.00	0.50	0.00		4	0.125	100.00	12.50						0	0.000		0.00
9	0.00	0.30	0.40	0.50		4	0.300	100.00	30.00						0	0.000		0.00
10	0.30	0.00	0.00	0.40		4	0.175	43.55	7.62	0.00	0.70	0.30	0.00		4	0.250	56.45	14.11
11	0.00	0.70	0.00			3	0.233	11.68	2.73	0.70	0.00	0.00	0.80	0.30	5	0.360	88.32	31.80
12	0.70	1.30	0.40	0.00	0.00	5	0.480	82.22	39.47	0.00	0.80	0.00			3	0.267	17.78	4.74
13	1.30	1.50	1.80	0.40		4	1.250	100.00	125.00						0	0.000		0.00
14	1.50	0.00	0.00	1.80		4	0.825	100.00	82.50						0	0.000		0.00
15	0.00	0.50	0.70	0.00		4	0.300	100.00	30.00						0	0.000		0.00
16	0.50	0.40	0.30	0.70		4	0.475	100.00	47.50						0	0.000		0.00
17	0.40	0.00	0.00	0.10	0.30	5	0.160	83.91	13.43	0.00	0.30	0.00			3	0.100	16.09	1.61
18	0.00	0.10	0.00			3	0.033	1.38	0.05	0.30	0.80	0.80	0.00	0.00	5	0.380	98.62	37.48
19	0.00	0.40	0.00			3	0.133	16.65	2.22	0.80	0.00	0.00	0.80		4	0.400	83.35	33.34
20	0.40	1.80	1.60	0.00		4	0.950	100.00	95.00						0	0.000		0.00
21	1.80	0.00	0.00	1.60		4	0.850	100.00	85.00						0	0.000		0.00
22	0.00	0.70	0.90	0.00		4	0.400	100.00	40.00						0	0.000		0.00
23	0.70	0.30	1.00	0.90		4	0.725	100.00	72.50						0	0.000		0.00
24	0.30	0.10	0.30	1.00		4	0.425	100.00	42.50						0	0.000		0.00
25	0.00	0.10	0.00	0.30		4	0.100	24.30	2.43	0.00	0.80	0.50	0.00		4	0.325	75.70	24.60
26						0	0.000		0.00	0.80	0.00	0.50	0.50		4	0.450	100.00	45.00
27	0.00	1.60	0.90	0.00		4	0.625	82.15	51.34	0.00	0.00	0.50			3	0.167	17.85	2.98
28	1.60	0.00	0.00	0.90		4	0.625	100.00	62.50						0	0.000		0.00
29	0.00	0.90	1.00	0.00		4	0.475	100.00	47.50						0	0.000		0.00
30	0.90	1.00	1.10	1.00		4	1.000	100.00	100.00						0	0.000		0.00
31	1.00	0.30	0.60	1.10		4	0.750	100.00	75.00						0	0.000		0.00
32	0.30	0.00	0.00	0.60		4	0.225	61.60	13.86	0.00	0.50	0.10	0.00		4	0.150	38.40	5.76
33						0	0.000		0.00	0.50	0.50	0.80	0.10		4	0.475	100.00	47.50
34	0.00	0.90	0.50	0.00		4	0.350	51.40	17.99	0.50	0.00	0.00	0.80		4	0.325	48.60	15.80
35	0.90	0.00	0.00	0.50		4	0.350	100.00	35.00						0	0.000		0.00
36	0.00	1.00	0.00	0.00		4	0.250	100.00	25.00						0	0.000		0.00
37	1.00	1.10	0.00	0.00		4	0.525	100.00	52.50						0	0.000		0.00
38	1.10	0.60	0.00	0.00		4	0.425	100.00	42.50						0	0.000		0.00
39	0.60	0.00	0.00			3	0.200	42.85	8.57	0.10	0.00	0.00			3	0.033	7.15	0.24
40						0	0.000		0.00	0.10	0.80	0.00	0.00		4	0.225	100.00	22.50
41	0.00	0.50	0.00			3	0.167	19.25	3.21	0.80	0.00	0.00			3	0.267	30.75	8.20
42	0.50	0.00	0.00	0.00		4	0.125	100.00	12.50						0	0.000		0.00

TOTAL FILL - Cubic metres		1,450	TOTAL CUT - Cubic metres	310
Shrinkage Factor	0.95		Swell Factor	0.25
Required Cubic Metres of Fill		1,527	Cubic Metres of Cut to Haul	387
Net Cubic Metres to Import		1,140		

FIGURE 8.14 Completed Cut and Fill Work Sheet

In the previous examples, it was assumed that the finish grade was the point at which the earthwork took place; however, this is typically not true. In Figure 8.15 the planned contour lines on the parking lot represent the top of the asphalt. Therefore, the rough grading will be at an elevation different from the one shown on the site plan. In this scenario, the elevation for the rough grading needs to be reduced by the thickness of the asphalt and base material.

EXAMPLE **8.8** **Cut and Fill with Paving**

Using Figures 8.15 and 8.16, determine cut for grid 9.

continued

$$\frac{(1.05 + 1.05 + 0.85 + 0.95)}{} \times 100 \text{ m}^2 = 98 \text{ m}^3$$

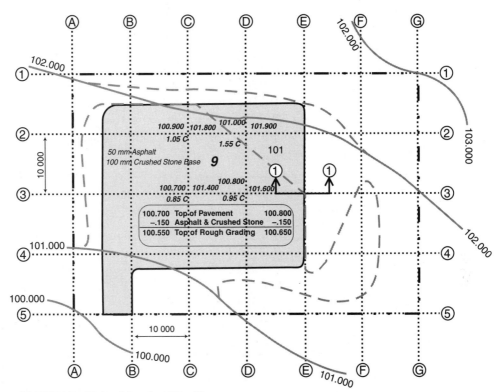

FIGURE 8.15 Parking Lot Site Plan

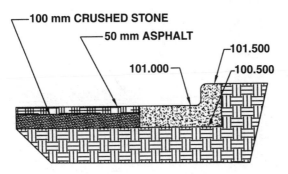

FIGURE 8.16 Cross-Section Through Pavement

Point	Top of Pavement	Top of Rough Grade	Existing Elevation	Cut (m)
C2	100.900	100.750	101.800	1.05
D2	101.000	100.850	101.900	1.05
C3	100.700	100.550	101.400	0.85
D3	100.800	100.650	101.600	0.95

$$\frac{(1.05 + 1.05 + 0.85 + 0.95)}{} \times 100 \text{ m}^2 = 98 \text{ m}^3$$

Average End Area

The average end area method of quantifying cut and fill is often used when dealing with long narrow tracts, such as for roads. In this method, the site is divided into stations. This labelling convention comes from plane surveying using 30-m-long measuring tapes. The first numbers are the number of tapes, and the last numbers are the number of metres on the partial tape. In Figure 8.17, station 00 + 00 is the beginning. Station 00 + 20 is 20 metres from the beginning station, and station 01 + 20 is 50 metres from the beginning station. The positioning of station lines is a function of the contour and requires accuracy. The closer the station lines, the greater is the accuracy.

The first step in determining the volume using the average-end-area method is to draw a profile at the station lines. Next, the cut and fill area for each of the profiles is calculated, and finally, the cut or fill area of two adjacent profiles is averaged and multiplied by the distance between the two stations to determine the cut and fill quantity between the stations.

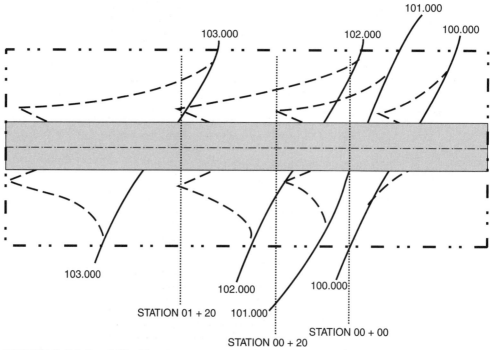

FIGURE 8.17 Road Site Plan

E X A M P L E   **8.9**   **Cut and Fill between Station 00 + 00 and 00 + 20**

The first step is to profile station 00 + 00 and 00 + 20. Figure 8.18 is the profile for all stations. A quick observation of these profiles shows that they only contain cut. The next step is to determine the cut area for each of the profiles. Because these profiles are drawn

continued

to scale, albeit different vertical and horizontal scales, they can be easily calculated by breaking up these profiles into rectangles and triangles to find the area. Figure 8.19 is an example of how the cut area for station 00 + 00 was calculated.

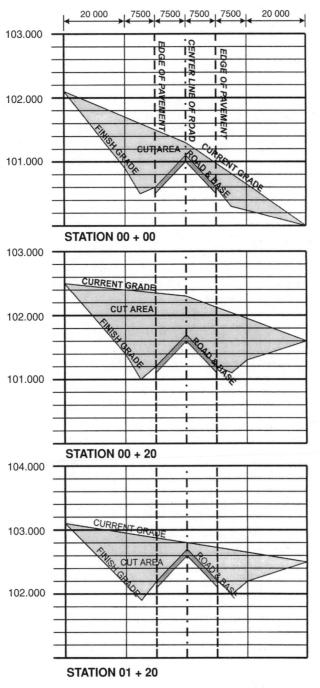

FIGURE 8.18 Example Profiles

continued

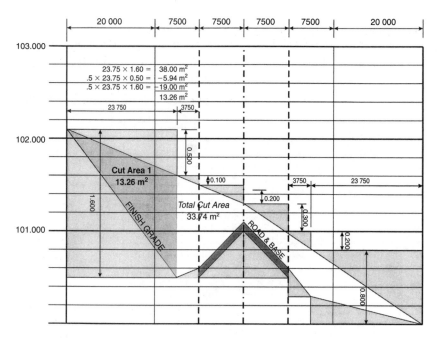

FIGURE 8.19 Cut Area Calculation at Station 00 + 00

Figure 8.20 shows that the cut area at station 00 + 00 is 33.74 m² and 43.75 m² at station 00 + 20. These amounts are averaged and then multiplied by the 20 metres between the two stations to find that the estimated cut is 775 m³.

Cut between Station 00 + 00 and 00 + 20

$$\frac{33.74 \text{ m}^2 + 43.75 \text{ m}^2}{2} \times 20 \text{ m} = 775 \text{ m}^3$$

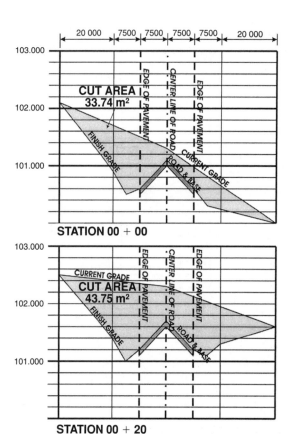

FIGURE 8.20 Road Cut Calculation

8.7   Perimeter and Area

Throughout the estimate, some basic information is used repeatedly. The perimeter of a building is one such basic dimension that must be calculated. The perimeter is the distance around the building; it is the total length around the building expressed in metres.

EXAMPLE **8.10**   **Perimeter Calculations**

Find the perimeter of the building shown in Figure 8.21. Starting in the upper left corner of the building and proceeding clockwise:

25.90 + 7.60 + 4.50 + 10.70 + 9.10 + 3.00 + 9.10 + 3.00 + 12.20 + 18.30 = 103.40 m

As an alternative to the above "piecemeal" approach, a faster and reliable way of measuring the perimeter of any rectangular building is to apply the following formula:

$$P = 2 (l + w + r)$$

where l is the extreme length, w is the extreme width and r is the depth of recess.

This "primary" perimeter can be used to calculate any "secondary" perimeter along the exterior girth of the building. Any perimeter (larger) outside the primary can be calculated by applying the formula $P_e = P + 8d$ where d is the horizontal distance from where P was calculated to where the new perimeter is to be calculated. Any perimeter (smaller) inside the primary perimeter can be calculated by applying the formula $P_i = P - 8d$.

The above calculation principle to obtain the primary perimeter or exterior face of the building shown in Figure 8.21 can be applied as follows:

Perimeter of exterior wall face = 2 [(25.90 + 4.50) + 18.30 + 3.00]
= 103.40 m

Assuming a 250 mm total wall thickness, the above perimeter can be used to calculate the perimeter (smaller) to the interior wall face.

Perimeter of interior wall face = 103.40 − (8 × 0.25)
= 101.40 m

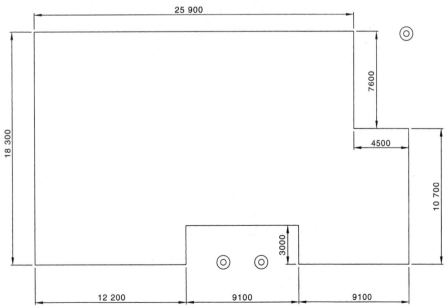

FIGURE 8.21 Sample Perimeter and Area Calculations

E X A M P L E **8.11** **Building Area**

Find the area of the building shown in Figure 8.21.

Gross area	30.40 × 18.30	556.32	m²
Area at set-back	9.10 × 3.00	−27.30	m²
Cutout	4.50 × 7.60	−34.20	m²
Net building area		494.82	m²

8.8 Topsoil Removal

The removal of topsoil to a designated area where it is to be stockpiled for finished grading and future use is included in Division 2 of the specifications. Thus, the estimator must determine the depth of topsoil, where it will be stockpiled, and what equipment should be used to strip the topsoil and move it to the stockpile area. Topsoil is generally removed from all building, walkway, roadway, and parking areas. The volume of topsoil is figured in cubic metres. A clearance around the entire basic plan must also be left to allow for the slope required for the general excavation; usually about 1.50 m is allowed on each side of a building and 0.30 to 0.60 m for walkways, roadways, and parking areas.

E X A M P L E **8.12** **Top Soil Removal**

In Figure 8.22, the "foot print" of the building has been enlarged by 1.50 m to compensate for accuracy and slope. Assume that the topsoil to be removed is 200 mm thick.

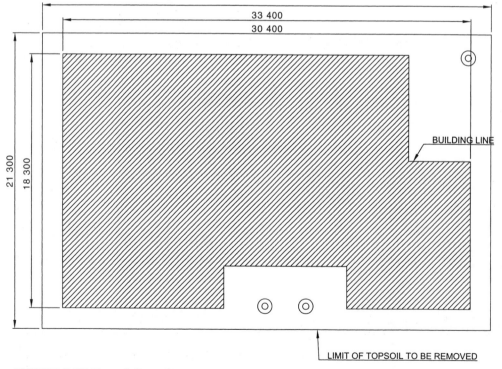

FIGURE 8.22 Topsoil Quantity

Quantity of Topsoil to be Removed (m³) = 33.40 × 21.30 × 0.20 = 142.28 m³

E X A M P L E **8.13**   **Equipment and Labour Cost**

Equipment selection for the removal of topsoil will probably be limited to either a bulldozer or front-end loader. Assume that a 0.76 m³ (1 c.y.) bucket front-end loader is selected (see Figures 8.23 and 8.24) and its production rate is estimated to be an average of 18.35 m³ per hour. Equipment move-in/move-out costs $400, the operating cost per hour for the equipment is estimated at $50.20, and the cost for an operator is $28.40 per hour. Estimate the number of hours and the cost.

Soil	Dozer				Tractor Shovel		Front-End Loader			Backhoe	
	15 m haul		30 m haul		No haul		15 m haul	30 m haul		No haul	
	50 hp (37 kW)	120 hp (89 kW)	50 hp (37 kW)	120 hp (89 kW)	0.76 m³ (1 c.y.)	1.72 m³ (2.25 c.y.)	0.76 m³ (1 c.y.)	1.72 m³ (2.25 c.y.)	0.38 m³ (.5 c.y.)	0.76 m³ (1 c.y.)	
Medium	31	76	23	57	31	54	18	23	19	42	
Soft, sand	34	84	27	65	34	69	23	31	19	46	
Heavy soil or stiff clay	11-15	31	8-11	23-27	11-15	27	8	9	8	11	

FIGURE 8.23 Equipment Capacity (cubic metres per hour)

Truck size	Load and Haul	
	Haul	m³
4.6 m³ (6 c.y.)	1.6 kilometres	9–12
4.6 m³ (6 c.y.)	3.2 kilometres	6–9
9.2 m³ (12 c.y.)	1.6 kilometres	14–17
9.2 m³ (12 c.y.)	3.2 kilometres	9–11

FIGURE 8.24 Truck Haul (cubic metres per hour)

First, the total work hours required to complete the topsoil removal (Example 8.12) must be calculated. Divide the total cubic metres to be excavated by the rate of work done per hour, and add the mobilization time; the answer is the total hours for this phase of work.

$$\frac{142.28 \text{ m}^3}{18.35 \text{ m}^3 \text{ per hour}} = 7.75 \text{ hours}$$

The total number of hours is then multiplied by the cost of operating the equipment per hour plus the cost of the crew for the period of time.

Equipment cost = $50.20 per hour × 7.75 hours = $387.50
Labour cost = $28.40 per work hour × 7.75 hours = $220.10
Move in/out = $400.00
Total = $1,007.60

Therefore, the total cost for 142.28 m³ is $1,007.60 at a unit rate of $7.08/m³.

8.9   General Excavation

Included under general (mass) excavation is the removal of all types of soil that can be handled in fairly large quantities, such as excavations required for a basement, mat footing, or a cut for a highway or parking area. Power equipment, such as power shovels, front-end loaders, bulldozers, and graders, is typically used in this type of excavation work.

When calculating the amount of excavation to be done for a project, the estimator must be certain that the dimensions used are the measurements of the outside face of the footings and not those of the outside of the building. The footings usually project beyond the wall. An extra 150 mm from the face of the footing or 600 mm from the face of the wall, whichever is greater, is added to all sides of the footing to allow the workers to install and remove forms. Sometimes, the excavation is cut to the face of the footing, and the concrete to the footing is poured without forms.

The *Regulations for Construction Projects* under the *Occupational Health and Safety Act* require that the sides of all earth embankments and trenches over 1.2 metres deep be adequately protected by a shoring system or by cutting back the sides of the excavations to a safe angle.

It is important that the estimator carefully studies the Excavation section of the Regulations.

The estimator, taking into consideration the depth of excavation, type of soil, and possible water conditions, must determine the amount of slope required. Some commonly used slopes, referred to as angle of repose (Figure 8.25) are given in Figure 8.26. Generally, for good soil conditions, a trench sloped at a 45-degree angle or 1-to-1 within 1.20 m of the trench bottom is usually adequate. For fairly good soils, the 45-degree or 1-to-1 angle should continue to the bottom of the trench. For poor or bad soils, the trench walls should be sloped at an angle of at least 1 to 3 (3 metres back for every 1 metre up from the trench bottom). If job conditions will not allow the sloping of soil, the estimator will have to consider using sheet piling (Figure 8.27) or some type of bracing to shore up the bank. Any building or column footing projection is a separate calculation, and the result is added to the amount of excavation for the main portion of the building.

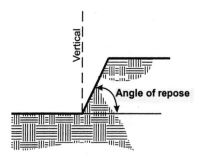

FIGURE 8.25 Angle of Repose

Material	Angle		
	Wet	**Moist**	**Dry**
Gravel	15-25	20-30	24-40
Clay	15-25	25-40	40-60
Sand	20-35	35-50	25-40

FIGURE 8.26 Earthwork Slopes

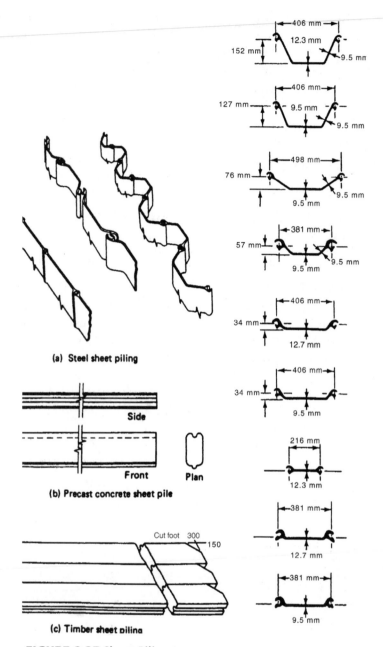

(a) Steel sheet piling

(b) Precast concrete sheet pile

(c) Timber sheet piling

FIGURE 8.27 Sheet Pilings

The actual depth of cut is the distance from the top of grade to the bottom of the fill material used under the concrete floor slab. If topsoil has been stripped, the average depth of topsoil is deducted from the depth of cut. If fill material, such as gravel, is not used under the concrete floor, the depth is then measured to the bottom of the floor slab. Because the footings usually extend below the fill material, a certain amount of excavation will be required to bring the excavation down to the proper elevation before footings can be placed. This would also be included under the heading of "general excavation," but kept separate from topsoil.

Before estimators can select equipment, they will have to determine what must be done with the excess excavation—whether it can be placed elsewhere on the site or whether it must be hauled away. If it must be hauled away, they should decide how far. The answers to these questions will help determine the types and amount of equipment required for the most economical completion of this phase of the work.

To determine the amount of general excavation, it is necessary to determine:

1. Size of building (building dimensions).

2. The distance the footing will project out beyond the wall.

3. The amount of working space required between the edge of the footing and the beginning of excavation.

4. The elevation of the existing land, by checking the existing contour lines on the site plan.

5. The type of soil that will be encountered. This is determined by first checking the soil borings (on the drawings), but this must be checked during the site investigation (Section 4.6). Almost every specification clearly states that the soil borings are for the contractor's information, but they are not guaranteed.

6. Whether the excavation will be sloped or shored. Slope angles (angles of repose) are given in Figure 8.26.

7. The required depth of the excavation. This is done by determining the bottom elevation of the cut to be made. Then, take the existing elevation, deduct any topsoil removed and subtract the bottom elevation of the cut. This will determine the depth of the general excavation.

When sloping sides are used for mass excavations, the volume of the earth that is removed is found by developing the average cut length in both dimensions and multiplying them by the depth of the cut. The average length of cut can be found as shown in Figure 8.28 or the top of cut and bottom of cut dimensions can be averaged. Either method will result in the same answer.

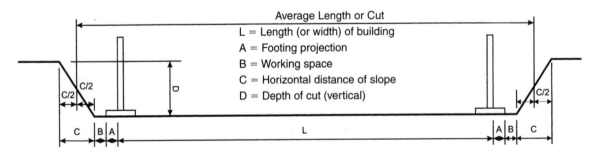

FIGURE 8.28 Typical Excavation

E X A M P L E   **8.14**   **Basement Excavation**

Determine the amount of general excavation required for the basement portion of the building shown in Figure 8.29.

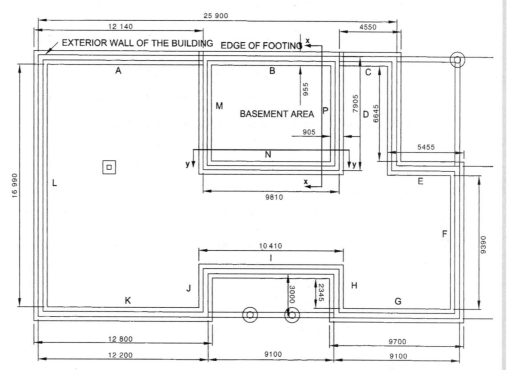

FIGURE 8.29 Building Plan

In order to estimate the excavation quantities for this building, one needs to look at the building cross-sections shown in the following sketches (Figures 8.30 and 8.31):

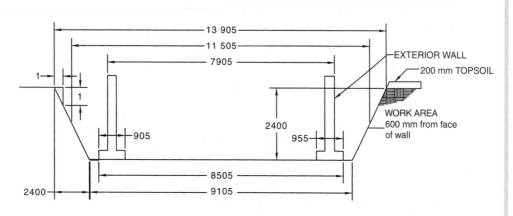

FIGURE 8.30 Basement Cross-Section

continued

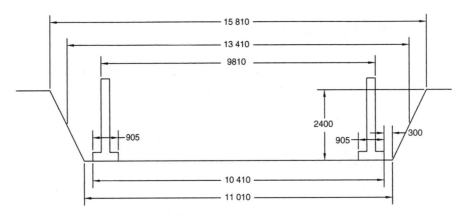

FIGURE 8.31 Basement Cross-Section

1. From the building plan, the exterior dimensions of the basement are 9,810 mm by 7,905 mm.
2. From the wall section, the footing projects out 300 mm from the foundation wall.
3. The workspace between the edge of the footing and the beginning of the excavation will be 300 mm or 600 mm from the face of the wall in this example.
4. The elevation of the existing land, by checking the existing contour lines on the plot (site) plan, is found and noted. In this example, the expected depth of the cut is 2,400 mm after deducting for the topsoil that would have already been removed.
5. Check the soil borings. For this example, a slope of 1:1 will be used (which means for every 1 metre of vertical depth an additional 1 metre of horizontal width is needed). Since the alternative is shoring or sheet piling on this project, the sloped excavation will be used.
6. The bottom elevation of the general excavation cut will be at the bottom of the gravel. Since this elevation is rarely given, it may have to be calculated. Generally, the drawings will give the elevation of the basement slab or bottom of the footing, and the depth of cut is calculated from these.

Average width of cut:
(9,105 mm + 13,905 mm) / 2 = 11,505 mm

Average length of cut:
(11,010 mm + 15,810 mm) / 2 = 13,410 mm

General excavation:
General excavation (m³) = 11.51 × 13.41 × 2.40
= 370 m³

Required equipment: Backhoe with 0.76 m³ (1 c.y.) bucket
Move-in/move-out costs: $400
Rate of work for backhoe: 60 m³ per hour for 100% efficiency. Assume 85% efficiency (51 m³/hour)
Equipment cost: $40.80 per hour
Operator cost: $31.60 per hour

Equipment hours = 370 m³ / 51 = 7.25 hours
Equipment $ = 7.25 hours × $40.80 per hour = $295.80
Labour $ = 7.25 hours × $31.60 per hour = $229.10
Move-in/move-out = $400
Total = $924.90

continued

If this were to be hauled off the site, assuming a 30 percent swell factor, it would take 61 loads using an 8 m³ truck.

$$\text{Required Haul (m}^3\text{)} = 370 \text{ m}^3 \times 1.3 = 481 \text{ m}^3$$
$$\text{Required loads} = 481 \text{ m}^3 / 8 \text{ m}^3 \text{ per load} = 61 \text{ loads}$$

EXAMPLE 8.15 Continuous Footing Excavation

Determine the amount of general excavation required for the continuous footings of the building shown in Figure 8.29. Figure 8.32 is a sketch of the continuous footing with dimensions. In this example, the slope is 1:1, which means that for every 1 metre of vertical rise there is 1 metre of horizontal run.

The simplest way to approximate the amount of cut is to multiply the average cut width times the perimeter of the building times the depth. From Example 8.10, the building perimeter is 103.40 metres. However, 9.81 m of that perimeter was included in the basement wall (refer to Figure 8.29). Therefore, the linear distance of continuous footing is 93.59 m (not taking into account the overlap of workspace allowances)

$$\text{General excavation (m}^3\text{)} = 93.59 \times 3.16 \times 1.60$$
$$= 473 \text{ m}^3$$

Required equipment: Backhoe with 0.38 m³ (.5 c.y.) bucket
Move-in/move-out costs: $400
Rate of work for backhoe: 30 m³ per hour for 100% efficiency.
Assume 85% efficiency (26 m³/hour)
Equipment cost: $40.80 per hour
Operator cost: $31.60 per hour

$$\text{Equipment hours} = 473 \text{ m}^3 / 26 = 18.19 \text{ hours}$$
$$\text{Equipment \$} = 18.19 \text{ hours} \times \$40.80 \text{ per hour} = \$742.15$$
$$\text{Labour \$} = 18.19 \text{ hours} \times \$31.60 \text{ per hour} = \$574.80$$
$$\text{Move-in/move-out} = \$400$$
$$\text{Total} = \$1,716.95$$

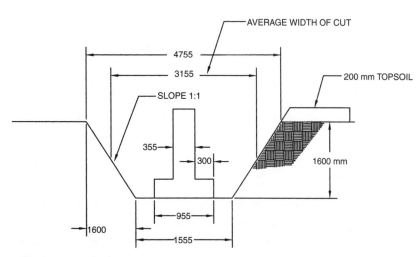

FIGURE 8.32 Continuous Footing Section

EXAMPLE **8.16** **Spread Footing Excavation**

There is one spread footing shown in Figure 8.29. Given a slope of 1:1, the details of this spread footing can be sketched as shown in Figure 8.33.

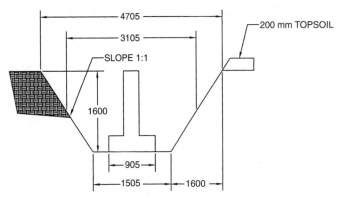

FIGURE 8.33 Sketch of Spread Footing

Since the spread footing is square, the general excavation can be found by squaring the average cut width and multiplying that by the depth.

$$\text{General excavation (m}^3) = 3.11 \times 3.11 \times 1.60$$
$$= 15 \text{ m}^3$$

8.10 Special Excavation

Usually, the special excavations are the portions of the work that require hand excavation, but any excavation that requires special equipment used for a particular portion other than general (mass) excavation may be included.

Portions of work most often included under this heading are footing holes, small trenches, and the trench-out below the general excavation for wall and column footings, if required. On a large project, a backhoe may be brought in to perform this work, but a certain amount of hand labour is required on almost every project.

The various types of excavation must be kept separately on the estimate, and if there is more than one type of special or general excavation involved, each should be considered separately and then grouped together under the headings "special excavation" or "general excavation."

In calculating the special excavation, the estimator must calculate the volume of excavation, select the method of excavation, and determine the cost.

8.11 Backfilling

Once the foundation of the building has been constructed, one of the next steps in construction is the backfilling required around the building. *Backfilling* is the placing back of excess soil that was removed from around the building during the general excavation. After the topsoil and general and special excavations have been estimated, it is customary to calculate the amount of backfill.

The material may be transported by wheelbarrows, scrapers, front-end loaders with scoops or buckets, bulldozers, and perhaps trucks, if the soil must be transported a long distance. The selection of equipment will depend on the type of soil, weather conditions, and distance the material must be moved. If tamping or compaction is required, special equipment will be needed, and the rate of work per hour will be considerably lower than if no tamping or compaction is required.

One method for calculating the amount of backfill to be moved is to determine the total volume of the building within the area of the excavation. This would be the total volume of the basement area, figured from the underside of fill material, and would include the volume of all footings, piers, and foundation walls. This volume is deducted from the volume of excavation that had been previously calculated. The volume of backfill required is the result of this subtraction. The figures should not include the data for topsoil, which should be calculated separately.

A second method for calculating backfill is to compute the actual volume of backfill required using the trench perimeter multiplied by the cross-sectional area of the trench.

The following examples illustrate how to calculate backfill quantities:

E X A M P L E **8.17** **Backfilling the Basement Walls**

Using the sketches in Figures 8.34, 8.35, and 8.36, the following volume calculations can be performed:

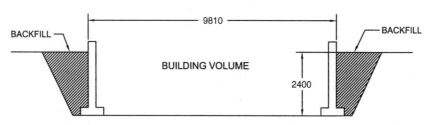

FIGURE 8.34 Backfill Section

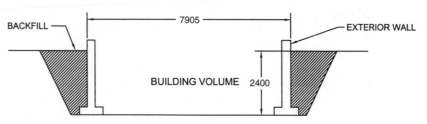

FIGURE 8.35 Backfill Section

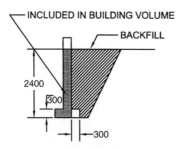

FIGURE 8.36 Footing Backfill

continued

$$\begin{aligned} \text{Building volume (m}^3\text{)} &= 9.81 \times 7.91 \times 2.40 \\ &= 186 \text{ m}^3 \end{aligned}$$

$$\begin{aligned} \text{Footing volume (m}^3\text{)} &= 0.30 \times 0.30 \times 7.91 \times 2 \\ &= 1.42 \text{ m}^3 \end{aligned}$$

$$\begin{aligned} \text{Footing volume (m}^3\text{)} &= 0.30 \times 0.30 \times 9.81 \times 2 \\ &= 1.77 \text{ m}^3 \end{aligned}$$

$$\begin{aligned} \text{Total footing volume (m}^3\text{)} &= 1.42 \text{ m}^3 + 1.77 \text{ m} \\ &= 3 \text{ m}^3 \end{aligned}$$

From Example 8.14, the basement excavation is 370 m^3

$$\begin{aligned} \text{Backfill (m}^3\text{)} &= 370 \text{ m}^3 \text{ (gen. excav.)} - 186 \text{ m}^3 \text{ (bldg. vol.)} \\ &\quad - 3 \text{ m}^3 \text{ (footing vol.)} \\ \text{Backfill} &= 181 \text{ m}^3 \end{aligned}$$

Equipment: Dozer (120 hp) at $100 per hour
Dozer work rate: 76 m^3 per hour for 100% efficiency. Assume 65% efficiency (49 m^3/hour)
Operator: $29.30 per work hour
Labourer: $18.50 per work hour
Move-in/move-out costs: $500

$$\text{Backfill time (hrs.)} = 181 \text{ m}^3 / 49 \text{ m}^3 \text{ per hour} = 3.69 \text{ hours}$$
$$\text{Operator cost (\$)} = 3.69 \text{ hours} \times \$29.30 \text{ per hour} = \$108.12$$
$$\text{Labourer cost (\$)} = 3.69 \text{ hours} \times \$18.50 \text{ per work hour} = \$68.27$$
$$\text{Equipment cost (\$)} = 3.69 \text{ Hours} \times \$100.00 \text{ per hour} = \$369$$
$$\text{Move in/move out} = \$500$$
$$\text{Total} = \$1,045.39$$

E X A M P L E 8.18   Backfilling the Foundation Walls

There are two ways in which the quantity of backfill can be determined. Both will yield virtually the same answer. The first is to subtract the area of the footing from the area backfill and multiply that by the length of the footing. Figure 8.37 is a sketch of the backfill requirements. The excavation for the footing was discussed in Example 8.15.

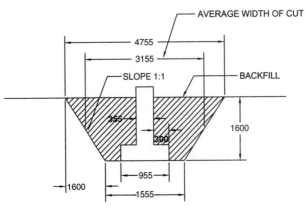

FIGURE 8.37 Continuous Footing Backfill

continued

$$\text{Volume of footing (m}^3) = 0.96 \times 0.30 \times 93.59$$
$$= 26.95 \text{ m}^3$$

$$\text{Volume in foundation wall (m}^3) = 1.30 \times 0.36 \times 93.59$$
$$= 43.80 \text{ m}^3$$

$$\text{Volume in continuous footing (m}^3) = 26.95 \text{ m}^3 + 43.80 \text{ m}^3$$
$$= 71 \text{ m}^3$$

$$\text{Volume of backfill} = 473 \text{ m}^3 - 71 \text{ m}^3$$
$$= 402 \text{ m}^3$$

Alternatively, figuring the area and multiplying that amount by the length can figure the area of backfill as shown in Figure 8.38.

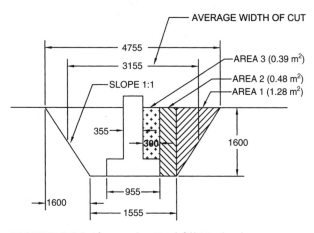

FIGURE 8.38 Alternative Backfill Method

$$\text{Fill area (m}^2) = (0.39 \text{ m}^2 + 0.48 \text{ m}^2 + 1.28 \text{ m}^2) \times 2 = 4.30 \text{ m}^2$$
$$4.30 \text{ m}^2 \times 93.59 \text{ m} = 402 \text{ m}^3$$

8.12 Excess and Borrow

Once the excavation has been calculated in terms of excavation, backfill, and grading, the estimator must compare the total amounts of cut and fill required and determine if there will be an *excess* of materials that must be discarded, or if there is a shortage of materials and some must be brought in *(borrow)*. Topsoil is not included in the comparison at this time; topsoil must be compared separately because it is much more expensive than the other fill that might be required, and excess topsoil is easily sold.

The specifications must be checked for what must be done with the excess material. Some specifications state that it may be placed in a particular location on the site, but many times, they direct the removal of excess materials from the site. If the material is to be hauled away, the first thing the estimator must know is how many cubic metres are required; then, the estimator must find a place to haul the material. Remember that soil swells; if the estimator calculated a haul of 100 m³ and the swell is estimated at 15 percent, then 115 m³ must be hauled away. Finding where the excess material can be hauled is not always a simple matter because it is desirable to keep the distance from the site as short as possible. If the haul distance is far and/or there is a restriction on the number of available dump trucks, the haul time may dictate the excavation equipment cost, rather than the actual amount of earth to be removed. If the backhoe or front-end loader has to wait for the truck to load, the equipment and operator costs are incurred for the wait time, even though no productive work is being performed. The estimator should check into this when visiting the site. If material must be brought in, the estimator must first calculate the amount required and then set out to find a supply of material as close to the job site as is practical. Check the specifications for any special requirements pertaining to the type of soil that may be used. Keep in mind that the material being imported is loose and will be compacted on the job. If it is calculated that 100 m³ are required, the contractor will have to haul in at least 106 to 114 m³ of soil—even more if it is clay or loam.

The next step is to select equipment for the work to be done which will depend on the amount of material, type of soil, and the distance it must be hauled.

8.13 Spreading Topsoil, Finish Grade

Many specifications call for topsoil to be placed over the rough grade to make it a *finish grade.* The topsoil that was stockpiled may be used for this purpose, but if there is no topsoil on the site, it will have to be purchased and hauled in—a rather costly proposition. The estimator must be certain to check the specification requirements and the soil at the site to see if the existing soil can be used as topsoil. On most projects, the equipment needed for finish grading consists of a bulldozer or scraper. The quantity of topsoil is usually calculated by multiplying the area used for rough grading by the depth of topsoil required.

After the topsoil has been spread throughout the site in certain areas, such as in the courtyards, around the buildings, or along the parking areas, it is necessary to hand rake the topsoil to a finish grade. The volume of soil to be moved must be calculated and entered on the estimate.

8.14 Landscaping

Most specifications require at least some landscaping work. If the work is seeding and fertilizing, the general contractor may do it or subcontract it out. The estimator should check the specifications for the type of fertilizer required. The specifications may also state the number of kilograms per square metre (kg/m²) that must be spread. The seed type will be included in the specifications, which will also state the number of kilograms to be spread per 10 square metres. Who is responsible for the growth of grass? If the contractor is, be certain that it will receive adequate water and that the soil will not erode before the grass begins to grow. Often, the seeded area is covered with straw or special cloth, which helps keeps in moisture and reduces erosion. The estimator will have to calculate the area to be fertilized and seeded, determine what equipment will be used, arrange covering and water, and then arrange the removal of the covering if it is straw. Sodding is often required when the owner wants an "instant lawn." The specifications will state the type of sod required, and the estimator must determine the area to be filled. The general contractor or a subcontractor may handle this phase of work. The estimator, in calculating labour costs, must consider how close to the actual area being sodded the truck bringing the sod can get, or whether the sod will have to be transferred from the truck to wheelbarrows or lift trucks and brought in.

A subcontractor who specializes in landscaping will generally handle trees and shrubs required for the project. The estimator should note the number, size, and species required. The landscape contractor may submit a proposal to perform all the landscape work; however, it is up to the estimator to make a preliminary decision about how it will be handled.

In the estimate, there must also be an allowance for the maintenance of the landscaping for whatever period is required in the specifications. Many specifications also require a warranty period during which the contractor must replace landscaping plants that fail to grow or if they die.

8.15 Pumping (Dewatering)

Almost all specifications state that the contractor provides all pumping and dewatering required. The estimator should examine any soil investigation reports made on the property for the possibility of a high water table. Also to be considered is the time of year that the soil investigation was made, as the water table varies throughout the year, depending on the particular area in which the project is located.

Water can present a problem to the contractor in almost any location. Even if the groundwater table presents no problem, the possibility exists of rainfall affecting the construction and requiring pumps for the removal of water.

Some projects require constant pumping to keep the excavation dry, whereas others require a small pump simply to remove excess groundwater. The variation in costs is extreme, and the estimator must rely on past experience in a given area as a guide. If the area is unfamiliar, then the problem should be discussed with people who are familiar with the locale. Who does the estimator ask? If the contacts in a given area are limited, the following may be good sources of information: the local building department, all retired construction managers, superintendents, surveyors (who are the most readily available source of information)—generally, these people will be most honest in their appraisal of a situation and will be delighted to be of help. Other sources might be local sales representatives, the consultant, and any local subcontractor, such as an excavator.

The problem of water will vary depending on the season of the year, location, type of work, general topography, and weather. The superintendent should be certain that during construction, the general slope of the land is away from the excavation so that in case of a sudden rainfall, it is not "washed out."

8.16 Rock Excavation

The excavation of rock differs from the excavation of ordinary soils. Rock is generally classified as soft, medium, or hard, depending on the difficulty of drilling it. Almost all rock excavation requires drilling holes into the rock and then blasting the rock into smaller pieces so that it may be handled on the jobsite. Types of drills include rotary and core jackhammers. The drill bits may be detachable, carbide, or have diamond-cutting edges, solid or hollow. Among the types of explosives used are blasting powder and dynamite. When a contractor has had no experience in blasting, a specialist is called to perform the required work. The estimator working for a contractor who has had blasting experience will make use of the cost information from past projects.

Factors that affect rock excavation include the type of rock encountered, the amount of rock to be excavated, whether bracing is required, the manner of loosening the rock, equipment required, length of haul, delays, and special safety requirements. Also, the estimator must decide whether it is to be an open cut or a tunnel in which water might be encountered. Once loosened, a cubic metre of rock will require as much as 50 percent more space than it occupied initially, depending on the type of rock encountered.

The type of rock, equipment, explosive, and depth of drilling will affect the cost of blasting. Only the most experienced personnel should be used and all precautions taken. Mats are often used to control the possibility of flying debris.

8.17 Subcontractors

Contractors who specialize in excavation are available in most areas. Specialized subcontractors have certain advantages over many general contractors. They own a large variety of equipment, are familiar with the soil encountered throughout a given area, and know where fill can be obtained and excess cut can be hauled. The subcontractor may bid the project as a lump sum or by the cubic metre. Either way, the estimator must still prepare a complete estimate—first, to check the subcontractor's price to be certain it is neither too high nor too low and, second, because the estimator will need the quantities to arrive at a bid price. All subcontractors' bids must be checked regardless.

The estimator should always discuss with subcontractors exactly what will be included in their proposals, and put this in writing so that both parties agree on what is being bid. It is customary for the general contractor to perform all hand excavation and sometimes the trenching. Selection of a subcontractor is most often based on cost, but also to be considered are the equipment owned by the subcontractor and a reputation for speed and dependability.

If the subcontractor does not meet the construction schedule, it will probably cost far more than the couple of hundred dollars that may have been saved on initial cost.

8.18 Excavation Checklist

Clearing site:
  removing trees and stumps
  clearing underbrush
  removing old materials from the
    project site
  removing fences and rails
  removing boulders
  demolishing old buildings

Foundations:
  underpinning existing buildings
  disconnecting existing utilities
  clearing shrubs

Excavation (including backfilling):
  basement

footings
foundation walls
sheet pilings
pumping
inspection chamber
catch basins
backfilling
tamping
blasting
grading (rough and fine)
utility trenches
grading and seeding lawns
trees
shrubs
topsoil removal
topsoil-imported

8.19 Piles

Piles used to support loads are referred to as *bearing piles.* They may be wood, steel "H" sections, poured-in-place concrete with metal casing removed, poured-in-place with metal casing left in place, wood and concrete, or precast concrete.

The wide variations in design, conditions of use, types of soil, and depth to be driven make accurate cost details difficult to determine, unless the details of each particular job are known. The various types and shapes available in piles also add to the difficulty, particularly in concrete piles. It is suggested that estimators approach subcontractors who specialize in this type of work until they gain the necessary experience. Figure 8.39 gives approximate requirements for labour when placing bearing piles.

Sheet piling may be wood, steel, or concrete. They are used when the excavation adjoins a property line or when the soil is not self-supporting. Sheet piling is taken off by the surface area to be braced or sheet-installed.

Wood sheet piling is purchased by the length or board feet, steel piling by weight (kg or tonne), and concrete piling by the piece. Depending on the type used, the quantity of materials required must also be determined.

The sheet piling may be placed in the excavation and braced to hold it erect or it may be driven into the soil. No one way is less expensive. Often, after the work is done, the piles are removed *(extracted).* Extraction may require the employment of a pile extractor for difficult jobs, or hand tools for the simple jobs. Figure 8.40 gives approximate requirements for labour when placing sheet piling.

Estimating. The first step is to determine exactly what type and shape of pile is required; next, the number of piles required; and third, decide the depth that the piles will have to be driven. With this information, the estimator can determine material quantities and their cost.

If the contractor's crew will drive the piles, the next determination should be the type of equipment required, how many hours the work will take, and cost per hour. Then, the crew size, along with the hours and cost per hour required, are calculated. To arrive

Type and Size of Pile	Approximate No. of Metres Driven per Hour
Wood	25–40
Steel	
10–30	15–30
30–60	10–20
Concrete	
30–60	20–30
60–120	3–12

FIGURE 8.39 Driving Bearing Piles

Type of Sheet Pile	Place, Drive By Hand	Place, Drive By Power Hammer	Brace	Pulling
Wood	13–30	11–26	7–10	10–16
Steel	10–30	10–25	7–10	10–16
Concrete	–	13–33	7–15	7–26

FIGURE 8.40 Sheet Piles, Approximate Work Hours Required Per Metre

at this, an estimate of the length (metres) placed per hour (including all cutting off, moving around, and so on) must be made so that the total hours required for the completion of the pile driving is determined. In dealing with sheet piling, if the sheets are to be extracted at the end of construction, the estimator must also include a figure for this. Costs must also comprise mobilization of workers, equipment, and material for pile driving, and demobilization when the work has been completed.

Check the soil borings for the kind of soil recorded and the approximate depth the piles must be driven. The accuracy of this information is quite important.

Specifications. Check the type of piles required and under what conditions sheet piling is required. Any requirements regarding soil conditions, equipment, or other special items should be noted. Many specifications require that only firms experienced in driving piles be allowed to do this type of work.

8.20 Pile Checklist

Type:
  wood
  wood and concrete
  H piles
  casing and poured in place
  poured in place
  precast
  wood sheets

  steel sheets
  precast sheets
  bracing

Equipment:
  pile driver
  compressors
  derricks
  cranes

8.21 Asphalt Paving

The asphalt paving required on the project is generally subcontracted out to someone specializing in paving. The general contractor's estimator will make an estimate to check the subcontractor's price.

Asphalt paving will most commonly be hot-mix and is generally classified by traffic (heavy, medium, or light) and use (walkways, driveways, courts, streets, driveways, and so on).

The estimator will be concerned with subgrade preparation, subdrains, soil sterilization, insulation course, sub-base course, base courses, prime coats, and the asphalt paving required. Not all items are required on any given project so the estimator should determine which items will be required, the material and equipment necessary for each portion of the work, and the requisite thickness and amount of compaction.

Specifications. Check the requirements for compaction, thickness of layers, total thickness, and materials required for each portion of the work. The drawings will also have to be checked for some of these items. The drawings will show the location of most of the work to be completed, but the specifications should also be checked. The specifications and drawings will list different requirements for the various uses. (These are called traffic requirements.)

Estimate. The number of square metres of surface area to be covered is determined, and the thickness (compacted) of each course and type of materials required are noted. Base courses and the asphalt paving are often measured by the tonne, as this is the unit in which these materials must be bought. The type of asphalt and aggregate size required must also be noted. Two layers of asphalt paving are required on some projects: a coarse base mix may be used with a fine topping mix. Equipment required may include a steel-wheel roller, trailers to transport equipment, dump trucks, paving machines, and various small tools.

To estimate the tonnes of material required per 100 m² of surface area, refer to Figure 8.41. Different requirements will be listed for the various uses (walkway, driveway, and so on), and the different spaces must be kept separately.

In many climates, the asphalt paving has a cutoff date in cold weather and the paving that is not placed when the mixing plants shut down will not be laid until the startup time in the spring. The plants may be shut down for as long as four months or more, depending on the region.

Compacted Thickness	Asphalt[1] Paving	Granular[2] Material	Subgrade[3] Material
25 mm	6.16	4.84	4.18
50 mm	12.32	9.68	8.36
75 mm	18.48	14.52	12.54
100 mm	24.64	19.36	16.72
125 mm	30.80	24.20	20.90
150 mm	36.96	29.04	25.08
200 mm	49.28	38.72	33.44
250 mm	61.60	48.40	41.80
300 mm	73.92	58.08	50.16

Per 100 m² of surface area, figures include 10 percent waste.
1. Asphalt paving, 2240 kg/m³
2. Granular material, 1760 kg/m³
3. Subgrade material, 1520 kg/m³

FIGURE 8.41 Approximate Asphalt Tonnage

E X A M P L E 8.19 Asphalt Takeoff

Determine the quantity of asphalt and subgrade material required for the parking lot in Figure 8.42. Figure 8.43 is a cross-section of the parking lot. Assume that the asphalt required is 75 mm thick and the subgrade is 150 mm thick.

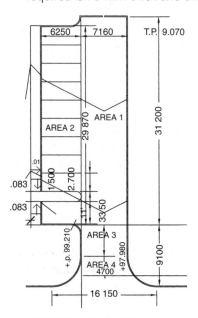

FIGURE 8.42 Parking Lot

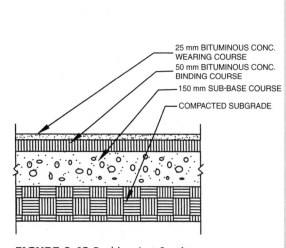

FIGURE 8.43 Parking Lot Section

continued

$$\text{Area 1} = 7.16 \times 31.20 = 223.39 \text{ m}^2$$
$$\text{Area 2} = 6.25 \times 29.87 = 186.69 \text{ m}^2$$
$$\text{Area 3} = 7.16 \times 4.70 = 33.65 \text{ m}^2$$
$$\text{Area 4} = [16.15 + 7.16) \times 4.40] / 2 = 51.28 \text{ m}^2$$
$$\text{Total area} = 495 \text{ m}^2$$

$$\text{Asphalt} = 18.48 \text{ tonnes per } 100 \text{ m}^2, 75 \text{ mm thick} \times 4.95 \text{ hundred m}^2$$
$$= 91.5 \text{ tonnes of asphalt}$$
$$\text{Sub-base material} = 29.04 \text{ tonnes per } 100 \text{ m}^2, 150 \text{ mm thick} \times 4.95 \text{ hundred m}^2$$
$$= 144 \text{ tonnes of granular material}$$

Due to the special equipment required, the installation is virtually always subcontracted.

Web Resources

www.worldsafety.com
www.unitedrentals.com
www.dirtpile.com
www.caterpillar.com
www.customer.deere.com
www.equipmentcentral.com

Review Questions

1. What type of information about the excavation can the estimator learn from the specifications?

2. How does the type of soil to be excavated affect the estimate?

3. What is the unit of measure in excavation?

4. How will the type of soil, shape of the excavation and amount of work to be done affect the equipment selection?

5. What is meant by cut and fill?

6. What does the estimator have to consider if there is a substantial amount of cut on the job? What if there is a substantial amount of fill?

7. How can the estimator get an estimate of the depth of topsoil on a specific project?

8. What type of excavation is considered to be general excavation?

9. How does general excavation differ from special excavation?

10. What is excess and borrow, and how are each considered in the estimate?

11. How will the possibility of a high water table or underground stream affect the bid?

12. What are piles, and under what conditions might they be required on the project?

13. Determine the quantities for the foundation excavation for the residential building shown in Appendix C (assuming that the existing grade is the same as the finished grade).

14. Determine the quantities for the foundation excavation for the commercial building shown in Appendix D (assuming that the building site grade was reduced to the underside of the granular base prior to any foundation excavations).

15. (Computer Exercise) Create an estimate for the foundation excavation works using Timberline Precision Estimating Basic by entering the data from Question 13. Change the prices to match prevailing market rates.

16. (Computer Exercise) Create an estimate for the foundation excavation works using Timberline Precision Estimating Basic by entering the data from Question 14. Change the prices to match prevailing market rates.

CHAPTER **9**

Concrete

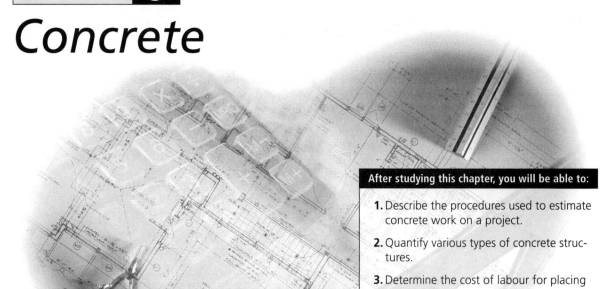

After studying this chapter, you will be able to:

1. Describe the procedures used to estimate concrete work on a project.

2. Quantify various types of concrete structures.

3. Determine the cost of labour for placing concrete.

4. Estimate the reinforcement requirements for concrete work.

5. Identify items classified as accessories in concrete work.

6. List the factors involved in formwork relative to costs.

9.1 Concrete Work

The concrete for a project may be either ready mixed or mixed on the project site. Most of the concrete used on commercial and residential work is ready mixed and delivered to the site by a ready-mix company. These ready-mix producers easily obtain quality control, proper gradation, water, and design mixes. When ready-mix is used, the estimator must determine the amount of concrete required and the type and amount of cement, aggregates, and admixtures. These are discussed with the supplier, who then gives a quotation for supplying the specific concrete.

Concrete for large projects or those in remote locations will typically be mixed at the project site and will require a field batch plant. To successfully estimate a project of this magnitude, the estimator must have a thorough understanding of the design of concrete mixes. The basic materials required for concrete are cement, aggregates (fine and coarse), and water. Various admixtures, which modify the properties of the concrete, may also be required. If field batching is desired, the estimator will have to compute the amounts of each material required, evaluate the local availability, and determine if these ingredients will be used or if materials will be shipped to the site from somewhere else. If the materials to be shipped are bulky and heavy, they are typically transported by rail. If the site is not adjacent to a rail line, provisions will need to be made to move the materials from the rail siding to the project batch plant. No attempt will be made here to show the design of concrete mixes. It is suggested that the estimator who is unfamiliar with mix design and control consult with a professional engineer.

9.2 Estimating Concrete

Concrete is estimated by the cubic metre (m³)/cubic yard (c.y.) The cubic metre is the pricing unit of the ready-mix companies, and most tables and charts available relate to the cubic metre.

Roof and floor slabs, slabs on grade, pavements, and sidewalks are most commonly measured and taken off in length, width, and thickness and converted to cubic metres. Often, irregularly shaped plans are broken down into smaller areas for more accurate and convenient manipulation.

When estimating footings, columns, beams, and girders, their volumes are determined by taking the length of each item in metres times its cross-sectional area. The volumes of the various items may then be tabulated.

With rectangular plan shapes, if an estimator calculates the wall length along the centreline, the same length can be used for the footing calculation (if the footings are all the same size). When estimating footings for buildings with irregular shapes and jogs, the estimator must be careful to include the corners only once. It is a good practice for the estimator to highlight on the plans which portions of the footings have been figured. When taking measurements, keep in mind that the footings extend out from the foundation wall, and therefore, this projection must be taken into account.

In completing quantities, the estimator makes no deductions for concrete displaced by other materials (reinforcing bars or other miscellaneous accessories) cast into the concrete nor for openings not exceeding 0.05 m³ in volume. Waste allowance ranges from 5 percent for footings, columns, and beams to 8 percent for slabs. Some estimators will reflect the waste allowance in the unit price of the concrete. For example, 30 m³ of concrete would be priced at $105 / m³ instead of 32 m³ of concrete at $100 / m³.

The procedure to be used to estimate the concrete on a project would be as follows:

1. Review the specifications to determine the requirements for each area in which concrete is used separately (such as footings, floor slabs, and walkways) and list the following:

   a. Type of concrete

   b. Strength of concrete

   c. Colour of concrete

   d. Any special curing or testing

2. Review the drawings to be certain that all concrete items shown on the drawings are covered in the specifications. If not, a call will have to be made to the consultant so that an addendum can be issued.

3. List each of the concrete items required on the project.

4. Determine the quantities required from the working drawings. Footing sizes are checked on the wall sections and foundation plans. Watch for different-sized footings under different walls.

Concrete slab information will most commonly be found on wall sections, floor plans, and structural details. Exterior walkways and driveways will most likely be identified on the site plan and in sections and details.

E X A M P L E 9.1 **Continuous Footings**

The objective of this example is to determine the quantity of concrete in the 955-mm-wide continuous footings. From the plan and section shown in Figures 9.1 and 9.2, it can be discerned that there are two different sizes of continuous footings. The continuous footings on the perimeter of the building are 955 mm wide, and the ones found on the interior of the building are 905 mm wide. The following steps should be taken:

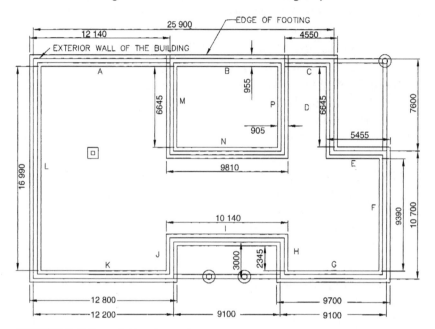

FIGURE 9.1 Foundation Layout

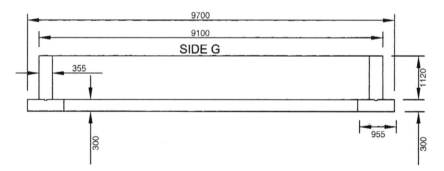

FIGURE 9.2 Footing and Foundation Wall Detail

1. Determine the length of footing for each width.
2. Determine the cross-sectional area for each of the differing sizes.
3. Determine the volume in cubic metres.

Determining the length of continuous footing typically requires some minor calculations. The dimensions (in millimetres) listed on the drawings typically reference the exterior face of the building or the centerline of the structural framing. Neither of these dimensions is appropriate for finding the length of footing. In addition, the overlap that would occur in the corners by taking the measurement along the exterior face of the footing needs to be compensated. Therefore, the best approach is to determine the dimensions along the exterior edge and dimension so that there is no overlap. Figure 9.1 shows the exterior dimensions and footing dimensions. Figure 9.2 is a sketch showing how

continued

the footing dimensions for side G were determined. Figure 9.3 is a tabulation of all dimensions for the 955-mm-wide footings.

$$\text{Cross-sectional area } (m^2) = \text{Width } (m) \times \text{Height } (m)$$
$$= 0.96 \times 0.30$$

Side	Length
A	12.14
B	9.81
C	4.55
D	6.65
E	5.46
F	9.39
G	9.70
H	2.35
I	10.14
J	2.35
K	12.80
L	16.99
TOTAL	102.33

FIGURE 9.3 Length of 955-mm-wide Continuous Footing

$$= 0.29 \ m^2$$
$$\text{Volume of concrete } (m^3) = \text{Cross-sectional area } (m^2) \times \text{Length } (m)$$
$$\text{Volume of concrete } (m^3) = 0.29 \ m^2 \times 102.33 \ m$$
$$= 29.68 \ m^3$$
Add 5% for waste and round off
$$29.68 \ m^3 \times 1.05 = 31 \ m^3$$

EXAMPLE 9.2 Spread Footing

Figure 9.4 details the one spread footing shown on the plan (Figure 9.1)

The concrete contained in this footing is found in virtually the same fashion as was the continuous footing.

$$\text{Volume of concrete } (m^3) = \text{Length } (m) \times \text{Width } (m) \times \text{Height } (m)$$
$$= 0.91 \times 0.91 \times 0.30$$
$$= 0.25 \ m^3$$
Add 5% for waste and round off
$$0.25 \ m^3 \times 1.05 = 0.26 \ m^3 - \text{Use } 1 \ m^3$$

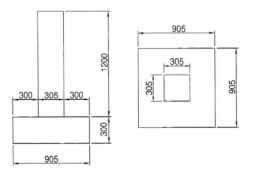

FIGURE 9.4 Spread Footing Detail

Caisson Piles. When dealing with caisson piles, some information about the soil is needed. If the soil is loose or the water table is high, it may be necessary to case the piers. The casing prevents the sides from caving in and water from seeping in. The casing, if steel, is typically removed as the concrete is placed; however, heavy-duty cardboard tubes can be used as casing. If this type of material is used, it is left in place and will eventually deteriorate. In clay soil, which is cohesive and impermeable, casings are rarely required.

To quantify a caisson pile, as shown in Figure 9.5, the shaft diameter, bell diameter, and angle of the bell must be known. The volume of the caisson pile can be calculated or a table similar to the one found in Figure 9.6 could be used to find the volume. If the volume is to be manually calculated, the bell diameter and angle can be used to find the height of the bell by using Formula 9.1 on the next page. Then, the volume of the bell is found using Formula 9.2, which is added to the volume of the band (150 mm thick) and volume of the shaft (Formula 9.3) to determine the total volume for a specific caisson pile.

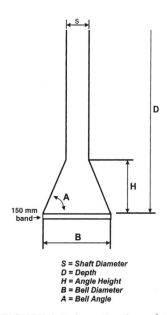

S = Shaft Diameter
D = Depth
H = Angle Height
B = Bell Diameter
A = Bell Angle

FIGURE 9.5 Cross-Section of Drilled Pier

VOLUME OF BELLS IN DRILLED PIERS											
BELL ANGLE		60									
SHAFT DIAMETER (mm)		BELL DIAMETER (mm)									
		450	600	750	900	1050	1200	1350	1500	1650	1800
300	h (metre)	0.13	0.26	0.39	0.52	0.65	0.78	0.91	1.04	1.17	1.30
	m³	0.01	0.05	0.09	0.18	0.30	0.47	0.68	0.95	1.29	1.70
450	h (metre)		0.13	0.26	0.39	0.52	0.65	0.78	0.91	1.04	1.17
	m³		0.03	0.08	0.15	0.27	0.42	0.62	0.88	1.19	1.58
600	h (metre)			0.13	0.26	0.39	0.52	0.65	0.78	0.91	1.04
	m³			0.05	0.12	0.22	0.37	0.56	0.80	1.10	1.47
750	h (metre)				0.13	0.26	0.39	0.52	0.65	0.78	0.91
	m³				0.07	0.17	0.31	0.49	0.72	1.00	1.36
900	h (metre)					0.13	0.26	0.39	0.52	0.65	0.78
	m³					0.10	0.23	0.40	0.62	0.90	1.24
1050	h (metre)						0.13	0.26	0.39	0.52	0.65
	m³						0.13	0.30	0.51	0.78	1.11

FIGURE 9.6 Volume in Bells 60-Degree Angle

E X A M P L E   9.3   Caisson Pile

Assume that the plan shown in Figure 9.1 calls for three identical caisson piles with the following dimensions:

Shaft diameter = 450 mm

Bell diameter = 1,050 mm

Angle of bell = 60 degrees

Depth = 4.80 m

Formula 9.1 Height of Bell on Reamed Footing

$$H = \tan (A) \times \left(\frac{B - S}{} \right)$$

$$H = \tan (60°) \times \left(\frac{1.05 - 0.45}{} \right)$$

$$= 1.73 \times 0.30$$

$$H = 0.52 \text{ m}$$

Once the bell height (H) is found, Formula 9.2 can be used to determine the volume of the belled portion of the footing.

Formula 9.2 Volume of Bell

$$\text{Volume of bell} = \left(\frac{\text{(Area of shaft)} + \text{(Area of band)}}{} \right) \times H$$

$$\text{Volume of bell} = \frac{\pi \left(\frac{S}{2} \right)^2 + \pi \left(\frac{B}{2} \right)^2}{2} \times H$$

$$= \frac{\pi (0.225)^2 + \pi (0.525)^2}{} \times 0.52$$

$$\text{Volume of bell} = 0.27 \text{ m}^3$$

The remaining element of the drilled piers is to determine the volume of the shaft, which is done by subtracting the height of the bell from the pier depth.

Length of shaft = 4.80 − 0.52

Length of shaft = 4.28 m

This length of shaft is then multiplied by its cross-sectional area to determine its volume.

continued

Formula 9.3 Shaft Volume

$$\text{Shaft volume} = \pi \left(\frac{S}{2}\right)^2 \times (\text{Length of Shaft})$$

$$\text{Shaft volume} = \pi\,(0.225)^2 \times 4.28$$
$$\text{Volume of shaft} = 0.68\ \text{m}^3$$

Formula 9.4 Band Volume

$$\text{Band volume} = \pi \left(\frac{B}{2}\right)^2 \times (\text{Depth of band})$$

$$\text{Band volume} = \pi\,(0.525)^2 \times 0.15$$
$$\text{Volume of band} = 0.13\ \text{m}^3$$

The volumes of the shaft, bell, and band can then be added together and multiplied times the number of caissons to determine the total volume.

$$\text{Volume of concrete in caissons} = \text{Number} \times (\text{volume in shaft} +$$
$$\text{volume in bell} + \text{volume in band})$$
$$\text{Volume of concrete in caissons} = 3 \times (0.68 + 0.27 + 0.13)$$
$$\text{Volume of concrete in caissons} = 3.24\ \text{m}^3$$
$$\text{Add 10\% for waste}$$

$$3.24\ \text{m}^3 \times 1.10 = 3.56\ \text{m}^3 - \text{Use 4 m}^3$$

The bell volume can be verified using the 60-degree table in Figure 9.6. From this table, the bell height is 0.52 m and the volume is 0.27 m³ per bell, which matches the previous calculations. Figure 9.7 is the workup sheet for the spread and continuous footings for the Foundation Layout in Figure 9.1.

E X A M P L E   9.4   Formed Piers

Spread footings typically have projecting formed piers that support the building structure. The foundation plan in Figure 9.1 has one formed pier. This pier, as shown in Figure 9.4, is 305 mm square and 1,200 mm high. The volume of the formed pier is found by multiplying the cross-sectional area by its height.

$$\text{Volume in formed piers} = \text{Number} \times \text{Cross-sectional area} \times \text{Height}$$
$$\text{Volume in formed piers} = 0.31 \times 0.31 \times 1.20$$
$$\text{Volume in formed piers} = 0.12\ \text{m}^3$$
$$\text{Use 1 m}^3$$

ESTIMATE WORK SHEET

Project:	Little Office Building							Estimate No.	1234	
Location	Mountainville, BC							Sheet No.	1 of 1	
Architect	C.K. Architects			**FOOTING CONCRETE**				Date	23/12/20xx	
Items	Foundation Concrete							By	DCK	**Checked** SRS

Cost Code	Description	Dimensions					Volume m³	Quantity	Unit
		Length m	Width m	Height m					
	955 mm Wide Footing (25 MPa)	102.25	0.96	0.30			29.47		
	Side A 12.14								
	Side B 9.81								
	Side C 4.55								
	Side D 6.65								
	Side E 5.46								
	Side F 9.39								
	Side G 9.70								
	Side H 2.35								
	Side I 10.14								
	Side J 2.35								
	Side K 12.80								
	Side L 16.99								
	102.33						29.47		
	Add 5% for Waste						1.47		
	Total Concrete in 955 mm wide Footings						30.94	**31**	m³
	905 mm Wide Footing (25 MPa)	23.44	0.91	0.30			6.40		
	Side M 6.65								
	Side N 10.14								
	Side P 6.65								
	23.44						6.40		
	Add 5% for Waste						0.32		
	Total Concrete in 905 mm wide Footings						6.72	**7**	m³
	Spread Footing								
	905 mm Square Column Footing	0.91	0.91	0.30			0.25		
	Add 5% for Waste						0.01		
	Total Concrete in 905 mm wide Footings						0.26	**1**	m³
	Use 1 m³								

FIGURE 9.7 Concrete Footing Takeoff

E X A M P L E   **9.5**   **Foundation Walls**

The concrete for the foundation walls is done in substantially the same manner as the spread footings. In the small example, the building perimeter foundation walls are 355 mm thick and the interior is 305 mm thick. Figures 9.8 and the table in Figure 9.9 detail the length of 355-mm-thick foundation walls. In addition, side B is 2.50 m high, as compared with 1.12 m high for the remaining 355-mm thick walls. This is why side B is not included in Figure 9.9.

continued

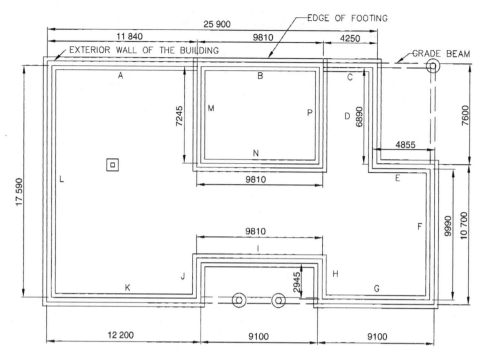

FIGURE 9.8 Foundation Wall Dimensions

Side	Foundation Wall Length (m)
A	11.84
C	4.25
D	6.89
E	4.86
F	9.99
G	9.10
H	2.95
I	9.81
J	2.95
K	12.20
L	17.59
TOTAL	92.43

FIGURE 9.9 Exterior Foundation Walls 1120 mm High

Foundation wall concrete (m³) = Length (m) × Height (m) × Thick (m)

1.12-m-High Wall

Quantity of foundation wall concrete (m³) = 92.43 × 1.12 × 0.36
$$= 37 \text{ m}^3$$

2.50-m-High Wall

Quantity of concrete in foundation wall (m³) = 9.81 × 2.50 × 0.36
$$= 9 \text{ m}^3$$

Quantity of concrete in exterior foundation walls = (37 + 9)
$$= 46 \text{ m}^3$$

+ 5% waste − Use 48 m³

E X A M P L E   9.6   Grade Beams

Grade beams are located in the front and right rear corners of the building (refer to Figures 9.10 and 9.11). These grade beams are required to tie the caisson piles to the remainder of the building foundation. The volume of concrete in the grade beam is found by multiplying the cross-sectional area of the grade beams by their length. From Figure 9.10, it can be discerned that the grade beams have different cross-sectional areas.

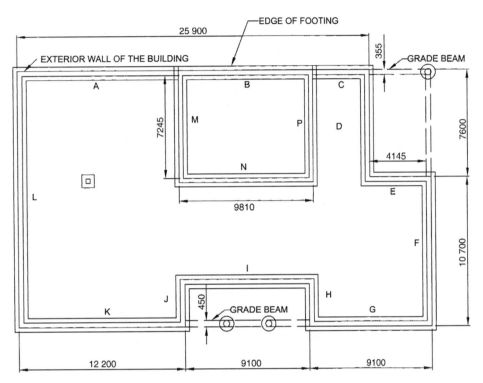

FIGURE 9.10 Grade Beam Location

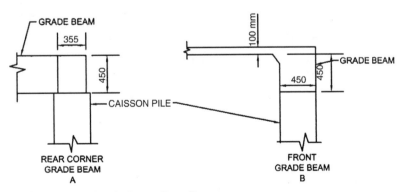

FIGURE 9.11 Grade Beam Details

11.75 m of 355-mm-wide grade beams (cross-sectional area = 0.16 m²)
9.10 m of 450-mm-wide grade beams (cross-sectional area = 0.20 m²)
Volume of concrete in 355 mm grade beams = 11.75 m × 0.16 m²
= 1.88 m³

continued

Volume of concrete in 450 mm grade beams = 9.10 m × 0.20 m²
= 1.82 m³
Total concrete in grade beams = 1.88 m³ + 1.82 m³ = 3.70 m³
Add 8% for waste − Use 4 m³

Figure 9.12 is a workup sheet for the concrete in the foundation walls and grade beams.

ESTIMATE WORK SHEET

Project:	Little Office Building						Estimate No.	1234	
Location	Mountainville, BC						Sheet No.	1 of 1	
Architect	C.K. Architects		Foundation Wall &				Date	23/12/20xx	
Items	Foundation Concrete		Grade Beams				By DCK	Checked	SRS

Cost Code	Description		Dimensions					Volume m³	Quantity	Unit
			Length m	Width m	Height m					
	1.12 m High Foundation Walls									
	355 mm Wide (25 MPa)		92.43	0.36	1.12			37.27		
	Side A	11.84								
	Side C	4.25								
	Side D	6.89								
	Side E	4.86								
	Side F	9.99								
	Side G	9.10								
	Side H	2.95								
	Side I	9.81								
	Side J	2.95								
	Side K	12.20								
	Side L	17.59								
		92.43								
	2.50 m High Foundation Walls									
	355 mm Wide (25 MPa)									
	Side B		9.81	0.36	2.50			8.83		
	2.50 m High Foundation Walls									
	305 mm Wide (25 MPa)		24.31	0.31	2.50			18.84		
	Side M	7.25								
	Side N	9.81								
	Side P	7.25								
		24.31								
	Total Foundation Walls							64.94		
	Add 5% for Waste							3.25		
								68.19	**68**	**m³**
	Grade Beams									
	355 mm x 450 mm Rear of Building		7.60	0.36	0.45			1.23		
	355 mm x 450 mm Rear of Building		4.15	0.36	0.45			0.67		
	450 mm x 450 mm Front of Building		9.10	0.45	0.45			1.84		
	Total Grade Beams							3.74		
	Add 8% for Waste							0.30		
								3.93	**4**	**m³**
	1 Only Formed Pier		0.31	0.31	1.20			0.12		
	Add 5% for Waste							0.01	**1**	**m³**

FIGURE 9.12 Foundation Wall and Grade Beam Takeoff

E X A M P L E   **9.7**   **Concrete Slabs**

The volume of a reinforced slab is found by taking the area and multiplying it by the depth of the slab. From the drawings in Figures 9.13, 9.14, and 9.15, there are four unique types of slabs. First, there is the 69 m² (9.51 m × 7.70 m less 4 m² for stair opening) of 50-mm-thick topping.

continued

Description	Dimensions	Square Metres
Whole Building Area	30.20 m × 18.10 m	547
Covered Patio	4.50 m × 7.60 m	-34
50 mm Topping		-73
100 mm Slab on Grab (Front)		-30
Total		410

FIGURE 9.13 Slab Areas

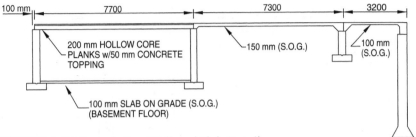

FIGURE 9.14 Foundation Wall and Slab Detail

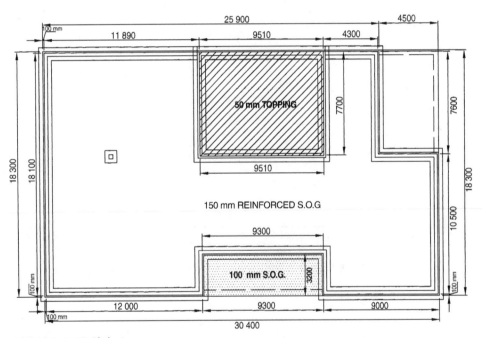

FIGURE 9.15 Slab Area

$$\text{Quantity of concrete (m}^3) = 69 \text{ m}^2 \times 0.05 \text{ m} = 3.45 \text{ m}^3$$
$$+ 5\% \text{ waste} - \text{Use 4 m}^3$$

The second slab is the 100-mm-thick slab that is in the front of the building. This slab is on grade and is 28 m² (9.10 m × 3.10 m) and 100 mm thick. Because the edges of this slab are thickened and it is on grade, a larger waste factor will be used to compensate for this situation.

$$\text{Quantity of concrete (m}^3) = 28 \text{ m}^2 \times 0.10 \text{ m} = 2.80 \text{ m}^3$$
$$+ 8\% \text{ waste} - \text{Use 3 m}^3$$

continued

Third is the slab on grade for the remaining portion of the building that is at grade. This area is found by determining the area for the whole slab and subtracting the 50 mm topping and 100 mm slab by the front entrance.

$$\text{Quantity of concrete (m}^3) = 410 \text{ m}^2 \times 0.15 \text{ m} = 61.50 \text{ m}^3$$
$$+8\% \text{ waste} - \text{Use 66 m}^3$$

The other remaining slab is the floor for the portion of the building that is below grade. That takeoff is similar to the three previous examples and can be found in Figure 9.16 with all the slab takeoffs.

ESTIMATE WORK SHEET

Project:	Little Office Building						Estimate No.	1234	
Location	Mountainville, BC						Sheet No.	1 of 1	
Architect	C.K. Architects		Foundation Wall & Grade Beams				Date	23/12/20xx	
Items	Foundation Concrete						By DCK	Checked SRS	

Cost Code	Description	Dimensions					Volume m³	Quantity	Unit
		Length m	Width m	Thickness m					
	All Concrete is 25MPa								
	50 mm thick topping over 25 MPa								
	pre-cast hollow core planks	9.51	7.70	0.05			3.66		
	Add 5% for waste						0.18		
							3.84	4	m³
	100 mm thick slab on grade								
	(front entrance)	9.10	3.10	0.10			2.82		
	Add 8% for waste & thickening						0.23		
							3.84	4	m³
	150 mm thick slab on grade								
	(whole area)	30.20	18.10	0.15			81.99		
	(Less Front Entrance)	-9.10	3.10	0.15			-4.23		
	(Less 50 mm topping)	-9.51	7.70	0.05			-3.66		
	(Less covered patio)	-4.50	7.60	0.05			-1.71		
	Total Volume						72.39		
	Add 8% for waste						5.79		
							68.19	68	m³
	100 mm thick slab on grade								
	(below finish floor)	9.15	7.29	0.10			6.67		
	Add 8% for waste						0.53		
							7.20	7	m³

FIGURE 9.16 Slab Takeoff

Labour Costs. The labour costs are found by multiplying the quantity takeoff by the appropriate productivity rate. Figure 9.17 shows various productivity rates for placing concrete in various situations.

Type of Placement	Productivity Rate (Work Hours/ m³)
Continuous Footings – Direct Chute	0.31
Spread Footings – Direct Chute	0.67
Drilled Piers	0.24
Formed Piers	0.67
Foundation Walls – Direct Chute	0.38
Grade Beams	0.31
Slab – Direct Chute	0.24

FIGURE 9.17 Labour Productivity Rates—Placing Concrete

EXAMPLE **9.8** **Labour Costs**

Find the labour cost for placing concrete in the 955-mm-wide continuous footings. From the concrete takeoff in Figure 9.7, there are 31 m³ of concrete. Using that quantity and the labour productivity information from Figure 9.17, the following calculations can be performed:

$$\text{Work hours} = \text{Quantity (m}^3) \times \text{Productivity rate (work hours / m}^3)$$
$$\text{Work hours} = 31 \text{ m}^3 \times 0.31 \text{ work hours / m}^3$$
$$\text{Work hours} = 10$$
$$\text{Basic labour costs} = \text{work hours} \times \text{Wage rate}$$
$$\text{Basic labour costs (\$)} = 10 \text{ work hours} \times \$21.50 \text{ / work hour}$$
$$\text{Basic labour cost} = \$215$$

Figure 9.18 is the priced-out estimate for all the concrete in the foundation of the small commercial building.

9.3 Reinforcing

The reinforcing used in concrete may be reinforcing bars, welded wire mesh (WWM), or a combination of the two. Reinforcing bars are listed (noted) by bar designation (bar size), which does not correspond to the bar diameter. For example, a 25M bar has a 25.2 mm nominal diameter. The nominal dimensions of a deformed bar are equivalent to those of a plain round bar having the same mass per metre as the deformed bar. The bar designation, diameters, areas, and mass are given in Figure 9.19.

Generally, reinforcing steel is estimated by the weight, obtained by listing all bars of different sizes and lengths and extending the total to kilograms. The takeoff (workup) sheet should be set up to include the number of the bars, pieces, lengths, and bends.

ESTIMATE SUMMARY SHEET

Project:	Little Office Building												Estimate No.	1234		
Location	Mountainville, BC												Sheet No.	1 of 1		
Architect	C.K. Architects												Date	23/12/20xx		
Items	Foundation Concrete												By	DCK	Checked SRS	

Cost Code	Description	Qty.	Waste Factor	Purch. Quan.	Unit	Crew	Prod. Rate	Wage Rate	Work Hours	Unit Cost Lab.	Unit Cost Mat.	Unit Cost Equip.	Lab.	Mat.	Equip.	Total
	Caisson Pile 25 MPa															
	450 mm diameter caisson	3.24	10%	4	m³		0.24	21.50	0.96	5.16	85.00		20.64	340.00		**360.64**
	Continuous Footings 25 MPa															
	905 mm Wide	6.41	5%	7	m³		0.31	21.50	2.17	6.67	85.00		46.66	595.00		**641.66**
	955 mm Wide	29.48	5%	31	m³		0.31	21.50	9.61	6.67	85.00		206.62	2,635.00		**2,841.62**
	Spread Footing 25 MPa															
	905 mm column footing	0.25	5%	1	m³		0.30	21.50	0.30	6.45	85.00		6.45	85.00		**91.45**
	Foundation Walls 25 MPa															
	Walls	64.94	5%	68	m³		0.38	21.50	25.84	8.17	85.00		555.56	5,780.00		**6,335.56**
	Grade Beams 25 MPa															
	Grade Beams	3.74	8%	4	m³		0.31	21.50	1.24	6.67	85.00		26.66	340.00		**366.66**
	Formed Pier 25 MPa															
	Piers	0.12	5%	1	m³		0.67	21.50	0.67	14.41	85.00		14.41	85.00		**99.41**
	Topping 25 MPa															
	50 mm thick topping	3.66	5%	4	m³		0.24	21.50	0.96	5.16	85.00		20.64	340.00		**360.64**
	Slab-on-grade 25 MPa															
	100 mm thick - front entrance	2.82	8%	4	m³		0.24	21.50	0.96	5.16	85.00		20.64	340.00		**360.64**
	100 mm thick - basement floor	6.67	8%	7	m³		0.24	21.50	1.68	5.16	85.00		36.12	595.00		**631.12**
	150 mm thick	72.39	8%	68	m³		0.24	21.50	16.32	5.16	85.00		350.88	5,780.00		**6,130.88**
	TOTAL								60.71				$1,305	$16,915		**$18,220**

FIGURE 9.18 Cast-in-Place Concrete Estimate Summary

Because reinforcing bars are usually priced by the tonne (1,000 kilograms), the weight of reinforcing required must be calculated. The steel can be bought from the mill or main warehouses, and the required bars will be cut, bundled, and tied. Bars will also be bent to job requirements at these central points. Bars purchased at smaller local warehouses are generally bought in 6 m lengths and cut and bent in the field. This process is usually more expensive and involves more waste. When time permits, the reinforcing bars should be ordered from the rebar suppliers and be shop fabricated. Often, the fabricators will provide the required shop drawings. Check the specifications to determine the type of steel required and whether it is plain or must be zinc coated or galvanized. Zinc coating and galvanizing can increase material cost by as much as 150 percent and often delay delivery by many weeks.

Allowance for splicing (lapping) the bars (Figure 9.20) must also be included (lap splicing costs may range from 5 to 15 percent, depending on the size of the bar and yield strength of steel used). Waste may range from less than 1 percent for precut and preformed bars to 10 percent when the bars are cut and bent on the job site.

Mesh reinforcing may be welded wire mesh or expanded metal. The former is an economical reinforcing for floor and driveways, and is commonly used as temperature reinforcing and beam and column wrapping. It is usually furnished in flat sheets or by the roll. The rolls are 1.83 m wide and 60.96 m long. Wire mesh is designated on

Bar Designation (Bar Size)	Mass kg/m	Nominal Dimensions	
		Diameter (mm)	Cross-Sectional Area (mm²)
10M	0.785	11.3	100
15M	1.570	16.0	200
20M	2.355	19.5	300
25M	3.925	25.2	500
30M	5.495	29.9	700
35M	7.850	35.7	1000
45M	11.775	43.7	1500
55M	19.625	56.4	2500

FIGURE 9.19 Areas and Mass of Reinforcing Bars

Bar Size	Splice required when specified as a number of bar diameters	
	24 d	30 d
10M	300 mm	351 mm
15M	450 mm	480 mm
20M	600 mm	585 mm
25M	750 mm	756 mm
30M	900 mm	897 mm
35M	1050 mm	1071 mm

FIGURE 9.20 Splice Requirements (Minimum splice is 305 mm)

the drawings by wire spacing and wire gauge in the following manner: 152 × 152 MW 9.1 × MW 9.1. This designation shows that the longitudinal and transverse wires are spaced 152 mm on centre, while both wires are MW9.1 × MW 9.1 gauge. Another example, 102 × 102 MW 13.3 × MW13.3, means that the longitudinal wires are spaced 102 mm on centre and the transverse wires are also spaced 102 mm on centre; the longitudinal wire is MW13.3 gauge, while the transverse is also MW13.3 gauge. The takeoff must be broken up into the various sizes required and the number of square metres required of each type. It is commonly specified that wire mesh have a lap of one square and an allowance of 10 percent extra for laps must be carried. The mesh may be either plain or galvanized; this information is included in the specifications. Galvanized mesh may require special ordering and delivery times of two to three weeks.

The 6.35 mm rib lath is designed primarily as reinforcement for concrete floor and roof construction. The 6.35 mm ribs are spaced 152 mm on centre, and this reinforcement is available only in coated copper alloy steel, 9.61 and 12.01 kilograms per square metre. The width available is 610 mm with lengths of 2,438, 3,048, and 3,658 millimetres, packed six sheets to a bundle (8.92, 11.15, and 13.38 square metres, respectively). Allow for any required lapping and 3 percent for waste (for rectangular spaces).

The reinforcing steel must be elevated into the concrete to some specified distance. This can be accomplished by using concrete blocks, bar chairs, spacers, or bolsters, or it may be suspended with wires. The supports may be plastic, galvanized or zinc-coated steel, steel with plastic-coated legs, and other materials. If the finished concrete will be exposed to view and the supports are touching the portion to be exposed, consideration should be given to using non-corrodible supports. Steel, even zinc-coated or galvanized steel, has a tendency to rust, and this rust may show through on the exposed finish. A wide selection of accessories and supports are available.

When the reinforcing must be fabricated into a special shape—perhaps round, spiral, or rectangular—it is usually cheaper to have this done at a fabricating shop, which has the equipment for bending these shapes. Thus, there would be a charge in addition to the base cost of the reinforcing, but the process provides speed and economy in most cases.

Corrugated steel subfloor systems are also used for reinforcing concrete. When corrugated steel floor deck material is used as reinforcing for the concrete, it also acts as a form for the concrete that is to be poured on top of it. The system may simply be corrugated deck with concrete or may be as elaborate as supplying in-floor distribution of electricity, hot air, and telephone requirements. The more elaborate the system, the more coordination between trades is required.

Steel deck subfloors are taken off by the area required (also discussed in Chapter 11). Available in a variety of heights and widths, the type used will depend on the span and loading requirements of the job. Finishes include galvanized, galvanized with primer on the underside, and phosphate treated on upper surfaces with primer on the underside.

Wire mesh is sometimes specified for use as temperature steel, in conjunction with the steel decking. The estimator must include it in the takeoff, when it is required.

Estimating Reinforcing Bars. The length of rebar can most often be worked up from the concrete calculations. The sections and details must be checked to determine the reinforcing requirements of the various footings. The various footing sizes can generally be taken from the concrete calculations and adapted to the reinforcing takeoff.

EXAMPLE **9.9** **Foundation Reinforcing**

For this example, determine the quantity of reinforcing steel required for the side A continuous footing of the building shown in Figure 9.3 and which was used to quantify the concrete. The dimensions for all the sides can also be found in Figure 9.3. From that table, side A is 12.14 m. This footing has reinforcing bars that run both the long and short dimensions. Figure 9.21 details the requirements for the rebar.

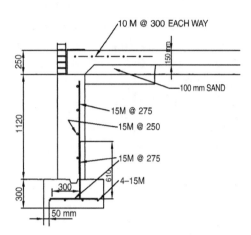

FIGURE 9.21 Footing Detail

Continuous Footing Short Bars:

Number of short bar spaces = 12.14 / 0.275 spacing = 45 spaces
(Round off to the nearest whole number)
Add 1 to get the number of bars – Use 46 bars
The bar length is 0.86 m, which is derived by subtracting the cover
from the footing width.
Total bar length (m) = Number of bars × Length of individual bars
Total bar length (m) = 46 bars × 0.86 m per bar = 39.56 m
15M bars weigh 1.570 kg/m
Total weight (kg) = 39.56 m × 1.570 kg/m = 62 kilograms

Continuous Footing Long Bars:

Length of long bars = 12.04 m
Total bar length (m) = 12.04 m × 4 bars
Total weight (kg) = 48.16 × 1.570 kg/m = 76 kilograms
Total footing rebar weight = 62 + 76 = 138 kilograms
Add 10% for waste and lap – Use 152 kilograms

The reinforcing for the foundation wall is found in virtually the same fashion. From Figure 9.8, side A is 11.84 m. Figure 9.21 specifies the vertical bar spacing at 275 mm on centre. Therefore, the quantity of vertical bars would be found in the following fashion:

continued

Foundation Wall Vertical Bars (Side A):

Number of vertical bar spaces = 11.84 / 0.275 = 44 spaces

Add 1 to get the number of bars − Use 45 bars

Bar length = Wall height − bar cover

Bar length = 1.12 − 0.10 = 1.02 m bar length

Total bar length (m) = Number of bars × Length of individual bars

Total bar length (m) = 45 × 1.02 = 45.90 m

Total weight (kg) = 45.90 × 1.570 kg/m = 76 kilograms

Add 10% for waste and lap − Use 84 kilograms

Number of horizontal spaces = 1.12 / 0.25 = 4.5 spaces − Use 5

Number of bars = Spaces + 1 − Use 6 bars

Bar length = 11.84 m − 0.10 = 11.74 m

Total bar length (m) = 11.74 × 6 = 70.44 m

Total weight (kg) = 70.44 × 1.570 kg/m = 111 kilograms

Add 10% for waste and lap − Use 122 kilograms

Dowels:

Length of each dowel = 910 mm

Number of dowels = 45

Length of dowels = 45 × 0.91 = 40.95 m

40.95 m × 1.570 kg/m = 64 kilograms

Add 5% for lap and waste − Use 67 kilograms

Figure 9.22 is a detailed takeoff of all reinforcing bars found in the foundation footings and walls.

ESTIMATE WORK SHEET
REINFORCING STEEL

Project:	Little Office Building													Estimate No.	1234
Location	Mountainville, BC													Sheet No.	1 of 2
Architect	C.K. Architects													Date	23/12/20xx
Items	Reinforcing Steel													By	Checked SRS

		Length/ Width	Bar Spacing		Bar Length	Cover-age	Bar Length		Bar Size				Bar Weights					
													10M	15M	20M	25M		
Cost Code	Description	mm	mm	Pcs.	m	mm	Ea	Total	10M	15M	20M	25M	0.785	1.570	2.355	3.925	Quantity	Unit
	Continuous Footing - Long Bars																	
	955 mm wide footing	0.96	400	4	102.33	50	102.23	408.9		X				642.00				
	905 mm wide footing	0.96	400	4	23.44	50	23.34	93.36		X				146.58				
	Continuous Footing - Short Bars																	
	955 mm wide footing	102.33	275	374	0.96	50	0.86	321.6		X				504.97				
	905 mm wide footing	23.44	275	87	0.96	50	0.86	74.82		X				117.47				
	Continuous Footing - Dowels	102.33	275	374	0.96	0	0.96	359		X				563.69				
	Total Continuous Footing																1975	kg
	Foundation Walls																	
	Short walls (Short Bars)	92.43	275	338	1.12	50	1.02	344.8		X				541.27				
	Short walls (Long Bars)	1.12	250	6	92.43	50	92.33	554		X				869.75				
	Tall walls (Short Bars)	34.12	275	126	2.50	50	2.40	302.40		X				474.77				
	Tall walls (Long Bars)	2.50	250	11	34.12	50	34.02	374.2		X				587.53				
	Total Foundation Walls																2473	kg
	Grade Beams																	
	Front (Long Bars)			4	9.10	0	9.10	36.40		X				57.15				
	Stirrups	9.10	300	32	1.40	0	1.40	44.80	X				35.17					
	Rear (Long Bars)			4	11.75	0	11.75	47.00		X				73.19				
	Stirrups	11.75	300	41	1.21	0	1.21	49.61	X				38.94					
	Total Grade Beams																205	kg
	Total Weight												74	4579			4653	kg

FIGURE 9.22 Reinforcing Steel Quantity Takeoff

continued

ESTIMATE WORK SHEET
REINFORCING STEEL

Project: Little Office Building   Estimate No. 1234
Location Mountainville, BC   Sheet No. 2 of 2
Architect C.K. Architects   Date 23/12/20xx
Items Reinforcing Steel   By Checked SRS

Cost Code	Description	Length /Width mm	Bar Spacing m	Pcs.	Bar Length mm	Cover m	Bar Length Ea	Bar Length Total	10M	15M	20M	25M	10M 0.785	15M 1.570	20M 2.355	25M 3.925	Quantity	Unit
	S.O.G.																	
	A Long Bars	11.89	300	41	7.70	50	7.60	311.60	X				244.61					
	A Short Bars	7.70	300	27	11.89	50	11.79	318.33	X				249.89					
	B Long Bars	12.00	300	41	10.40	50	10.30	422.30	X				331.51					
	B Short Bars	10.40	300	36	12.00	50	11.90	428.40	X				336.29					
	C Long Bars	9.30	300	32	7.20	50	7.10	227.20	X				178.35					
	C Short Bars	7.20	300	25	9.30	50	9.20	230.00	X				180.55					
	D Long Bars	7.60	300	27	4.30	50	4.20	113.40	X				89.02					
	D Short Bars	4.30	300	16	7.60	50	7.50	120.00	X				94.20					
	E Long Bars	10.50	300	36	9.00	50	8.90	320.40	X				251.51					
	E Short Bars	9.00	300	31	10.50	50	10.40	322.40	X				253.08					
	Total Slab-on- grade																2209	kg
	Caisson Piles - Vertical Bars			12	4.80	75	4.65	55.80		X				87.61				
	Caisson Piles - Dowels			12	1.09	0	1.09	13.08		X				20.54				
	Caisson Piles - Stirrups	14.40	300	49	1.10	0	1.10	53.90	X				42.31					
	Total Caisson Piles																150	kg
	Spread Footings (Long Bars)	0.91	300	5	0.91	50	0.81	4.05		X				6.36				
	Spread Footings (Short Bars)	0.91	300	5	0.91	50	0.81	4.05		X				6.36				
	Pier - Dowels				0.96	0	0.96	3.84		X				6.03				
	Pier - Vertical Bars			4	1.20	50	1.10	4.40		X				6.91				
	Pier - Stirrups	1.20	300	5	0.82	0	0.82	4.10	X				3.22					
	Total Spread Footings																29	kg
	Front Entrance - Long Bars	9.30	300	32	3.20	50	3.10	99.20	X				77.87					
	Front Entrance - Short Bars	3.20	300	12	9.30	50	9.20	110.40	X				86.66					
	Total Exterior Slab-on-grade																165	kg
	Total Weight												2419	134			2553	kg

FIGURE 9.22 Reinforcing Steel Quantity Takeoff

E X A M P L E **9.10** Slab Reinforcing

If the reinforcing in a slab is done with sized deformed bars, the bars are quantified in the exact manner as the footings and foundation walls. Once again, the quantities of long and short bars need to be determined. Figure 9.23 is an example of how the slab can be divided into unique areas so the quantity of reinforcing bars can be determined. Using area A as an example, the long bars will be (11.89 m − 0.10 m) = 11.79 m and the short bars would be (7.70 m − 0.10 m) = 7.60 m.

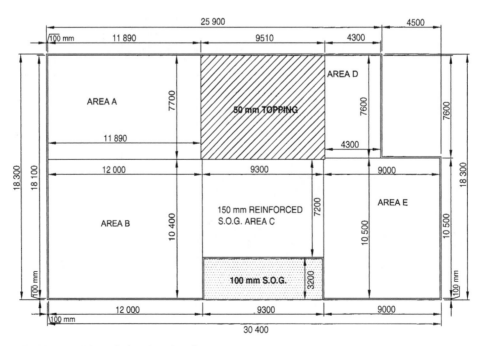

FIGURE 9.23 Reinforcing Steel Layout

continued

Long bar spaces = 7.70 / 0.30 − Use 26 spaces
Add 1 to convert to bar count − Use 27 bars
Short bar spaces = 11.89 / 0.30 = 40 spaces
Add 1 to convert to bar count − Use 41 bars
Total length of long bars (m) = 11.79 × 27 bars = 318.33 m
Total length of short bars (m) = 7.60 m × 41 bars = 311.60 m
Total length of bars (m) = 318.33 m + 311.60 m = 629.93 m
Total weight of bars (kilograms) = 629.93 m × 0.785 kg/m
Total weight = 495 kilograms
Add 10% for lap and waste − Use 544 kilograms

E X A M P L E **9.11** **Reinforcing Caisson Piles**

Estimating the reinforcing in the caisson piles consists of counting the number of vertical bars and determining their length. Since there are three caisson piles, results will be multiplied by three to determine the total.

From the caisson detail shown in Figure 9.24, there are four 15M vertical bars and 10M stirrup bars at 300 mm on centre. Stirrups are usually kept separate in the rebar estimate due to the higher cost of fabrication.

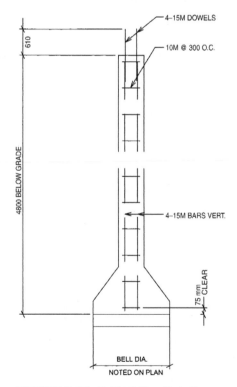

FIGURE 9.24 Drilled Pier Detail

<u>Vertical bars − 15M</u>
Length of vertical bars = Caisson length − Coverage
Vertical bars (m) = 4.65 × 4 bars = 18.60 m
Vertical dowels (m) = (610 mm vertical extension +

continued

480 mm into pier) × 4 dowels = 4.36 m
Total quantity (m) = (18.60 + 4.36) × 3 piles = 68.88 m
Total weight of bars (kg) = 68.88 × 1.57 kg/m
Total weight = 108.14 kg
Add 10% for lap and waste − Use 119 kilograms
<u>Stirrup bars − 10 M</u>
Pile diameter = 450 mm
Stirrup diameter = Pier diameter − Cover = 0.45 − 0.10 = 0.35
Stirrup length (m) = 2πr = 2 × π × 0.175 = 1.10 m per stirrup
Number of vertical spaces = 4.80 / 0.30
Add 1 to convert to number of stirrups − Use 17
Total length of stirrups (m) = 17 stirrups × 1.10 m per stirrup × 3 piers
Total length of stirrup (m) = 56.10 m
Total weight of bars (kg) = 56.10 × 0.785 kg/m
Total weight = 44.04 kg

E X A M P L E 9.12 Grade Beams

The grade beam in the rear of the building is 11.75 m long. The specifics of this grade beam are shown in Figure 9.25. There are four 15M horizontal bars, and 10M bars are used for stirrups at 300 mm on centre (o.c.).

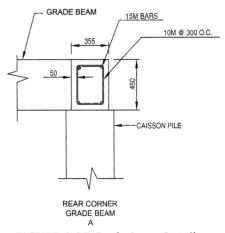

FIGURE 9.25 Grade Beam Details

Total quantity of long bars 15M (m) = 4 bars × 11.75 m per bar = 47 m
Total weight of bars (kg) = 47.00 × 1.570 kg/m
Total weight = 73.79 kg
Add 10% for lap and waste − Use 81 kilograms
Stirrup length = (0.255 + 0.350) m × 2
Stirrup length (m) = 1.21 m per stirrup
Stirrup spaces = 11.75 / 0.30 stirrup spacing = 40 spaces − Use 41 stirrups
Total quantity of stirrups 10M (m) = 41 stirrups × 1.21 m per stirrup
Total of 10M stirrups = 49.61 m
Total weight of bars (kg) = 49.61 m × 0.785 kg/m
Total weight = 38.94 kg
Add 10% for lap and waste − Use 43 kilograms

Estimating Wire Mesh. The area of floor to be covered may be taken from the slab concrete calculations. Check the sections and details for the size of the mesh required. To determine the number of rolls required, add the lap required to the area to be covered and divide by 111 m² (the coverage in a roll). Waste averages about 5 percent, unless much cutting is required; only full rolls may be purchased in most cases.

E X A M P L E   **9.13**   **Wire Mesh Reinforcing**

Using the plan shown on Figure 9.15 and details in Figure 9.14, the wire mesh is used for the basement floor and over the precast hollow core planks. Both these areas are roughly the same. From Figure 9.23, this area is 9.30 m × 7.70 m

$$\text{m}^2 \text{ of concrete requiring mesh} = 9.30 \times 7.70 \times 2 = 143.22 \text{ m}^2$$
$$\text{Rolls of mesh required} = (143.22 \text{ m}^2 \times 1.05 \text{ waste \%}) / 111 \text{ m}^2 \text{ per roll}$$
$$\text{Order 2 rolls of mesh}$$

Figure 9.22 (see page 139–140) is the workup sheet for all reinforcing found in the building foundation.

Labour. The following table (Figure 9.26) contains sample productivity rates for tying and placing the reinforcing steel. In addition, Figure 9.27 is the priced estimate for the reinforcing steel.

Type of Placement	Productivity Rate (Work Hrs./Tonne)
Beams 10M–25M	22
Columns 10M–25M	34
Footings10M–25M	15
Walls 10M–25M	11
Slab on Grade 10M–25M	13
Wire Mesh	2/Roll

FIGURE 9.26 Reinforcing Productivity Rates

ESTIMATE SUMMARY SHEET

Project: Little Office Building   Estimate No. 1234
Location Mountainville, BC   Sheet No. 1 of 1
Architect C.K. Architects   Date 23/12/20xx
Items Reinforcing Steel   By DCK Checked SRS

Cost Code	Description	QTY.	Waste Factor	Purch. Quan.	Unit	Crew	Prod. Rate	Wage Rate	Work Hours	Unit Cost						
										Lab.	Mat.	Equip.	Lab.	Mat.	Equip.	Total
	Continuous Footings	1,975	10%	2.17	tonne		15	22.50	32.59	337.50	950.00		733.22	2,063.88		2,797.09
	Foundation Walls	2,473	10%	2.72	tonne		11	22.50	29.92	247.50	950.00		673.27	2,584.29		3,257.56
	Grade Beams	205	10%	0.23	tonne		22	22.50	4.96	495.00	950.00		111.62	214.23		325.85
	Slab-on-grade	2,209	10%	2.43	tonne		13	22.50	31.59	292.50	950.00		710.75	2,308.41		3,019.15
	Caisson Piles	150	10%	0.17	tonne		24	22.50	3.96	540.00	950.00		89.10	156.75		245.85
	Spread Footing	29	10%	0.03	tonne		15	22.50	0.48	337.50	950.00		10.77	30.31		41.07
	Exterior Slab-on-grade	165	10%	0.18	tonne		13	22.50	2.36	292.50	950.00		53.09	172.43		225.51
	Wire Mesh	20	2%		roll		2	22.50	4.00	45.00	210.00		90.00	420.00		510.00
	TOTAL								109.86				$2,472	$7,950		$10,422

FIGURE 9.27 Reinforcing Steel Estimate

EXAMPLE **9.14** **Labour Costs for Continuous Footings**

Determine the labour cost for placing the reinforcing steel for the continuous footings (refer to Figure 9.22).

$$\text{Work hrs.} = (1{,}975 \text{ kg} / 1{,}000 \text{ kg} / \text{tonne}) \times 15 \text{ work hours/tonne}$$
$$= 29.6 \text{ work hours}$$
$$\$666.00 \text{ basic labour cost} = 29.6 \text{ work hours} \times \$22.50/\text{work hours}$$

Figure 9.27 is a complete estimate for the reinforcing found in the building in Appendix C.

9.4 Vapour Barrier (Retarder)

The vapour barrier or retarder is placed between the gravel and the slab poured on it and is usually included in the concrete portion of the takeoff. This vapour barrier most commonly consists of polyethylene films or kraft papers. The polyethylene films are designated by the required mil thickness (usually 4 or 6 mil). The material used should be lapped about 150 mm, so an allowance for this must be made depending on the widths available. Polyethylene rolls are generally available in widths of 2.44 and 3.05 m, and are 15, 30 and 45 m long. Careful planning can significantly cut down on waste, which should average 5 percent plus lapping. Two workers can place 100 m² of vapour barrier on the gravel in about one hour, including the time required to get the material from storage and place and secure it. Large areas can be covered in proportionately less time.

Vapour barriers are part of Division 7, Thermal and Moisture Protection (Chapter 13). They are included here since they are part of the concrete slab takeoff.

Estimating Vapour Barrier. To determine the number of rolls of vapour barrier required, calculate the area of the slab, allow for a 10 percent waste factor (cutting and overlapping) and divide by the area of the roll.

EXAMPLE **9.15** **Vapour Barrier (Retarder)**

Slab area (main Level) from Figure 9.13 = 410 m²
Slab area (basement) = 9.15 m × 7.29 m = 67 m²
(410 + 67) / 137.25 = 3.48
Add 10 % for waste — Use 4 rolls (3.05 m × 45 m)

Labour. The time required for the installation of the vapour barrier has been estimated from the data presented in Figure 9.28. For this example the wage rate is $15 per hour.

m² of Vapour Barrier	Work Hours per 100 m²
Up to 100	1.50
100 to 500	1.25
Over 500	1.00

FIGURE 9.28 Vapour Barrier Labour Productivity Rates

Work hours = 4.77 hundred m² (actual) × 1.25 work hr. per hundred m²
= 7.2 work hours
Basic labour = 6 work hours × $15 per work hour = $90

9.5 Accessories

Any item cast into the concrete should be included in the concrete takeoff. The list of items that might be included is extensive; the materials vary depending on the item and intended usage. The accessory items may include the following:

Expansion Joint Covers. Made of aluminum and bronze, expansion joint covers are available in a wide variety of shapes for various uses. These are usually specified in Division 10—Miscellaneous Specialties. Takeoff by the length (m) is required.

Expansion Joint Fillers. Materials commonly used as fillers are asphalt, fibre, sponge rubber, cork, and asphalt-impregnated fibre. These are available in thicknesses of 6, 9.5, 12.7, 19, and 25 mm; widths of 50 to 200 mm are most common. Sheets of filler are available and may be cut to the desired width on the job. Whenever possible, filler of the width to be used should be ordered to save labour costs and reduce waste. Lengths of filler strips may be up to 3 m, and the filler should be taken off by length (m) plus 5 percent waste.

Waterstops. Used to seal construction and expansion joints in poured concrete structures against leakage caused by hydrostatic pressure, waterstops are commonly composed of polyvinyl chloride, rubber, and neoprene in a variety of widths and shapes. The takeoff should be in length (m) and the estimator must check the roll size in which the specified waterstop is available and add 5 percent for waste. Full rolls only may be purchased (usually 15 m long).

Manhole Frames and Covers. Manhole frames and covers are available in round or square shapes. The frame is cast in the concrete and the cover put in the frame later. The materials used are aluminum (lighter duty) and cast iron (light and heavy duty). The covers may be recessed to receive tile; have a surface that is plain or abrasive; or have holes put in them, depending on the intended usage and desired appearance. The size, material, type, and installation appear in the specifications and on the drawings. They are taken off by the actual number of each type required. The various size frames and types of covers cost varying amounts, so different items are kept separate.

Trench Frames and Covers. The trench frame is cast in the concrete. The length of frames required must be determined and the number of inside and outside corners noted as well as the frame type. Cast-iron frames are available in 910 mm lengths, and aluminum frames are available up to 6 m long. Covers may also be cast iron or aluminum in a variety of finishes (perforated, abrasive, plain, recessed).

The takeoff is in length (m) with the widths, material, and finish all noted. Cast-iron covers are most commonly available in 610 mm lengths while aluminum frames are available in 3, 4, and 6 m lengths, depending on the finish required. Many cast-iron trench manufacturers will supply fractional sizes of covers and frames to fit whatever size trench length is required. Unless the aluminum frames and covers can be purchased in such a manner, sizable waste may be incurred on small jobs.

Miscellaneous accessories, such as anchor bolts, bar supports, screed chairs, screw anchors, screw anchor bolts, plugs, inserts of all types (to receive screws and bolts), anchors, and splices for reinforcing bars, must be included in the takeoff. These accessories are taken off by the number required; they may be priced individually or in 10s, 25s, 50s, or 100s, depending on the type and manufacturer. The estimator must carefully note the size required, since so many sizes are available. The material (usually steel, cast iron, or plastic) should be listed on the drawings or in the specifications. Some inserts are also available in bronze and stainless steel.

Such accessories as reglets, dovetail anchor slots, and slotted inserts are taken off by the length required. Available in various widths, thicknesses, and lengths, they may be made out of cold rolled steel, galvanized or zinc-coated steel, bronze, stainless steel, copper, or aluminum.

Estimate Expansion Joint Filler. Determine from the plans, details, and sections exactly the length required. A 12.5-mm-thick expansion joint filler is required where any concrete slab abuts a vertical surface. This would mean that filler would be required around the outside of the slab, between the slab and the wall.

9.6 Concrete Finishing

All exposed concrete surfaces require some type of finishing. Basically, finishing consists of the patchup work after the removal of forms and the dressing up of the surface by trowelling, sandblasting, and other methods.

Patchup work may include patching voids and stone pockets and removing fins and patching chips. Except for some floor slabs (on grade), there is always a certain amount of this type of work on exposed surfaces. It varies considerably from job to job and can be kept to a minimum with good quality concrete, with use of forms that are tight and in good repair and with careful workmanship, especially in stripping the forms. This may be included with the form stripping costs, or it may be a separate item. As a separate item, it is much easier to get cost figures and keep a cost control on the particular item rather than "bury" it with stripping costs. Small patches are usually made with a cement-sand grout mix of 1:2; be certain the type of cement (even the brand name) is the same used in the pour because different cements are varying shades of grey. The work hours required will depend on the type of surface, number of blemishes, and the quality of the patch job required. Scaffolding will be required for work above 1.80 metres.

The finishes required on the concrete surfaces will vary throughout the project. The finishes are included in the specifications and finish schedules; sections and details should also be checked. Finishes commonly required for floors include hand or machine trowelled carborundum rubbing (machine or hand), wood float, broom, floor hardeners, and sealers. Walls and ceilings may also be trowelled, but they often receive decorative surfaces, such as bush hammered, exposed aggregate, rubbed, sandblasted, ground, and lightly sanded. Finishes, such as trowelled, ground, sanded, wood float, broom finishes, and bush hammered, require no materials to get the desired finish, only labour and equipment. Exposed aggregate finishes may be of two types. In the first, a retarder is used on the form liner, and then the retarder is sprayed off and the surface cleaned. This finish requires the purchase of a retarding agent, spray equipment to coat the liner, and a hose with water and brushes to clean the surface. (These must be added to materials costs.) The second method is to spray or trowel an exposed aggregate finish on the concrete; it may be a two- or three-coat process, and both materials and equipment are required. For best results, it is recommended that only experienced workers place this finish. Subcontractors should price this application by the square metre.

Rubbed finishes, either with burlap and grout or with float, require both materials and work hours plus a few hand tools. The burlap and grout rubbed finish requires less material and more labour than the float finish. A mixer may be required to mix the grout.

Sandblasting requires equipment, labour, and the grit to sandblast the surface. It may be a light, medium, or heavy sandblasting job, with best results usually occurring with green to partially cured concrete.

Bush hammering, a surface finish technique, is done to expose portions of the aggregates and concrete. It may be done by hand, with a chisel and hammer, or with pneumatic hammers. The hammers are commonly used, but hand chiseling is not uncommon. Obviously hand chiselling will raise the cost of finishing considerably.

Other surface finishes may also be encountered. For each finish, analyze thoroughly the operations involved, material and equipment required, and work hours needed to do the work.

The finishing of concrete surfaces is estimated by the area (m²), except bases, curbs, and sills, which are estimated by length (m). Since various finishes will be required throughout, keep the takeoff for each one separately. The work hours required (approximate) for the various types of finishes are shown in Figure 9.29. Charts of this type should never take the place of experience and field observations and are included only as a guide.

Materials for most operations (except exposed aggregate or other coating) will cost only 10 to 20 percent of labour. The equipment required will depend on the type of finishing done. Trowels (hand and machine), floats, burlap, sandblasting equipment, sprayers, small mixers, scaffolding, and small hand tools must be included with the costs of their respective items of finishing.

Estimating Concrete Finishing. Areas to be finished may be taken from other concrete calculations, either for the actual concrete required or the area (m²) of forms required. Roof and floor slabs, and slabs on grade, pavements, and sidewalk areas can most easily be taken from actual concrete required. Be careful to separate each area requiring a different finish. Footing, column, walls, beam, and girder areas are most commonly found in the form calculations.

Finish	Method / Location	Work Hours / 10 m²
Trowelling	Machine – Slab on Grade	1
	Hand – Slab on Grade	2
Broom Finish	Slabs	.75
	Stairs	4
Float Finish	Slab on Grade	3.5
	Slabs Suspended	4.0
	Walls	3.5
	Stairs	3.0
	Curbs	8.5
Bush Hammer	Machine – Green Concrete	3
	Machine – Cured Concrete	6
	Hand – Green Concrete	9
Rubbed	With Burlap and Grout	2.5
	With Float Finish	4

FIGURE 9.29 Concrete Finish Labour Productivity

E X A M P L E   **9.16**   **Concrete Finishing**

Using the concrete takeoff from Example 9.7, the needed information for determining the quantity of concrete to be finished can be found. From that takeoff the following information can be gleaned:

Area of slab (m²) = 410 m² of 150 mm slab + 69 m² of 50 mm topping = 479 m²

Labour. The time required for finishing the concrete can be estimated from the information provided in Figure 9.29. From this table, the productivity rate for machine trowelling concrete is 1 work hour per 10 m². Using the locally prevailing wage rate of $21.20 per work hour for concrete finishers, the labour costs can be determined.

479 m² × 1 work hour per 10 m² = 48 work hours
$1,017.60 basic labour cost = 48 work hours × $21.20 per work hour

9.7 Curing

Proper *curing* is an important factor in obtaining good concrete. The concrete cures with the chemical combinations between the cement and water. This process is referred to as *hydration*. Hydration (chemical combination) requires time, favourable temperatures, and moisture; the period during which the concrete is subjected to favourable temperature and moisture conditions is referred to as the *curing period*. The specified curing period usually ranges from three to 14 days. Curing is the final step in the production of good concrete.

Moisture can be retained by the following methods: leaving the forms in place, sprinkling, ponding, spray mists, moisture retention covers, and seal coats. Sometimes, combinations of methods may be used and the forms left on the sides and bottom of the concrete while moisture retention covers or a seal coat is placed over the top. If the forms are left in place in dry weather, wetting of the forms may be required because they have a tendency to dry out; heat may be required in cold weather.

Sprinkling fine sprays of water on the concrete helps keep it moist, but use of caution is advised. If the surface dries out between sprinklings, it will have a tendency to crack.

Temperature must also be maintained on the concrete. With cold weather construction it may be necessary to begin with heated materials and provide a heated enclosure. Although the hydration of cement causes a certain amount of heat (referred to as the heat of hydration) and is of some help, additional heat is usually required. The forms may be heated by use of steam, and the enclosed space itself may be heated by steam pipes or unit heaters of the natural gas or LP gas type. The number of heaters required varies on the size.

During mild weather (5 to 10°C), simply heating the mixing water may be sufficient for the placement of the concrete, but if temperatures are expected to drop, it may be necessary to apply some external heat. Specifications usually call for a temperature of 15 to 20°C for the first three days or above 10°C for the first five days.

Hot weather concreting also has its considerations. High temperatures affect the strength and other qualities of the concrete. To keep the temperature down, some concrete plants will use ice in lieu of water when mixing the concrete. Wood forms should be watered to avoid the absorption of the moisture, needed for hydration, by the forms.

A continuous spray mist is one of the best methods of keeping an exposed surface moist. There is no tendency for the surface to crack with a continuous spray mist. Special equipment can be set up to maintain the spray mist so it will not require an excessive number of work hours.

Moisture-retaining covers, such as burlap, are also used. They should thoroughly cover the exposed surfaces and be kept moist continuously because they dry out (sometimes rapidly). Straw and canvas are also used as moisture retaining covers. Burlap and canvas may be reused on many jobs, thus spreading out their initial cost considerably.

Watertight covers may be used on horizontal areas, such as floors. Materials employed are usually papers and polyethylene films. Seams should be overlapped and taped, and the papers used should be nonstaining. Often, these materials may be reused several times with careful handling and storage.

Sealing compounds are usually applied immediately after the concrete surface has been finished. Both the one- and two-coat applications are colourless, black, or white. If other materials must be bonded to the concrete surface at a later date, they must bond to the type of compound used.

Estimating the cost of curing means that first a determination must be made as to what type of curing will be required and in what type of weather the concrete will be poured. Many large projects have concrete poured throughout the year, and thus, the estimators must consider the problems involved in cold, mild, and hot weather. Smaller projects have a tendency to fall into one or perhaps two of the seasons, but there is a general tendency to avoid the very coldest of weather.

If a temporary enclosure is required (perhaps of wood and polyethylene film), the size and shape of the enclosure must be determined, a material takeoff made, and the number of work hours determined. For simple enclosures, two workers can erect 10 m^2 in 30 to 60 minutes, once the materials are assembled in one place. If the wood may be reused, only a portion of the material cost is charged to this portion of the work. The enclosure must be erected and possibly moved during the construction phases and taken down afterwards—and each step costs money.

Heating of water and aggregate raises the cost of the concrete by 3 to 10 percent, depending on how much heating is needed and whether ready-mix or job-mix concrete is being used. Most ready-mix plants already have the heating facilities and equipment, with the cost spread over their entire production. Job mixing involves the purchase and installation of equipment on the job site, as well as its operation in terms of fuel, work hours, and upkeep.

The cost of portable heaters used to heat the space is usually based on the volume of the space to be heated. One worker (not an operator) can service the heaters for 75 m^3 in about one to two work hours, depending on the type of heater and fuel being used. Fuel and equipment costs are also based on the type used.

The continuous spray requires purchase of equipment (which is reusable) and the employment of labour to set it up. It will require between one-half and one work hour to run the hoses and set up the equipment for an area of 10 m^2. The equipment should be taken down and stored when it is not in use.

Moisture-retaining and watertight covers are estimated by the area of the surface to be covered and separated into slabs or walls and beams. The initial cost of the materials is estimated and is divided by the number of uses expected of them. Covers over slabs may be placed at the rate of 100 m^2 per one to three work hours, and as many as five uses of the material can be expected (except in the case of canvas, which lasts much longer); wall and beam covers may be placed at the rate of 100 m^2 per two to six hours. The sealing of watertight covers takes from three-quarters to one work hour per 30 m.

Sealing compounds are estimated by the area (m^2) to be covered divided by the coverage (in square metres) per litre to determine the number of litres required. Note whether one or two coats are required. If the two-coat application is to be used, be certain to allow for material to do both coats. For two-coat applications, the first coat coverage varies from 20 to 50 m^2, while the second coat coverage varies from 30 m^2 to 60 m^2. However, always check the manufacturer's recommendations. Equipment required will vary according to the size of the project. Small areas may demand the use of only a paint roller, while medium-size areas may require a pressure-type hand tank or backpack sprayer. Large, expansive areas can best be covered with special mobile equipment. Except for the roller, the equipment has much reuse, usually a life expectancy of five to eight years with reasonable care. The cost of the equipment is spread over the time of its estimated usage. Using the pressure-type sprayer, the estimator figures from one and a half to three hours will be required per 10 m^2. In estimating labour for mobile equipment, the estimator should depend more on the methods in which the equipment is mobilized than on anything else, but labour costs can usually be cut by 50 to 80 percent if the mobile equipment is used.

Specifications. The requirements of the consultant regarding curing of the concrete are spelled out in the specifications. Sometimes, they detail almost exactly what must be done for each situation, but often they simply state that the concrete must be kept at a certain temperature for a given number of days. The responsibility of protecting the concrete during the curing period is that of the general contractor (who may, in turn, delegate it to the concrete subcontractor).

9.8   Transporting Concrete

The methods used to transport the concrete from its point of delivery by the ready-mix trucks or the field-batching plant include truck chutes, buggies (with and without power), crane and bucket, crane and hoppers, tower and buggies, forklifts, conveyors, and pumps. Several combinations of methods are available, and no one system is the answer. It is likely that several methods will be used on any one project. Selection of the transporting method depends on the type of pour (floor, wall, curb, and so on), the total volume to be poured in each phase (in cubic metres), the distance above or below grade, equipment already owned, equipment available, distance from point of delivery to point of use, and possible methods of getting the concrete to that point.

Once a decision on the method of transporting the concrete has been made, the cost of equipment must be determined, as well as the anticipated amount of transporting that can be done using that equipment and a given crew of workers. From these, the cost of transporting the concrete may be estimated. Generally speaking, the higher the building and the further the point of use from the point of delivery, the higher is the cost of transporting.

9.9   Forms

This portion of the chapter is not a course in form design but identifies the factors involved in formwork relative to costs. No one design or system will work for all types of formwork. In general, the formwork must be true to grade and alignment, braced against displacement, resistant to all vertical and horizontal loads, resistant to leaking through tight joints, and of a surface finish that produces the desired texture. The pressure on the forms is the biggest consideration in the actual design of the forms.

In the design of wall and column forms, the two most important factors are the rate of placement of the concrete (m³ per hour) and the temperature of the concrete in the forms. From these two variables, the lateral pressure (MPa) may be determined. Floor slab forms are governed primarily by the actual live and dead loads that will be carried.

Actual construction field experience is a big factor in visualizing exactly what is required in forming and should help in the selection of the form type to be used. The types of forms, form liners, supports, and methods are many; preliminary selections must be made during the bidding period. This is one of the phases in which the proposed job superintendent should be included in the discussions of the methods and types of forms being considered, as well as in the consideration of what extra equipment and work power may be required.

Engineering data relative to forms and the design of forms are available from the American Concrete Institute (ACI), Portland Cement Association (PCA), and most manufacturers. These reference manuals should definitely be included in the estimator's reference library.

The forms for concrete footings, foundations, retaining walls, and floors are estimated by the area (in m²) of the concrete that comes in contact with the form. The plans should be studied carefully to determine if it is possible to reuse the form lumber on the building and the number of times it may be reused. It may be possible to use the entire form on a repetitive pour item, or the form may have to be taken apart and reworked into a new form.

On higher buildings (eight stories and more), the cost of forms may be reduced and the speed of construction increased if high early strength cements are used instead of ordinary Portland cement. The forms could then be stripped in two or three days instead of the usual seven to 10 days; however, not all consultants will permit this.

Many types of forms and form liners may be rented. The rental firms often provide engineering services as well as the forms themselves. Often, the cost of the forms required for the concrete work can be reduced substantially.

Wood Forms. Wood is one of the most common materials used to build forms. The advantages of wood are that it is readily accessible and easy to work with and that once used, it may be taken apart and reworked into other shapes. Once it has been decided to use wood, the estimator must determine the quantity of lumber required and the number of uses. This means the construction of the forms must be decided upon with regard to plywood sheathing, wales, studs or joists, bracing, and ties. The estimator can easily determine all of this if the height of the fresh concrete pour (for columns and walls), the temperature of the placed concrete, and the thickness of the slab (for floors) are known. The manufacturer's brochures or ACI formwork engineering data may be used.

Metal Forms. Prebuilt systems of metal forms are used extensively on poured concrete not only on large projects but also for foundation walls in homes. Advantages are that these systems are reusable several times, easily adaptable to the various required shapes, interchangeable, and require a minimum of hardware and a minimum of wales and ties, which are easily placed. They may be purchased or rented, and several time-saving methods are employed. Curved and battered weights are easily obtained, and while the plastic-coated plywood face liner is most commonly used, other liners are available. Heavy-duty forms are available for heavy construction jobs in which a high rate of placement is desired.

Engineering data and other information pertaining to the uses of steel forms should be obtained from the metal form supplier. The supplier can give information regarding costs (rental and purchase), tie spacing, number of forms required for the project, and labour requirements.

Miscellaneous Forms. Column forms are available in steel and laminated plies of fibre for round, square, and rectangular columns. Many manufacturers will design custom forms of steel, fibre, and fibreglass to meet project requirements. These would include tapered, fluted, triangular, and half-rounded shapes. Fibre tubes are available to form voids in cast-in-place (or precast) concrete; various sizes are available. Most of these forms are sold by the length (m) required of a given size. The fibre forms are not reusable, but the steel forms may be used repeatedly.

Estimate. The unit of measurement used for forms is the actual contact area (in m^2) of the concrete against the forms (with the exception of mouldings, cornices, sills, and copings, which are taken off by the metre). The forms required throughout the project must be listed and described separately. There should be no deductions in the area for openings of less than 10 m^2.

Materials in the estimate should include everything required for the construction of the forms, except stagings and bridging. Materials that should appear are struts, posts, bracing, bolts, wire, ties, form liners (unless they are special), and equipment for repairing, cleaning, oiling, and removing.

Items affecting the cost of concrete wall forms are the height of the wall (since the higher the wall, the more lumber that will be required per square metre of contact surface) and the shape of the building, including pilasters.

Items affecting the cost of concrete floor forms include the floor-to-floor height, reusability of the forms, length of time the forms must stay in place, type of shoring and supports used, and the number of drop beams required.

The various possibilities of renting or purchasing forms, using gang forms built on ground and lifted into place, slip forming, and so on should be considered during this phase.

Although approximate quantities of materials are given (Figure 9.30) for work forms, a complete takeoff of materials should be made. The information contained in Figure 9.30 is approximate and should be used only as a check. In addition, Figure 9.31 is a sample of the labour productivity to build different configurations of formwork.

Estimating Wood Footing Forms. The estimator must first determine if the entire footing will be poured at one time or if it will be poured in segments, which would permit the reuse of forms.

Type of Form	Lumber BF	Work Hours			
		Assemble	Erect	Strip & Clean	Repair
Footings	200-400	2-6	2-4	2-5	1-4
Walls	200-300	6-12	3-6	1-3	2-4
Floors	170-300	2-12	2-5	1-3	2-5
Columns	170-350	3-7	2-6	2-4	2-4
Beams	250-700	3-8	3-5	2-4	2-4
Stairs	300-800	8-14	3-8	2-4	3-6
Mouldings	170-700	4-14	3-8	2-6	3-6
Sill, Coping	150-600	3-12	2-6	2-4	2-6

FIGURE 9.30 Wood Forms, Approximate Quantity of Materials, and Work Hours

Application	Productivity Rate Work Hrs / SMCA
Foundation Wall (Plywood)	
1 Use	4.74
2 Use	3.01
3 Use	2.69
4 Use	2.37
Footings	
1 Use	1.51
2 Use	0.97
3 Use	0.86
4 Use	0.75
Slab On Grade	
1 Use	12.92
2 Use	11.84
3 Use	10.76
4 Use	8.61
Columns	
1 Use	1.94
2 Use	1.57
3 Use	1.54
4 Use	1.51
Beams	
1 Use	2.15
2 Use	1.94
3 Use	1.61
4 Use	1.51

FIGURE 9.31 Approximate Quantity of Work Hours

EXAMPLE **9.17** Foundation Wall Forms

Determine the labour costs associated with placing the 1.12-m-high foundation wall form shown in Figures 9.1 and 9.2. From Example 9.5 and Figure 9.9, the quantity of formwork can easily be found. From that example, there are 92.43 m of 1.12-m-high foundation wall. The contact area is found by multiplying these two dimensions and then doubling that quantity to compensate for both sides being formed.

Contact area (SMCA) = Length (m) × width (m) × sides formed
Contact area (SMCA) = 92.43 × 1.12 × 2
Contact area (SMCA) = 207 m²

The time required for forms has been estimated from Figure 9.30. Assuming that the forms will be used twice, the following calculation can be performed.

Work hours = Quantity × productivity rate
Work hours = 207 SMCA × 3.01 work hours per SMCA = 623 work hours
Basic labour cost = Work hours × wage rate
623 work hours × $21.20 per work hour = $13,207.60

9.10 Form Liners

The type of liner used with the form will determine the texture or pattern obtained on the surface of the concrete. Depending on the specified finish, formed concrete surfaces requiring little or no additional treatment can be easily obtained. Using various materials as liners may produce a variety of patterns and textures. Fibreglass liners, plastic-coated plywood, and steel are among the most commonly used. Different textures, such as wood grain, rough sawn wood, corrugations (of various sizes and shapes), and all types of specialized designs are available. They have liners that will leave a finish resembling sandblast, acid etch, and bush hammered, as well as others. These textures may be used on floors or walls.

Liners are also used to form waffle slabs and tee beam floor systems; they may be fibreglass, steel, or fibre core. When liners of this type are used, the amount of void must be known so that the quantity of concrete may be determined. This information is given in Figure 9.32. Complete information should be obtained from whichever company supplies the forms for a particular project.

Form liners are often available on a rental or purchase arrangement; specially made form liners of a particular design may have to be purchased. When special designs are required, be certain to get a firm proposal from a manufacturer.

Specifications. Check the specifications for requirements concerning textures and patterns, liner materials, thicknesses, configurations, and any other liners. It is not uncommon for the specifications to state the type of material from which the liner must be made. The drawings should be checked for types of texture, patterns, or other requirements of the form liner. The form liners required may be already in stock at the manufacturer or may need to be a special order, often requiring months of work, including shop drawing approvals.

Estimating.   Estimators take off the area (m^2) of the surface requiring a particular type of liner and decide how many liners can be used effectively on the job—this will be the total area of liner or number of pieces required. Being able to use them several times is what reduces the cost. Dividing the total cost of the liners by the area of surface provides a cost per square metre for liners. The same approach is used for rental of liners. Different textures and patterns may be required throughout, so details must be checked carefully.

Waffle Slab Form Liners	
Void Size and Depth	**Concrete Voided m^2**
483 mm × 483 mm × 100 mm	0.02
150 mm	0.03
200 mm	0.04
250 mm	0.05
300 mm	0.06
762 mm × 762 mm × 200 mm	0.11
250 mm	0.14
300 mm	0.16
355 mm	0.19
406 mm	0.21
508 mm	0.26

FIGURE 9.32 Void Area in Concrete

9.11 Checklist

Forms for:
  Footings, walls, and
    columns
  Floors
  Piers
  Beams
  Columns
  Girders/Stairs
  Platforms
  Ramps/Miscellaneous

Forms:
  Erection
  Removal
  Repair
  Ties
  Clamps
  Braces
  Cleaning
  Oiling
  Repairs
  Liners

Concrete:
  Footings, walls, and
    columns
  Floors
  Toppings
  Piers

Beams
Columns
Girders
Stairs
Platforms
Ramps
Curbing
Coping
Walkways
Driveways
Architectural
Slabs

Materials, Mixes:
  Cement
  Aggregates
  Water
  Colour
  Air entraining
  Other admixtures
  Strength requirements
  Ready-mix
  Heated concrete
  Cooled concrete

Finishes:
  Hand trowel
  Machine trowel

Bush hammer
Wood float
Cork float
Broom
Sand
Rubbed
Grouted
Removing fins

Curing:
  Admixtures
  Ponding
  Spraying
  Straw
  Canvas
  Vapour barrier
  Heat

Reinforcing:
  Bars
  Wire mesh
  Steel grade
  Bends
  Hooks
  Ties
  Stirrups
  Chairs
  Cutting

9.12 Precast Concrete

The term *precast concrete* is applied to individual concrete members that are cast in separate forms and then placed in the structure. The general contractor or supplier may cast them in a manufacturing plant or on the jobsite. The most common types of structural precast concrete are double and single tees, floor planks, columns, beams, and wall panels (Figure 9.33).

Depending on the requirements of the project, precast concrete is available reinforced or prestressed. Reinforced concrete utilizes reinforcing bars encased in concrete; it is limited in its spans, with 12 m being about the maximum for a roof. When longer spans are required, prestressed concrete is used, either pretensioned or post-tensioned. Prestressing generally utilizes high-strength steel or wire or wire strands as the reinforcing. In pretensioning, the longitudinal reinforcing is put in tension before the concrete is cast. The reinforcing is stretched between anchors and held in this state as the concrete is poured around the steel and cured. With post-tensioning, the longitudinal reinforcement is not bonded with the concrete. The reinforcement may be greased or wrapped to avoid bonding with the concrete, or conduits of some type (tubes, hoses) may be cast in the concrete and the reinforcement added later. The prestressing steel is

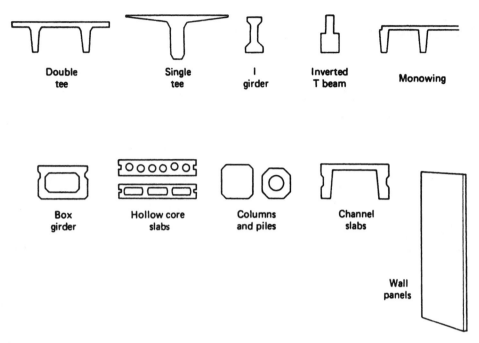

FIGURE 9.33 Precast Shapes

then stressed by jacking against the concrete. In this way the concrete is compressed as the steel is stretched.

Most precast concrete is priced by the area (m²), length (m) or in a lump sum. It is important to determine exactly what is included in the price. Most suppliers of precast concrete items price them delivered to the jobsite and installed, especially if they are structural items. When the specifications permit, some contractors will precast and install the pieces required themselves. Unless experienced personnel are available, doing your own precasting may cost more than subcontracting the work to others.

The aggregates used for precast concrete may be heavyweight or lightweight; however, it should be noted that some types of lightweight concrete are not recommended by some consulting engineers for use on post-tensioning work, and care should be taken regarding all materials used in the concrete.

Hole cutting in tees is subject to the distance between the structural tee portions of the members. The lengths of holes are not as rigidly controlled but should be approved by the structural engineer.

9.13 Specifications

The specifications must be checked to determine if a particular manufacturer is specified. All manufacturers who can supply materials and meet the specifications should be encouraged to bid the project. Often, there is only limited competition in bidding on precast concrete items. The estimator must also determine the strength, reinforcing, and inserts as well as any special accessories required.

If the project entails any special finish, such as sandblasting, filling air holes, coloured concrete, special aggregates, or sand finish, this should be noted on the workup sheets. Be certain that there is a clear understanding of exactly what the manufacturer is proposing to furnish. For example, it should be understood who would supply any required anchor bolts, welding, cutting of required holes, and filling of joints. Always check to see if the manufacturer (or subcontractor) will install the precast. If not, the estimator will have to calculate the cost of the installation.

9.14 Estimating

Floor, wall, and ceiling precast concrete are most commonly taken off in square metres (m^2) with the thickness noted. Also, be certain to note special requirements, such as insulation cast in the concrete, anchorage details, and installation problems.

Beams and columns are taken off in metres (m) with each required size kept separately. Anchorage devices, inserts, and any other special requirements must be noted.

Determine exactly what the suppliers are proposing. If they are not including the cutting of all holes, finishing of concrete, welding, or caulking, then all of these items will have to be figured by the estimator.

9.15 Precast Tees

Precast tees are available as simply reinforced or prestressed. The shapes available are double and single tee. For the double tee, the most common widths available are 1.20 metres and 1.80 metres, with depths of 200 to 650 mm. Spans range up to 23 m depending on the type of reinforcing and size of the unit. Single-tee widths vary from 1.80 m to 3.60 m with lengths up to about 30 m. The ends of the tees must be filled with some type of filler. Fillers may be concrete, glass, plastic, and so on. Refer to Figure 9.34.

Most manufacturers who bid this item, bid it delivered and installed either on a square-metre or lump-sum basis. One accessory item often overlooked is the special filler block required to seal the end of the tees.

Concrete fill is often specified as a topping for tees; it may be used for floors or sloped on a roof to direct rainwater. The fill is usually a minimum of 50 mm thick and should have at least light reinforcing mesh placed in it. Due to the camber in tees, it is sometimes difficult to pour a uniform 50-mm-thick topping and end up with a level floor.

Specifications. The estimator determines which manufacturers are specified, the strength of concrete, type and size of reinforcing, type and size of aggregate, the finish required, and whether topping is specified. For reinforcing bars they must determine the type of chairs to be used to hold the bars in place. The bars should be corrosion resistant. The estimator must determine who will cut the holes and caulk the joints, as well as what type of caulking will be used.

Estimating. The estimator must take off the area (m^2) required. If the supplier made a lump-sum bid, check the square-metre price. The area required should have all openings deducted. If the project is bid by the square metre, call the supplier to check your area against their takeoff. This provides a check for your figures. (However, it is only a check. If they don't agree, recheck the drawings. Never use anyone else's quantities when working up an estimate.)

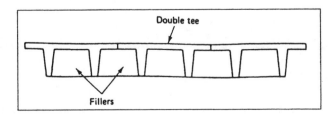

FIGURE 9.34 Double Tees Detail

When an interior finish is required, the shape of the tee is a factor: A tee 1.20 m wide and 12 m long does not have a bottom surface of 14.40 m² to be finished. The exact area involved varies, depending on the depth and design of the unit; it should be carefully checked.

The installation of mechanical and plumbing items sometimes takes some special planning with precast tees with regard to where the conduit, heating, and plumbing pipes will be located and how the fixtures will be attached to the concrete. All of these items must be checked.

The concrete fill *(topping)* is placed by the general contractor in most cases. The type of aggregate must be determined and the volume (m³) required must be taken off. The area (m²) of reinforcing mesh and the area of concrete to be finished must be determined. From the specifications, estimators determine the aggregate, strength, type of reinforcing, surface finish, and any special requirements. This operation will probably be done after the rest of the concrete work is completed. The decision must be made as to whether ready-mixed or field-mixed concrete will be used, how it will be moved to the floor, and the particular spot on which it is to be placed.

9.16 Precast Slabs

Precast slabs are available in hollow, cored, and solid varieties for use on floors, walls, and roofs. For short spans, various types of panel and channel slabs with reinforcing bars are available in both concrete and gypsum. Longer spans and heavy loads most commonly involve cored units with prestressed wire.

The solid panel and channel slabs are available in heavyweight and lightweight aggregates. The thicknesses and widths available vary considerably, but the maximum span is generally limited to about 3 m. Some slabs are available tongue-and-grooved and some with metal-edged tongue-and-groove. These types of slabs use reinforcing bars or reinforcing mesh for added tension strengths. These lightweight, easy-to-handle nail, drill, and saw pieces are easily installed on the job over the supporting members. A clip or other special fastener should be used in placing the slabs.

Cored units with prestressed wire are used on roof spans up to about 13 m. Thicknesses available range from 100 to 400 mm with various widths available, 1,016 mm and 1,219 mm being the most common. Each manufacturer must be contacted to determine the structural limitations of each product. The units generally have high fire resistance ratings and are available with an acoustical finish. Some types are available with exposed aggregate finishes for walls.

Specifications. The type of material used and the manufacturer specified are the first items to be checked. The materials used to manufacture the plank, type and size of reinforcing, and required fire rating and finish must be checked.

The estimator should also note who cuts required holes in the planks and who caulks the joints and the type of caulking to be used. If topping is required, the thickness, reinforcing, aggregates, and strength specified must also be noted. Any inserts, anchors, or special requirements must be noted as well.

Estimating. The precast slabs are generally quoted by the square metre or in a lump sum. Solid panels and channel slabs can often be purchased from the manufacturer and installed by the general contractor. The manufacturers of cored units generally furnish and install the planks themselves. When calculating the area, deduct all openings. Also, determine who will cut the holes, do any welding and special finishing, and provide anchors and inserts. Concrete topping is commonly used over the cored units, the most commonly specified thickness being 50 mm. Determine the volume of concrete, area of reinforcing mesh, and square metres of surface to be finished.

Check the drawings and specifications to determine how the planks will be held in place, special anchorage details, inserts, and any other items that may present a cost or problem on the job. Mechanical and electrical requirements should also be checked.

9.17 Precast Beams and Columns

Precast beams and columns are available in square, rectangular, T- and I-shaped sections. They are available simply reinforced, with reinforcing bars, or prestressed with high-strength strand. The sizes and spans depend on the engineering requirements of each project, and the beams and columns are not poured in any particular size or shape. Special forms can easily be made out of wood to form the size and shape required for a particular project.

Specifications. The manufacturers' specified strengths and materials, reinforcing, connection devices, anchors, inserts, and finishes required should all be noted. The different shapes needed throughout and any other special requirements should also be specified.

Estimating. If the contractor intends to precast the concrete, the costs involved are indicated in Section 9.19. Manufacturers who bid this item will bid it per metre or in a lump sum. In doing a takeoff, keep the various sizes separate. Take special note of the connection devices required.

Labour. Precast beams (lintels) for door and window openings are generally installed by the mason. If the weight of an individual lintel exceeds about 136 kg, it may be necessary to have a small lift or crane to put them in place.

9.18 Miscellaneous Precast

Precast panels for the exterior walls of homes, warehouses, apartment, and office buildings are available. Their sizes, thicknesses, shapes, designs, and finishes vary considerably. Each individual system must be analyzed carefully to determine the cost in place. Particular attention should be paid to the attachment details at the base, top, and midpoints of each panel, how attachments will be handled at the jobsite, how much space is required for erection, how much bracing is required for all panels to be securely attached, and how many workers are required.

Some of the various methods involve the use of panels 1.20 metres wide, panels the entire length and width of the house, and precast boxes that are completely furnished before they are installed on the job. A tremendous amount of research goes into precast modules. The higher the cost of labour in the field, the more research there will be to arrive at more economical building methods.

The estimator must carefully analyze each system, consider fabrication costs and time, space requirements, how mechanicals will relate and be installed, and try to determine any hidden costs. New systems require considerable thought and study.

9.19 Precast Costs

If the specifications allow the contractor to precast the concrete shapes required for a project, or if the contractor decides to at least estimate the cost for precasting and compare it with the proposals received, the following considerations will figure into the cost:

1. Precasting takes a lot of space. Is it available on the jobsite or will the material be precast off the site and transported to the jobsite? If precast off the site, whatever facilities are used must be charged off against the items being made.

2. Determine whether the types of forms will be steel, wood, fibreglass, or a combination of materials. Who will make the forms? How long will manufacture take? How much will it cost? The cost of forms must be charged to the precast items being made.

3. Will a specialist be required to supervise the manufacture of the items? Someone will have to coordinate the work and the preparation of shop drawings. This cost must also be included.

4. Materials required for the manufacture must be purchased:
   a. Reinforcing
   b. Coarse and fine aggregates
   c. Cement
   d. Water
   e. Anchors and inserts

5. Allowance must be made for the actual cost of labour required to:
   a. Clean the forms.
   b. Apply oil or retarders to the forms.
   c. Place the reinforcing (including pretensioning if required).
   d. Mix and pour the concrete (including trowelling the top off).
   e. Cover the concrete and apply curing method.
   f. Uncover the concrete after the initial curing.
   g. Strip from form and stockpiling to finish curing.
   h. Erect the concrete on the job. If poured off the site, it will also be necessary to load the precast concrete on trucks, transport it to the jobsite, unload it, and then erect it.

6. Equipment required may include mixers, lift trucks, and cranes. Special equipment is required to prestress concrete. Equipment to cure the concrete may be required and miscellaneous equipment, such as hoes, shovels, wheelbarrows, hammers, and vibrators, will also be required.

7. Shop drawings should be prepared either by a company draftsperson or a consultant.

9.20 Precast Checklist

Shapes	Girders
Bearing requirements	Lintels
Accessories	Strength requirements
Walls	Inserts
Floors	Attachment requirements

Ceilings	Finish
Beams	Colour
Joists	Special requirements

Web Resources

www.aci-int.org
www.concretenetwork.com
www.sweets.com
www.worldofconcrete.com
www.4specs.com
www.tilt-up.org
www.precast.org
www.crsi.org
www.pci.org

Review Questions

1. What is the difference between plant ready-mixed concrete and job-mixed concrete?

2. Under what circumstances might it be desirable to have a field batching plant for job-mixed concrete?

3. What is the unit of measure for concrete?

4. Why does the estimator have to keep the different places that the concrete will be used separate in the estimate (e.g., concrete sidewalks, floor slabs)?

5. Why should the different strengths of concrete be kept separate?

6. Where would the estimator most likely find the strength of the concrete required?

7. How are rebars taken off? In what unit of measure are large quantities ordered?

8. How does lap affect the rebar quantities?

9. How is wire mesh taken off? How is it ordered?

10. How is the vapour barrier taken off? How does the estimator determine the number of rolls required?

11. What unit of measure is used when taking off expansion joint fillers?

12. What unit of measure is used when taking off concrete finishes?

13. Why should each finish be listed separately on the estimate?

14. Where would the estimator look to determine if any curing of concrete is required on the project?

15. Why must the estimator consider how the concrete will be transported to the jobsite?

16. What unit of measure is used when taking off concrete forms? How can reuse of forms affect the estimate?

17. What unit of measure is used for form liners? When might it be more economical to rent instead of purchase them?

18. What two methods of pricing might a subcontractor use for precast concrete?

19. Many suppliers take the responsibility of installing the units. Why might this be desirable?

20. Under what conditions might it be desirable for a contractor to precast the concrete on the jobsite?

21. Determine the quantities of formwork, cast-in-place concrete and accessories required for the residential building shown in Appendix C.

22. Determine the quantities of formwork, reinforcement, cast-in-place concrete and accessories required for the commercial building shown in Appendix D.

23. (Computer Exercise) Create an estimate for all the concrete works using Timberline Precision Estimating Basic by entering the data from Question 21. Change the prices to match prevailing market rates.

24. (Computer Exercise) Create an estimate for all the concrete works using Timberline Precision Estimating Basic by entering the data from Question 22. Change the prices to match prevailing market rates.

CHAPTER 10

Masonry

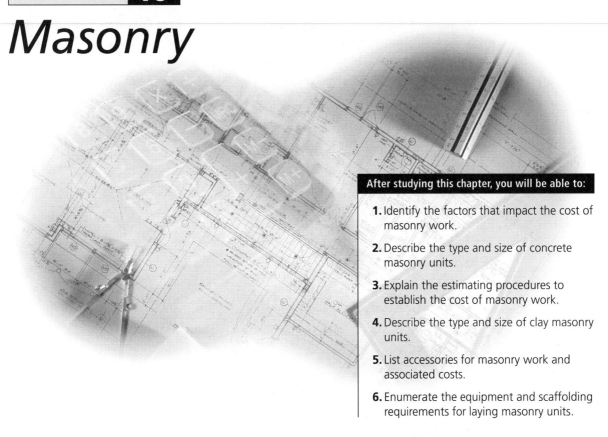

After studying this chapter, you will be able to:

1. Identify the factors that impact the cost of masonry work.

2. Describe the type and size of concrete masonry units.

3. Explain the estimating procedures to establish the cost of masonry work.

4. Describe the type and size of clay masonry units.

5. List accessories for masonry work and associated costs.

6. Enumerate the equipment and scaffolding requirements for laying masonry units.

10.1 General

The term *masonry* encompasses all the materials used by masons in a project, such as block, brick, clay, tile, or stone. The mason is also responsible for the installation of lintels, setting miscellaneous metals, integral flashings, joint wall reinforcing, ties, anchors, weep holes, control and expansion joints.

The tremendous amount of varied material available requires that estimators be certain they are bidding exactly what is required. The masonry trade associations and bid depositories publish trade scopes that define the work included for the masonry trade. Read the specifications, check the drawings, and call the manufacturers and builder suppliers to determine the exact availability, costs, and special requirements of the units needed.

10.2 Specifications

Specification sections for masonry include mortar, masonry reinforcement and connectors, unit masonry systems, such as concrete masonry units, brick units, and associated accessories. The specifications should be carefully checked for size of the unit, type of bond, colour, and shape. The estimator must check with the manufacturer and suppliers to ensure the availability of all items, their compliance with the specifications, and the standards established by the municipality building code. Prior to the inclusion of a particular masonry product in the specifications, the architect may consult with the local sales representative on the technical and availability aspects of that product. Bidders are then required to comply with the recommendations and stipulations of the manufacturer. Publications by the Canadian Masonry Research Institute are a reliable source of technical information for contractors on the latest masonry technology.

The masons must install many masonry accessories, such as lintels and flashing, which are built into the wall. The general contractor needs to verify if the masonry subcontractor has included these items in their quote. Large precast concrete lintels may require the use of special equipment, and steel angle lintels often require cutting of the masonry units. In addition, the type of joint "tooling" should be noted, as the different types impact labour costs.

10.3 Labour

The amount of time required for a mason (with the assistance of helpers) to lay a masonry unit varies with:

1. Size, weight, and shape of the unit

2. Bond (pattern)

3. Number of openings

4. Whether the walls are straight or have jogs in them

5. Distance the units must be moved (both horizontally and vertically)

6. The shape and colour of the mortar joint

The height of the walls becomes important in estimating labour for masonry units. The masonry work that can be laid up without the use of scaffolding is generally the least expensive; however, that is typically limited to 1.2 to 1.5 metres. Generally, masonry contractors will exclude the cost of hoisting over 7.5 metres from their bids. Labour costs arise from the erection, moving, and dismantling of the scaffolding as the building goes up. Estimating the cost of scaffolding is discussed in Section 10.18. The units and mortar have to be placed on the scaffold, which further adds to the labour costs.

The estimator should check union regulations in the locality in which the project will be constructed, since unions may require that two masons work together where the units weigh more than 15 kilograms each.

The weather conditions always affect labour costs because a mason will lay more brick on a clear, warm, dry day than on a damp, cold day. Winter construction requires the building and maintenance of temporary enclosures and heating.

10.4 Bonds (Patterns)

Some types of bonds (Figures 10.1, 10.2, and 10.3) required for masonry units can add tremendously to the labour cost of the project. The least expensive bond (pattern) is the *running bond*. Another popular bond is the *stacked bond*; this type of bond will increase labour costs by as much as 50 percent if used instead of the running bond. Various *ashlar patterns* may also be required; these may demand several sizes laid to create a certain effect. The estimator must study the drawings, check the specifications, and keep track of the different bonds that might be required on the project. When doing the quantity takeoff, the estimator must keep the amounts required for the various bonds separate. The types of bonds that may be required are limited only by the limits of the designer's imagination.

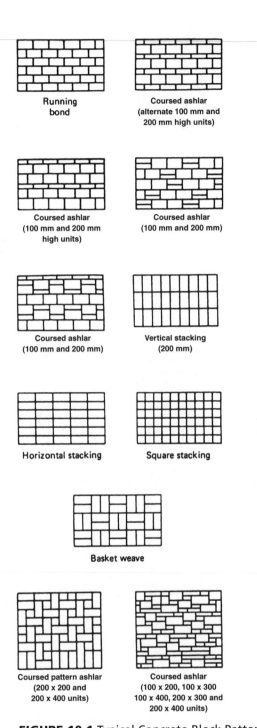

Running bond

Coursed ashlar
(alternate 100 mm and 200 mm high units)

Coursed ashlar
(100 mm and 200 mm high units)

Coursed ashlar
(100 mm and 200 mm)

Coursed ashlar
(100 mm and 200 mm)

Vertical stacking
(200 mm)

Horizontal stacking

Square stacking

Basket weave

Coursed pattern ashlar
(200 x 200 and 200 x 400 units)

Coursed ashlar
(100 x 200, 100 x 300 100 x 400, 200 x 300 and 200 x 400 units)

FIGURE 10.1 Typical Concrete Block Patterns

The most common shapes of mortar joints are illustrated in Figure 10.4. The joints are generally first struck off flush with the edge of the trowel. Once the mortar has partially set, the tooled joints are molded and compressed with a rounded or V-shaped joint tool. Some joints are formed with the edge of the trowel. The *raked joint* is formed by raking or scratching out the mortar to a given depth, which is generally accomplished with a tool made of an adjustable nail attached to two rollers.

FIGURE 10.2 Stone Patterns

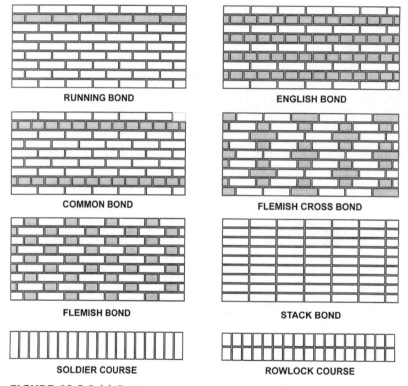

RUNNING BOND

ENGLISH BOND

COMMON BOND

FLEMISH CROSS BOND

FLEMISH BOND

STACK BOND

SOLDIER COURSE

ROWLOCK COURSE

FIGURE 10.3 Brick Patterns

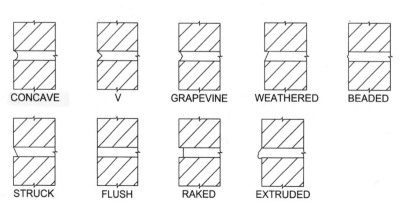

CONCAVE   V   GRAPEVINE   WEATHERED   BEADED

STRUCK   FLUSH   RAKED   EXTRUDED

FIGURE 10.4 Shapes of Mortar Joints

10.5 Concrete Masonry Units

Concrete masonry comprises all moulded concrete units used in the construction of a building and includes concrete brick, hollow and solid block, and decorative types of block. A wide variety of concrete blocks are manufactured and are produced in three major groups: solid load bearing, hollow load bearing and non-load-bearing units. Block manufacturers can produce almost any shape and size, provided a mold can be made to produce the unit.

EDCON   Reference

Concrete Block

Concrete block has no complete standard of sizes. The standard modular face dimensions of the units are 200 mm high and 400 mm long. Thicknesses available are 100, 150, 200, 250, and 300 mm. (These are nominal dimensions; actual dimensions are 10 mm less.) A 10 mm mortar joint provides face dimensions of 390 mm L × 190 mm H; it requires 12.5 blocks per square metre of wall. Some of the more common sizes and shapes are shown in Figure 10.5. If the order requires a large number of special units, it is very possible that manufacturers will produce the units needed for the project.

Concrete blocks are available either as heavyweight or lightweight units. The heavyweight unit is manufactured out of dense or normal weight aggregates, such as sand, gravel and crushed limestone, cement, and water. Lightweight units use aggregates, such as vermiculite, expanded slag, or pumice with the cement and water. The lightweight unit may weigh 30 percent less than the heavyweight unit, although it usually costs a few cents more per unit and usually has a slightly lower compressive strength.

The specifications will have to be checked to determine the size, shape, and colour of units, as well as the conformance with some set of standards, such as that of the American Society for Testing and Materials (ASTM), which sets up strength and absorption requirements and the Canadian Standards Association (CSA) Standards on Concrete Masonry Units. If a specific fire rating is designated for the units, it may require the

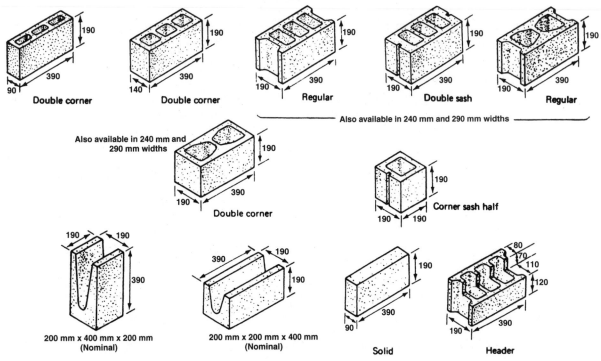

FIGURE 10.5 Typical Block Sizes (All sizes are actual unless otherwise noted.)

manufacture of units with thicker face shells than are generally provided by a particular supplier. Check also for the type of bond, mortar, and any other special demands. If the estimator is unfamiliar with what is specified, then he or she can call the builder's suppliers to discuss the requirements of the specifications with them to be certain that what is specified can be supplied.

10.6 Specifications—Concrete Masonry

The specifications will state exactly which types of units are required in each location. They give the size, shape, colour, and any requisite features, such as glazed units, strength, and fire ratings. The type, colour, thickness, and shape of the mortar joint must be determined (refer to the Canadian Standards Association, Mortar and Grout for Unit Masonry), as well as the style of bond required. Also to be checked is the reinforcing, control and expansion joints, wall ties, anchors, flashing, and weep holes needed.

If the specifications are not completely clear as to what is required, the estimator should call the consultant's office to check; they should never guess what the specifications mean.

10.7 Estimating—Concrete Masonry

Concrete masonry should be taken off from the drawings in square metres of wall required and divided into the different thickness of each wall. No deductions shall be made for openings not exceeding 1 m². The total area of each wall, of a given thickness, is then multiplied by the number of units required per square metre (Figure 10.6). For sizes other than those found in Figure 10.5, the following formula can be used:

Number of blocks required in a square metre of wall using a nominal size 200 mm × 200 mm × 400 mm block:

$$\text{Units per square metre} = 1 / (0.20 \times 0.40)$$
$$= 12.5 \text{ blocks per m}^2$$

When estimating the quantities of concrete masonry, use the exact dimensions shown. Corners should only be taken once, and deductions should be made for all openings in excess of 1 m². This area is then converted to units, and to this quantity an allowance for waste and breakage must be added.

While performing the takeoff, the estimator should note how much cutting of masonry units would be required. Cutting of the units is expensive and should be anticipated.

Materials	Materials (per 10 m² face area)			
	Concrete Block			Standard Modular Brick
	100 × 200 × 400	150 × 200 × 800	200 × 200 × 400	90 W × 57 H × 190 L
No. of Units	125	125	125	746
Mortar Cubic Metre				
Shell bedding	0.06	0.06	0.06	0.173
Full bedding	0.07	0.07	0.07	0.173

Values shown are net. Waste for block and brick ranges from 5 to 100%. Waste for mortar may range from 25 percent to 75 percent and actual job experience should be considered on this item. It is suggested that 100% waste be allowed by the inexperienced, and actual job figures will allow a downward revision. Figures are for 10-mm-thick joint.

FIGURE 10.6 Materials Required for 100 m² of Face Area

In working up the quantity takeoff, the estimator must separate masonry according to:

1. Size of the units
2. Shape of the units
3. Colours of the units
4. Type of bond (pattern)
5. Shape of the mortar joints
6. Colours of the mortar joints
7. Any other special requirements (such as fire rating)

In this manner, it is possible to make the estimate as accurate as possible. For mortar requirements, refer to Figure 10.7 and Section 10.14.

Materials	Quantities by Volume (m³) Mortar Type & Proportions by Volume				Quantities by Weight (kg) Mortar Type & Proportions by Volume			
	M 1:1/4: 3 1/2	S 1:1/2: 4 1/2	N 1: 1: 6	O 1: 2: 9	M 1:1/4: 3 1/2	S 1:1/2: 4 1/2	N 1: 1: 6	O 1: 2: 9
Cement	0.286	0.222	0.167	0.111	430	335	251	167
Lime	0.071	0.111	0.167	0.222	46	71	107	142
Sand	1.000	1.000	1.000	1.000	1,281	1,281	1,281	1,281

FIGURE 10.7 Mortar Mixes for 1 m³ of Mortar

EXAMPLE 10.1 Concrete Block

Determine the quantity of concrete block required for the wall shown in Figures 10.8 and 10.9.

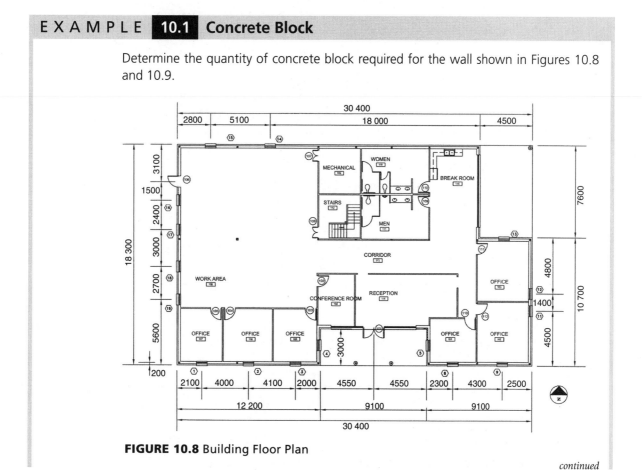

FIGURE 10.8 Building Floor Plan

continued

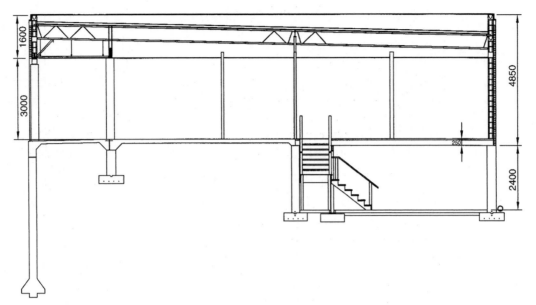

FIGURE 10.9 Wall Section

Wall height for concrete block = 4.85 − 0.25 = 4.60 m
West wall length = 18.30 − 0.20 (both end brick ledge) = 18.10 m
Gross wall area (m²) = 18.10 m × 4.60 m = 83.26 m²

Openings:

Windows 4 @ 0.91 × 2.13 = 7.75 m²
Doors 1 @ 0.91 × 2.13 = 1.94 m²
Area of openings = 7.75 m² + 1.94 m² = 9.69 m²
Net west wall area = 83.26 − 9.69 = 73.57 m²
Using a 200 mm × 200 mm × 400 mm = 12.5 blocks per m²
74 m² × 12.5 blocks per m² = 925 blocks
Waste @ 6% − Use 981 blocks

Header / Lintel blocks:

Top course = 18.10 / 0.40 = 45 blocks
Windows / doors = 4 per opening
5 openings = 20 blocks
Plain block = 981 − 45 − 20 = 916 blocks
For west wall purchase
916 − 200 × 200 × 400 blocks
65 header blocks

From Figure 10.10 the work hours can be determined.

continued

Work Hours per 10 m²										
Wall Thickness	100 mm		150 mm		200 mm		250 mm		300 mm	
Workers	Mason	Labourer	Mason	Labourer	Mason	Labourer	Mason	Labourer	Mason	Labourer
Type of Work										
Simple Foundation					4.5-6.0	4.0-7.5	6.0-9.0	7.0-10.5	7.0-10.0	8.0-11.5
Foundation with several corners, openings					5.0-7.5	5.0-7.5	6.5-10.0	7.5-12.0	7.5-10.0	8.5-12.0
Exterior walls 1.20 m high	3.5-5.5	3.5-6.0	4.0-6.0	4.0-6.5	4.5-6.0	5.0-7.5	6.0-9.0	7.0-10.5	7.0-10.0	8.0-11.5
Exterior walls, 1.20 to 2.40 m above ground or floor	3.5-6.0	4.5-7.5	4.0-6.5	4.5-7.0	4.5-6.5	6.0-9.0	6.5-10.0	7.5-12.0	7.5-10.0	8.5-12.0
Exterior walls, more than 2.40 m above ground or floor	4.5-8.0	6.0-9.5	5.0-9.0	7.0-10.0	5.0-7.0	7.0-10.0	7.0-10.5	7.5-12.0	7.5-10.0	8.5-12.0
Interior partitions	3.0-6.0	3.5-7.0	3.5-6.5	4.5-7.5	4.5-6.0	5.0-7.5				

Note:
1. The more corners and openings in the masonry wall, the more work hours it requires.
2. When lightweight units are used the work hours should be decreased by 10 percent.
3. Work hours include simple pointing and cleaning required.
4. Special bonds and patterns may increase the work hours by 20 to 50 percent.

FIGURE 10.10 Work Hours Required for Concrete Masonry

Using 6 mason work hours per 10 m² and 8 labourer work hours per 10 m²

7.4 ten m² × 6 = 44 mason work hours

7.4 ten m² × 8 = 59 labourer work hours

Assuming a basic labour rate of $22.50 for masons and $16 for labourers per work hour, the basic labour costs can be determined.

Mason labour cost = 44 work hours × $22.50 per work hour = $990

Labourer labour cost = 59 work hours × $16 per work hour = $944

Total labour cost = $990 + $944 = $1,934

10.8 Clay Masonry

Clay bricks are manufactured in a variety of types and sizes. It is sometimes necessary for architects to check with the local brick manufacturer to verify the size and colour of the units supplied. Commonly manufactured bricks are shown in Figure 10.11. Modular bricks are units in which the actual size plus a mortar joint can be assembled on a standard unit or module. The modular unit for metric brick sizes is 100 mm. The actual size plus the thickness of a mortar joint (10 mm) is the nominal size. Three dimensions—width, thickness and length, given in that sequence—specify the sizes for brick units. In the 100 mm module, three vertical courses of brick produce a 200 mm module that also corresponds to the height of one concrete block course. The actual sizes of metric bricks, number of bricks, and mortar requirements per 10 m² are given in Figure 10.12.

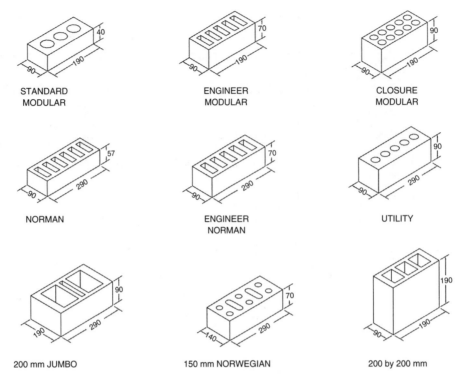

FIGURE 10.11 Metric Brick Sizes (Actual dimensions)

Brick Type	Metric Size Actual (mm)			Number of Bricks per 10 m²	Volume (m³) of Mortar per 10 m²
	W	H	L		
100 mm Standard Modular	90	57	190	746	0.173
Engineer	90	70	190	625	0.152
Roman (Miniature)	90	40	190	1000	0.216
100 mm Norman	90	70	290	417	0.139
100 mm Continental	90	90	190	500	0.131
100 mm Jumbo (Saxon)	90	90	290	333	0.117
Saturn (Solid)	90	140	390	167	0.081
85 mm Veneer	75	70	256	470	0.118

Quantities are net and based on running bond.
For bricks allow at least 5%, 10–25% for mortar.

FIGURE 10.12 Metric Brick and Mortar Requirements

10.9 Specifications—Brick

The specifications must be checked to determine exactly the type of material and the type of mortar as well as the shape, thickness, and colour of the joint itself. The style of bond must also be determined. From the specifications, the estimator also determines the types of lintels, flashing, reinforcing, and weep holes required, and who supplies and who installs each item.

Face brick with special shapes for stretchers, jambs, corners, sills, wall ends, and for use in other particular areas is available. Because these special units are relatively expensive, the estimator must allow for them. Also, the materials must meet all Canadian Standards Association (CSA) and related American Society for Testing and Materials (ASTM) Standards requirements of the specification.

Many specifications designate an amount, *cash allowance,* for the purchase of face brick. The estimator then carries this allowance in the bid. This practice allows the owner and consultant to determine the exact type of brick desired at a later date, at which time the contractor will be issued a change notice to expend this sum of money included in the bid amount.

10.10   Estimating Brick

The first thing to be determined in estimating the quantity of brick is the size of the brick and the width of the mortar joint. They are both necessary to determine the number of bricks per square metre of wall area and the quantity of mortar. Brick is sold by the thousand units, so the final estimate of materials required must be in the number of units required.

To determine the number of bricks required for a given project, the first step is to obtain the length and height of all walls to be faced with brick and then calculate the area of wall. Make deductions for all openings exceeding 1 m² so that the estimate will be as accurate as possible. The extra labour cost incurred in forming openings less than 1 m² is accounted for by the extra material quantified as a result of ignoring these smaller openings at the bid stage of the project. Check the jamb detail of the opening to determine if extra brick will be required for the reveal; generally, if the reveal is over 100 mm deep, extra bricks will be required. Special jambs and bullnose corner units may also be required.

Once the number of square metres has been determined, the number of bricks can be calculated. This calculation varies depending on the size of the brick, width of the mortar joint, and style of bond required. The figures must be extremely accurate, as actual quantities and costs must be determined. It is only in this manner that estimators will increase their chances of getting work at a profit.

Figure 10.13 shows the number of bricks required per square metre of wall surface for various patterns and bonds. Special bond patterns require that the estimator analyze the style of bond required and determine the number of bricks. One method of analyzing the amount of brick required is to make a drawing of several square metres of wall surface, determine the brick to be used, and divide that into the total area drawn. The estimator often makes sketches right on the workup sheets.

Labour costs will be affected by lengths of straight walls, number of indents in the wall, windows, piers, pilasters, expansion/control joints, and anything else that might slow the mason's work, such as weather conditions. Also to be calculated are the amount of mortar required and any lintels, flashing, reinforcing, and weep holes that may be specified. Any special requirements, such as coloured mortar, shape of joint, and type of flashing are noted on the workup sheet.

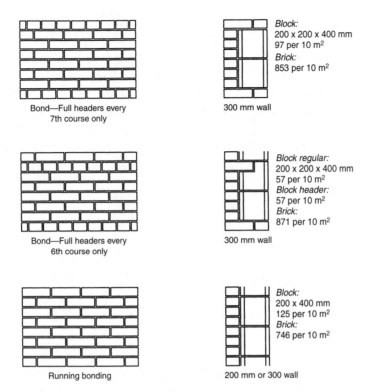

Block:
200 x 200 x 400 mm
97 per 10 m²
Brick:
853 per 10 m²

Bond—Full headers every
7th course only

300 mm wall

Block regular:
200 x 200 x 400 mm
57 per 10 m²
Block header:
57 per 10 m²
Brick:
871 per 10 m²

Bond—Full headers every
6th course only

300 mm wall

Block:
200 x 400 mm
125 per 10 m²
Brick:
746 per 10 m²

Running bonding

200 mm or 300 wall

FIGURE 10.13 Bricks (Standard Modular) Required per Square Metre of Wall Surface

E X A M P L E   10.2   Face Brick

Using the west wall that was quantified in the previous example, most of the dimensions can be reused. The gross area will be slightly different, as the exterior face dimensions will be used rather than the face dimensions of the concrete block.

Gross wall area = 4.85 × 18.30 = 88.76 m²
Area of openings = 9.69 m² (from Example 10.1)
Net wall area = 88.76 m² − 9.69 m² = 79 m²
Assuming modular brick, there are 746 bricks per 10 m² of wall
(refer to Figure 10.6)
7.9 ten m² of wall × 746 = 5,893 bricks
Waste @ 5% − Use 6,188 bricks

The work hours and basic labour costs can be determined using the productivity rates found in Figure 10.14.

Type Bond	Mason	Labourer
Common	10.0-15.0	11.0-15.5
Running	8.0-12.0	9.0-13.0
Stack	12.0-18.0	10.0-15.0
Flemish	11.0-16.0	12.0-16.0
English	11.0-16.0	12.0-16.0

Note:
1. The more corners and openings in the wall, the more work hours it will require.

Norman Brick	+30%
Roman Brick	+25%
Jumbo Brick	+25%
Modular Brick	Same as Standard
Jumbo Utility Brick	+30%
Spartan Brick	+25%

FIGURE 10.14 Work Hours Required to Lay 1,000 Standard-Size Bricks

10.11 Stone Masonry

Stone masonry is primarily used as a veneer for interior and exterior walls; it is also used for walkways, riprap, and trim on buildings. Stone masonry is usually divided into that which is laid up dry with no mortar being used—such as on some low walls, sloping walls, walkways, and riprap—and wet masonry in which mortar is used.

Stone is used in so many sizes, bonds, and shapes that a detailed estimate is required. The types of stone most commonly used are granite, sandstone, marble, slate, limestone, and trap. The finishes available include various split finishes and tooled, rubbed, machine, cross-broached, and brushed finishes. Stone is generally available in random, irregular sizes, sawed-bed stone, and cut stone, which consists of larger pieces of cut and finished stone pieces. Random, irregular-sized stone is used for rubble masonry and rustic and cobblestone work; it is often used for chimneys, rustic walls, and fences. Sawed-bed stones are used for veneer work on the interior and exterior of buildings. Patterns used include random and coursed ashlar. Coursed ashlar has regular courses, whereas random ashlar has irregular size pieces and generally will require fitting on the job.

10.12 Specifications—Stone

Check exactly the type of stone required, the coursing, thickness, type, and colour of mortar as well as wall ties, flashing, weep holes, and other special requirements. Not all types and shapes of stone are readily available. The supplier of the stone must be involved early in the bidding period. For large projects, the stone required for the job will actually be quarried and finished in accordance with the specifications. Otherwise, the required stone may be shipped in from other parts of the country or from other countries. In these special cases, estimators place a purchase order as soon as they are awarded the bid.

10.13 Estimating Stone

Stone is usually estimated by the area in square metres, with the thickness given. In this manner, the total may be converted to cubic metres, while still providing the estimator with the basic square metre measurement. Stone trim is usually estimated by the metre or enumerated.

Stone is sold in various ways—sometimes by the cubic metre, often by the tonne. Cut stone is often sold by the square metre; of course, the square metre price goes up as the thickness increases. Large blocks of stone are generally sold by the cubic metre. It is not unusual for suppliers to submit lump-sum proposals whereby they will supply all of a certain type of stone required for a given amount of money. This is especially true for cut-stone panels.

In calculating the quantities required, note that the length times height equals the area required; if the number must be in cubic metres, multiply the area by the thickness. Deduct all openings but usually not the corners. This calculation gives the volume of material required. The stone does not consume all the space; the volume of mortar must also be deducted. The pattern in which the stone is laid and the type of stone used will greatly affect the amount of mortar required. Cut stone may have 2 to 4 percent of the total volume as mortar, ashlar masonry 6 to 20 percent, and random rubble 15 to 25 percent. Waste is equally hard to anticipate. Cut dressed stone has virtually no waste, whereas ashlar patterns may have 10 to 15 percent.

Dressing a stone involves the labour required to provide a certain surface finish to the stone. Dressing and cutting stone require skill on the part of the mason, which varies considerably from person to person. There is an increased tendency to have all stone dressed at the quarry or supplier's plant rather than on the jobsite.

The mortar used should be nonstaining mortar cement mixed in accordance with the specifications. When cut stone is used, some specifications require that the mortar joints be raked out and that a specified thickness of caulking be used. The type and quantity of caulking must then be taken off. The type, thickness, and colour of the mortar joint must also be taken off.

Wall ties are often used for securing random, rubble, and ashlar masonry to the backup material. The type of wall tie specified must be noted, as must the number of wall ties per square metre. Divide this into the total area to determine the total number of ties required.

Cleaning must be allowed for as outlined in Section 10.17. Flashing should be taken off in square metres and lintels in metres. In each case note the type required, the supplier, and the installer.

Stone trim is used for door and windowsills, steps, copings, and mouldings. The supplier prices this item by the length (m) or as a lump sum. Some type of anchor or dowel arrangement is often required for setting the pieces. The supplier will know who is supplying the anchors and dowels and who will provide the anchor and dowel holes. The holes must be larger than the dowels being used.

Large pieces of cut stone may require cranes or other lifting devices to move them and set them in place. When large pieces of stone are being used as facing on a building, special inserts and attachments to hold the stone securely in place will be required.

All the aforementioned factors or items will also affect the work hours required to put the stone in place.

10.14 Mortar

The *mortar* used for masonry units may consist of Portland cement, mortar cement, sand, and hydrated lime; or of mortar cement, cement, and sand. The amounts of each material required vary depending on the proportions of the mix selected, the thickness of the mortar joint (10 mm is the common joint thickness), and the colour of the mortar. There has been an increase in the use of coloured mortar on masonry work; colours commonly used are red, brown, white, and black, but almost any colour may be specified. Pure white mortar may require the use of white cement and white mortar sand; the use of regular mortar sand will generally result in a creamy colour. The other colours are obtained by adding colour pigments to the standard mix. Considerable trial and error may be required before a colour acceptable to the owner and consultant is found.

When both coloured and grey mortars are required on the same project, the mixer used to mix the mortar must be thoroughly cleaned between mixings. If both types will be required, it may be most economical to use two mixers: one for coloured mortar, the other for grey. White mortar should never be mixed in the mixer that is used for coloured or grey mortar, unless the mixer has been thoroughly cleaned.

Labour. The amount of labor required for mortar is usually considered a part of the labour to lay the masonry units.

EXAMPLE **10.3** **Mortar Takeoff**

Example 10.1 contains 74 m² of block. Figure 10.6 details mortar at 0.07 m³ per 10 m² of face area. Therefore, the required mortar would be:

$$7.4 \times 0.07 = 0.52 \text{ m}^3$$
$$\text{Waste @ 40\%} = 0.73 \text{ m}^3$$

From Example 10.2, there were 79 m² of brick. Using Figure 10.6, the required mortar is 0.173 m³ per 10 m² of face wall area. Therefore, the required mortar for the brick would be:

$$7.9 \times 0.173 = 1.37 \text{ m}^3$$
$$\text{Waste @ 40\%} = 1.92 \text{ m}^3$$

10.15 Accessories

Masonry Wall Reinforcing. Steel reinforcing, which comes in a wide variety of styles and wire gauges, is placed continuously in the mortar joints (Figures 10.15 and 10.16). It is used primarily to minimize shrinkage, temperature, and settlement cracks in masonry, as well as to provide shear transfer to the steel. The reinforcing is generally available in lengths up to 6 m. The estimator must determine the length required. The drawings and specifications must be checked to determine the spacing required (sometimes every course, often every second or third course). The reinforcement is also used to tie the outer and inner wythes together in cavity wall construction. The reinforcing is available in plain or corrosion-resistant wire. The estimator should check the specifications to determine what is required and check with local suppliers to determine prices and availability of the specified material.

In calculating this item, the easiest method for larger quantities (over 150 m²) is to multiply the area by the factor given:

Reinforcement every course 1.50
Second course 0.75
Third course 0.50
To this amount, add 5 percent waste and 5 percent lap.

For small quantities, the courses involved and the length of each course must be figured. Deduct only openings in excess of 5 m².

Control Joints. A control joint is a straight vertical joint that cuts the masonry wall from top to bottom. The horizontal distance varies from 10 mm to about 50 mm. The joint must also be filled with some type of material; materials usually specified are caulking, neoprene/moulded rubber, and copper/aluminum. These materials are sold by the length and in a variety of shapes. Check the specifications to find the types required and check both the drawings and the specifications for the locations of control joints. Extra labour is involved in laying the masonry, since alternate courses utilizing half-size units will be required to make a straight vertical joint.

Wall Ties. Wall ties (Figure 10.17) are used to tie the outer wythe with the inner wythe. They allow the mason to construct one wythe of wall to a given height before

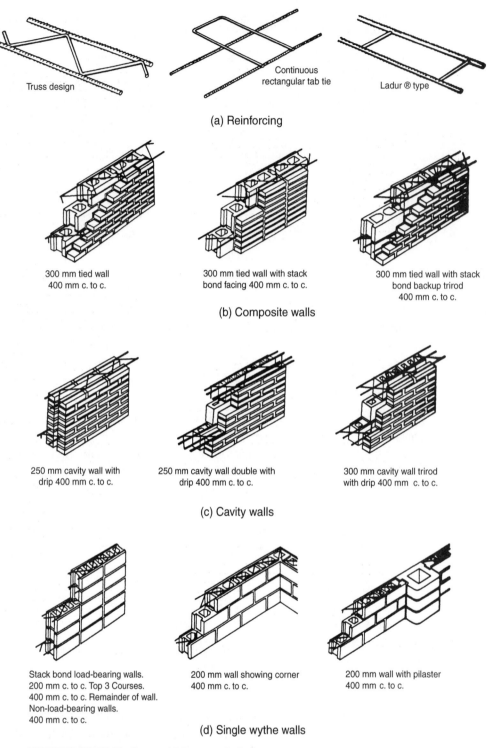

Truss design

Continuous
rectangular tab tie

Ladur ® type

(a) Reinforcing

300 mm tied wall
400 mm c. to c.

300 mm tied wall with stack
bond facing 400 mm c. to c.

300 mm tied wall with stack
bond backup trirod
400 mm c. to c.

(b) Composite walls

250 mm cavity wall with
drip 400 mm c. to c.

250 mm cavity wall double with
drip 400 mm c. to c.

300 mm cavity wall trirod
with drip 400 mm c. to c.

(c) Cavity walls

Stack bond load-bearing walls.
200 mm c. to c. Top 3 Courses.
400 mm c. to c. Remainder of wall.
Non-load-bearing walls.
400 mm c. to c.

200 mm wall showing corner
400 mm c. to c.

200 mm wall with pilaster
400 mm c. to c.

(d) Single wythe walls

FIGURE 10.15 Horizontal Masonry Reinforcement

(Courtesy of Dur-O-Wall)

working on the other wythe, resulting in increased productivity. Wall ties are available in a variety of sizes and shapes, including corrugated strips of metal 32 mm wide and 150 mm long, and wire bent to a variety of shapes. Adjustable wall ties are among the most popular, as they may be used where the coursing of the inner and outer wythes is not lined up. Non-corrodible metals or galvanized steel may be used. Check

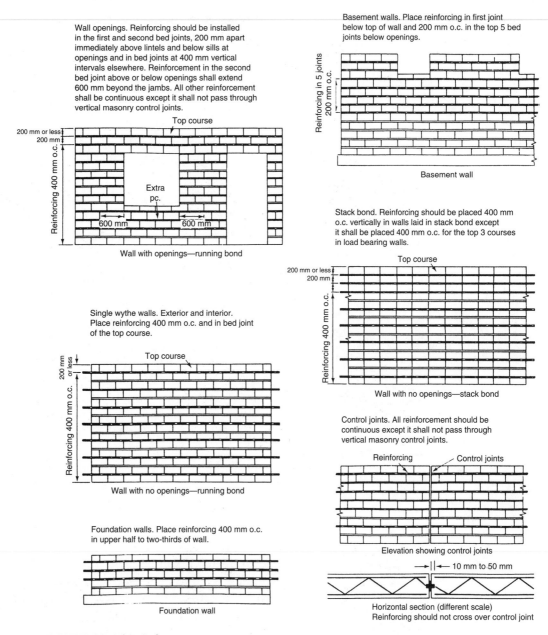

FIGURE 10.16 Reinforcement Layout

(Courtesy of Dur-O-Wall)

the specifications for the type of ties required and their spacing. To determine the amount required, take the total area of masonry and divide it by the spacing. A spacing of 400 mm vertically and 600 mm horizontally requires one tie for every 0.25 m^2; a spacing of 400 mm vertically and 900 mm horizontally requires one tie for every 0.37 m^2. Often, closer spacings are required. Also allow for extra ties at control joints, wall intersections, and vertical supports as specified.

Flashing. (Flashing is discussed in Chapter 13, Thermal and Moisture Protection, as well.) The flashing (Figure 10.18) built into the walls is generally installed by the mason. It is installed to keep moisture out and to divert any moisture that does get in back to the outside of the building. Flashing may be required under sills and copings,

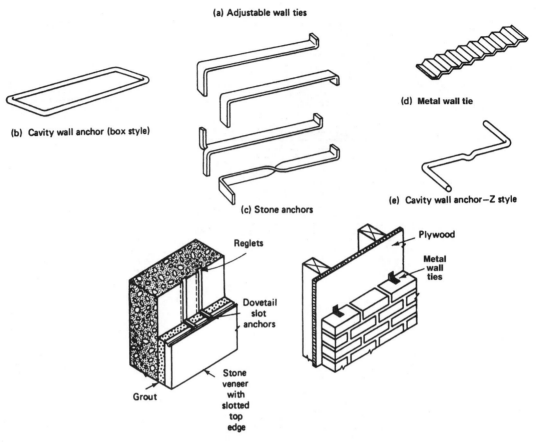

FIGURE 10.17 Typical Wall Ties and Wall Tie Installation

(Courtesy of Dur-O-Wall)

over openings for doors and windows, at intersections of roof and masonry walls, at floor lines, and at the bases of the buildings (a little above grade) to divert moisture. Materials used include copper, aluminum, copper-backed paper, copper and polyethylene, plastic sheeting (elastomeric compounds), wire and paper, and copper and fabric. Check the specifications to determine the type required. The drawings and specifications must also be checked to determine the locations in which the flashing must be used. Flashing is generally sold by the square metre or by the roll. A great deal of labour may be required to bend metal flashing into shape. Check carefully as to whether the flashing is to be purchased and installed under this section of the estimate, or if it is to be purchased under the roofing section and installed under the masonry section.

Weep Holes. In conjunction with the flashing at the base of the building (above grade level), weep holes are often provided to drain any moisture that might have gotten through the outer wythe. Weep holes may also be required at other locations in the construction. The maximum horizontal spacing for weep holes is about 900 mm, but specifications often require closer spacing. The holes may be formed by using short lengths of cord inserted by the mason or by well-oiled rubber tubing. The material used should extend upward into the cavity for several millimetres to provide a drainage channel through the mortar droppings that accumulate in the cavity.

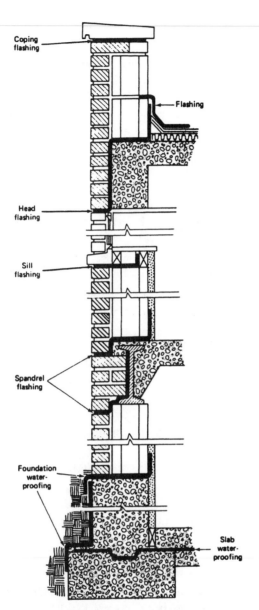

FIGURE 10.18 Flashing Location

Lintels. A *lintel* is the horizontal member that supports the masonry above a wall opening. In other words, it spans the opening. Materials used for lintels include steel-angle iron, composite steel sections, lintel block (shaped like a U) with reinforcing bars and filled with concrete, and precast concrete. The lintels are usually set in place by masons as they lay up the wall. Some specifications require that the lintel materials be supplied under this section, whereas other specifications require the steel angles and composite steel section to be supplied under "Structural Steel" or "Miscellaneous Accessories." Precast lintels may be supplied under "Concrete"; the lintel block will probably be included under "Masonry," as will the reinforcing bars and concrete used in conjunction with it.

It is not unusual for several types of lintels, in a variety of sizes, to be required on any one project. They must be separated into the types, sizes, and lengths for each material used. Steel lintels may require extra cutting on the job so that the masonry will

Accessories	Unit	Work Hours
Wall Reinforcing	300 m	1.0 to 1.5
Lintels – Precast Concrete (up to 140 kilograms)	30 m	4.5 to 7.0
Coping, Sills and Precast Concrete (up to 70 kilograms)	30 m	5.0 to 8.0
Wall Ties	100 pieces	1.0 to 1.5

FIGURE 10.19 Work Hours Required for Installing Accessories

be able to fit around them. If a lintel is heavy, it may be necessary to use equipment (such as a lift truck or a crane) to put it in place. In determining length, be certain to take the full masonry opening and add the required lintel bearing on each end. Lintel bearing for steel is generally a minimum of 150 mm, whereas lintel block and precast lintels are often required to bear 200 mm on each end. Steel is purchased by the kilogram, precast concrete by metre or piece, and lintel block by the unit (note the width, height, and length). Precast concrete lintels are covered in Section 9.18.

Sills. *Sills* are the members at the bottoms of window or door openings. Materials used are brick, stone, tile, and precast concrete. The mason installs these types of sills, although the precast concrete may be supplied under a different portion of the specifications. The brick and tile sills are priced by the number of units required, and the stone and precast concrete sills are sold by the metre. The estimator should check the maximum length of stone and precast concrete sills required and note it on the takeoff.

Also to be checked is the type of sill required: A *slip sill* is slightly smaller than the width of the opening and can be installed after the masonry is complete. A *lug sill*, which extends into the masonry at each end of the wall, must be built into the masonry as the job progresses. Some specifications require special finishes on the sill and will have to be checked. Also, if dowels or other inserts are required, that fact should be noted.

Coping. The *coping* covers the top course of a wall to protect it from the weather. It is most often used on parapet walls. Masonry materials used include coping block, stone, tile, and precast concrete. Check the specifications for the exact type required and who supplies it. The drawings will show the locations in which it is used, its shape, and how it is to be attached. The coping block and tile are sold by the unit, and the stone and precast coping are sold by the metre. Figure 10.19 gives the labour productivity rates for installing masonry accessories.

Special colours, finishes, dowels, dowel holes, and inserts may be required. Check the drawing and specifications for these items and note all requirements on the workup sheet.

10.16 Reinforcing Bars

Reinforcing bars are often used in masonry walls to create multistorey bearing wall construction; they are also used in conjunction with bond beam block and grout in bond beams used to tie the building together. Reinforcing bars are sold by weight: The various lengths required of each size are taken off, and the total weight of each size required is multiplied by the price.

The specifications should be checked to determine the type of steel required. If galvanized reinforcing bars are required, the cost for materials will easily double. Also galvanized reinforcing bars must be specially ordered, thus they are ordered quickly after the contract has been awarded (Section 9.3 also discusses reinforcing).

10.17 Cleaning

The specifications must be checked to determine the amount of cleaning required and the materials that must be used in the cleaning process. The materials exposed inside and outside of the building will probably require cleaning, while the concealed masonry, such as block used as a backup, generally receives no cleaning.

Clay Masonry. For brickwork, there should be no attempt to clean for a minimum of 48 hours after completion of the wall. After the minimum time, soap powder (or other mild solutions) with water and stiff brushes may be tried. When cleaning unglazed brick and tile, first use plain water and a stiff brush. If these solutions do not work, the surface should be thoroughly wetted with clear water, scrubbed with a solution of acid and water, and thoroughly rinsed. Always try the acid solution on an inconspicuous area prior to using it on the entire wall.

Concrete Masonry. Acid is not used on concrete masonry. If mortar droppings fall on the units, the droppings should be allowed to dry before removal to avoid smearing the face of the unit. When the droppings have dried, they can be removed with a trowel, and a final brushing will remove most of the mortar.

The estimator must determine the type of materials required for cleaning, the area of the surfaces to be cleaned, the equipment required, the amount of cleaning that will actually have to be done, and how many work hours will be required. The better the workmanship on the job, the less money has to be allowed for cleaning. When the colour of the mortar is different from that of the masonry unit, the cost for cleaning will be higher because all mortar droppings must be cleaned off to get an unblemished facing.

Stone Masonry. Clean stone masonry with a stiff fibre brush and clear water (soapy water may be used, if necessary). Then, rinse with clear water to remove construction and mortar stains. The stone supplier should approve machine-cleaning processes before they are used. Wire brushes, acids, and sandblasting are not permitted for cleaning stonework.

10.18 Equipment

The equipment required for laying masonry units includes the mason's hand tools, mortar boxes, mortar mixer, hoes, hoses, shovels, wheelbarrows, mortar boards (tubs), pails, scaffolding, power hoist, hand hoist, elevator tower, hoisting equipment, and lift trucks. On large commercial projects, the general contractor may provide cranes, forklift trucks, and other mechanical lifting devices. The masonry estimator must decide what equipment is required on the project, how much of each type is required, and the cost that must be allowed. The estimator must also remember to include the costs of ownership (or rental), operating the equipment, and mobilization to and from the project, erection, and dismantling. Items to be considered in determining the amount of equipment required are the height of the building, the number of times the scaffolding will be moved, the number of masons and helpers needed, and the type of units being handled.

The item for erection and dismantling of scaffold frames can be quantified on the basis of square metres of wall area (or number of masonry units) and priced by an appropriate unit rate. One approach is to measure the total length of each "lift" of scaffolding. For example, a wall 30 m long requiring two lifts of 1.50 m scaffolding is expressed as 60 m of scaffold in lifts of 1.50 m.

10.19 Cold Weather

Cold weather construction is more expensive than warm weather construction. Increased costs stem from the construction of temporary enclosures to enable the masons to work, higher frequency of equipment repair, thawing materials, and the need for temporary heat.

Masonry should not be laid if the temperature is 5 °C and falling or less than 0 °C and rising at the place where the work is in progress, unless adequate precautions against freezing are taken. The masonry must be protected from freezing for at least 48 hours after it is laid. Any ice on the masonry materials must be thawed before the masonry is used.

Mortar also has special requirements. Its temperature should be between 21 °C and 49 °C. During cold weather construction, it is common practice to heat the water used to raise the temperature of the mortar. Moisture present in the sand will freeze unless heated; upon freezing, it must be thawed before it can be used.

10.20 Subcontractors

In most localities, masonry subcontractors are available. The estimator will have to decide whether it is advantageous to use a subcontractor on each project. The decision to use a subcontractor does not mean the estimator does not have to prepare an estimate for that particular item; the subcontractor's bid must be checked to be certain it is neither too high nor too low. Even though a particular contractor does not ordinarily subcontract masonry work, it is possible that the subcontractor can do the work for less money. There may be a shortage of masons, or the contractor's masonry crews may be tied up on other projects.

If the decision is made to consider the use of subcontractors, the first thing the estimator should decide upon is which subcontractors he or she wants to submit a proposal for the project. The subcontractors should be notified as early in the bidding period as possible to allow them time to make a thorough and complete estimate. Often, the estimator will meet with the subcontractors to discuss the project in general and go over exactly which items are to be included in the proposal. Sometimes, the proposal is for materials and labour and other times for labour only. Both parties must clearly understand the items that are to be included.

E X A M P L E   **10.4   Masonry Takeoff**

Figures 10.20 and 10.21 are the completed masonry takeoffs for the building shown in Figures 10.8 and 10.9.

continued

ESTIMATE WORK SHEET

Project:	Little Office Building		Estimate No.	1234
Location	Mountainville, BC		Sheet No.	1 of 1
Architect	C.K. Architects	Gross Wall Area	Date	23/12/20xx
Items	Masonry		By	HYZ Checked DCK

Cost Code	Description	DIMENSIONS						Gross Length	Gross Brick Ledge	Gross Corners	Net Length	Wall Height	Gross Side Total	Quantity	Unit
		Gross Length	Brick Ledge		Corner										
		m	Width	No.	Thickness (m)	No.	m	mm		m	m	m		
	Concrete Block													
	(West Side)	18.30	0.11	2		0	18.30	0.22	0.00	18.08	4.60	83.17	83	m²
	(Front Side - North Wall)	12.20	0.11	2	200	2	12.20	0.22	0.40	11.58	4.60	53.27		
		3.00	0.11	1	200	1	3.00	0.11	0.20	2.69	4.60	12.37		
		0.60					0.60			0.60	4.60	2.76		
		0.60					0.60			0.60	4.60	2.76		
		3.00	0.11	1	200	1	3.00	0.11	0.20	2.69	4.60	12.37		
		9.10	0.11	2	200	2	9.10	0.22	0.40	8.48	4.60	39.01	123	m²
	(East Side)	10.70	0.11	2			10.70	0.22		10.48	4.60	48.21		
		4.50	0.11	1			4.50	0.11		4.39	4.60	20.19		
		0.60					0.60			0.60	4.60	2.76		
		0.30					0.30			0.30	4.60	1.38	73	m²
	(South Side)	30.40	0.11	2	200	1	30.40	0.22	0.20	29.98	4.60	137.91	138	m²
										90.47				
													417	m²
	Brick Veneer													
	(West Side)	18.30					18.30	0.00	0.00	18.30	4.85	88.76		
	(Front Side - North Wall)	12.20					12.20	0.00	0.00	12.20	4.85	59.17		
		3.00					3.00	0.00	0.00	3.00	4.85	14.55		
		0.60					0.60	0.00	0.00	0.60	4.85	2.91		
		0.60					0.60	0.00	0.00	0.60	4.85	2.91		
		3.00					3.00	0.00	0.00	3.00	4.85	14.55		
		9.10					9.10	0.00	0.00	9.10	4.85	44.14		
	(East Side)	10.70					10.70	0.00	0.00	10.70	4.85	51.90		
		4.50					4.50	0.00	0.00	4.50	4.85	21.83		
		0.60					0.60	0.00	0.00	0.60	4.85	2.91		
		0.30					0.30	0.00	0.00	0.30	4.85	1.46		
	(South Side)	25.90					25.90	0.00	0.00	25.90	4.85	125.62		
													431	m²

FIGURE 10.20 Gross Wall Area

ESTIMATE WORK SHEET

Project:	Little Office Building		Estimate No.	1234
Location	Mountainville, BC		Sheet No.	1 of 1
Architect	C.K. Architects	Masonry Quantities	Date	23/12/20xx
Items	Masonry		By	HYZ Checked DCK

Cost Code	Description	Comments/Calculations						Sub Totals	Quantity	Unit
	Lintel Blocks									
	(Top Course)	100.97 m / 0.40 m per block						252		
	(Front Side - North Wall)	12.20+9.10+4.50+25.90 = 51.70 m Divide by 0.40 m						129		
	(Openings)	4 per Opening = 17 Openings						68		
	Total Lintel Blocks							449		
	Required Lintel Blocks	Waste @ 5 %							471	Lintel Blocks
	Concrete Blocks									
	(Gross Area)							417.00		
	(Windows)	16 x (0.91m x 2.13 m)						-31.01		
	(Doors)	1 x (0.91m x 2.13 m)						-1.94		
	Net Wall Area							384.05		
	Gross Blocks	125 per 10 m²						4,800		
	Less Lintel Blocks							449		
								4,351		
	Required Blocks	Waste @ 5 %							4,569	Blocks
	Mortar	0.07 m³ per 10 m² (417 m²)						2.92		
		Waste @ 40 %							4	m³
	Horizontal Reinforcing	Every 3rd Course (0.50) Factor						208.50		
		Waste @ 40 %							219	m³
	Vertical Reinforcing	10M bars Every 1.80 m (90.47/1.80) Spaces						51		
		Nominal 4.50 m Ht.						230.00		
		Waste @ 20 % (High because of splice)							276	m³
	Face Brick									
	(Gross Area)							431.00		
	(Windows)	16 x (0.91m x 2.13 m)						-31.01		
	(Doors)	1 x (0.91m x 2.13 m)						-1.94		
								398.05		
		746 per 10 m²						29,694		Face Bricks
		Waste @ 5 %							31,179	
	Mortar	0.173 m³ per 10 m²						6.89		
		Waste @ 40 %							10	m³

FIGURE 10.21 Masonry Quantities

10.21 Checklist

Masonry:
type (concrete, brick, stone)
kind
size (face size and thickness)
load-bearing
non-load-bearing
bonds (patterns)
colours
special facings

Miscellaneous:
fire ratings
inserts
amount of cutting
anchors
copings
bolts
sills
dowels
steps
reglets
walkways
wall ties
flashing

Mortar:
cement
lime (if required)
fine aggregate
water
admixtures
colouring
shape of joint

Reinforcing:
lintels
bars
expansion joints
wall reinforcing
control joints
galvanizing (if required)
weep holes

Web Resources

www.masonrycanada.ca
www.bia.org
www.mca-canada.com
www.canmasonry.com
www.masonrysociety.org
www.ncma.org
www.brickinfo.org

Review Questions

1. What factors affect the costs of labour when estimating masonry?

2. How may the type of bond (pattern) affect the amount of materials required?

3. Why is high accuracy required with an item such as masonry?

4. Why should local suppliers be contacted early in the bidding process when special shapes or colours are required?

5. Why must the estimator separate the various sizes of masonry units in the estimate?

6. What is a cash allowance; and how does it work?

7. How is stone veneer quantities estimated?

8. How may cold weather affect the cost of a building?

9. Determine the quantities for masonry, mortar, accessories and any other related items required for the residential building shown in Appendix C.

10. Determine the quantities for masonry, mortar, accessories and any other related items required for the commercial building shown in Appendix D.

11. (Computer Exercise) Create an estimate for all the masonry works using Timberline Precision Estimating Basic by entering the data from Question 9. Change the prices to match prevailing market rates.

12. (Computer Exercise) Create an estimate for all the masonry works using Timberline Precision Estimating Basic by entering the data from Question 10. Change the prices to match prevailing market rates.

CHAPTER 11

Metals

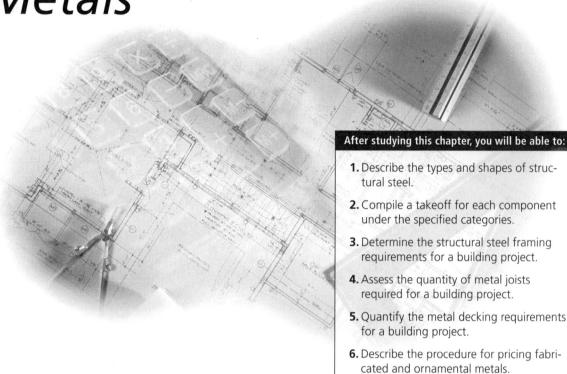

After studying this chapter, you will be able to:

1. Describe the types and shapes of structural steel.

2. Compile a takeoff for each component under the specified categories.

3. Determine the structural steel framing requirements for a building project.

4. Assess the quantity of metal joists required for a building project.

5. Quantify the metal decking requirements for a building project.

6. Describe the procedure for pricing fabricated and ornamental metals.

11.1 General

This chapter covers structural metal framing, metal joists, metal decking, cold-formed framing and metal fabrications, ornamental metal, and expansion control assemblies. The structural products are most commonly made of steel and aluminum. The rest of the products are available in a wide range of metals, including steel, aluminum, brass, and bronze.

11.2 Structural Steel

General contractors typically rely on subcontractors for the fabrication and erection of structural steelwork, which includes all columns, beams, lintels, trusses, joists, bearing plates, girts, purlins, decking, bracing, tension rods, and any other items required.

When estimating structural steelwork, the estimator should quantify each item, such as column bases, columns, trusses or lintels, separately. Structural steel is purchased by the tonne. The cost per tonne varies depending on the type and shape of metal required. Labour operations are different for each type.

The estimate of the field cost of erecting structural steel will vary depending on weather conditions, delivery of materials, equipment available, size of the building, and the amount of riveting, bolting, and welding required.

11.3 Structural Steel Framing

The metals used for the framing of a structure primarily include wide flange beams, (W series) light beams, I-beams, plates, channels, and angles. The wide flange shapes are the most commonly used today and are designated in Figure 11.1.

FIGURE 11.1 Wide Flange Beam Designation

Special shapes and composite members require customized fabrication and should be listed separately from the standard mill pieces. The estimator needs to be meticulous when quantifying each size and length. It is also a good practice to check not only the structural drawings but also the sections and details. The specifications should include the fastening method required and the drawings of the fastening details.

Estimating. Both aluminum and steel are sold by weight, so the takeoff is done in kilograms and converted into tonnes. The takeoff should first include a listing of all metal required for the structure. A definite sequence for the takeoff should be maintained. A commonly used sequence is columns and details, beams and details, and bracing and flooring (if required). Floor by floor, a complete takeoff is required.

Structural drawings, details, and specifications do not always show required items. Among the items that may not be shown specifically, but are required for a complete job, are various field connections, field bolts, ties, beam separators, anchors, rivets, bearing plates, welds, setting plates, and templates. The specification may require conformance with the Canadian Institute for Steel Construction (CISC) and the Canadian Standards Association (CSA) standards, with the exact methods to be determined by the fabricator and erector. When this situation occurs, a complete understanding is required of the CISC and Building Code requirements to do a complete estimate. Without this thorough understanding, the estimator can make an approximate estimate for a check on subcontractor prices but should not attempt to use those figures for compiling a bid price.

Once a complete takeoff of the structural steel has been done, the grouping should ⊹ FLOOR be according to the grade required and the shape of the structural piece. Special built-up shapes must be listed separately; in each area, the shapes should be broken down into small, medium, and large weights per metre. For standard mill pieces, the higher the weight per metre, the lower is the cost per kilogram. The weights of the various standard shapes may be obtained from any structural steel handbook or from the manufacturing company. To determine the weight of built-up members, the estimator adds each of the component shapes used to make the special shape.

E X A M P L E   **11.1**   **Structural Steel Framing Takeoff**

Determine the structural steel framing (columns and beams) requirements for the building shown in Figure 11.2, Roof Framing Plan, and Figure 11.3, Column Details.

From Figures 11.2 and 11.3 and details, calculations for structural steel can be found in Figure 11.4.

Weight of W310 × 45s that are 4.95 m long:
Weight (kilograms) = 4.95 × 45 kilograms per metre
4.95 × 45 × 3 = 668 kilograms

continued

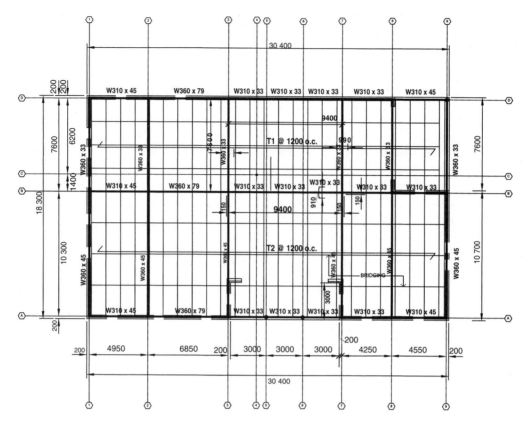

FIGURE 11.2 Roof Framing Plan

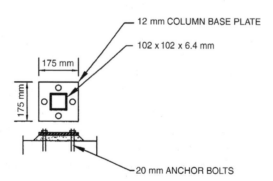

FIGURE 11.3 Column Details

Designation	Length (m)	Count
W310 x 45	4.95	3
W310 x 45	4.55	3
W310 x 33	3.00	9
W310 x 33	4.24	3
W360 x 33	7.60	6
W360 x 45	10.70	6
W360 x 79	6.65	3
102 x 102 x 6.4		22

FIGURE 11.4 Structural Metals Takeoff

continued

ESTIMATE WORK SHEET

		STRUCTURAL STEEL							

Project   Little Office Building   Estimate No. 1234
Location   Mountainville, BC   Sheet No. 1 of 1
Architect   C.K. Architects   Date 23/12/20xx
Items   Structural Steel   By HYZ   Checked DCK

Cost Code	Description	Designation	Kilogram/Metre	Length m	Count			Quantity	Unit
	Roof Framing (Beams)	W310 × 45	45.00	4.95	3			668	kg
	Roof Framing (Beams)	W310 × 45	45.00	4.55	3			614	kg
	Roof Framing (Beams)	W310 × 33	33.00	3.00	9			891	kg
	Roof Framing (Beams)	W310 × 33	33.00	4.25	3			421	kg
	Roof Framing (Beams)	W360 × 33	33.00	7.60	6			1,505	kg
	Roof Framing (Beams)	W360 × 45	45.00	10.70	6			2,889	kg
	Roof Framing (Beams)	W360 × 79	79.00	6.85	3			1,623	kg
	Columns	102 × 102 × 6.4	18.20	3.00	22			1,181	kg
								9,792	Kilograms

FIGURE 11.5 Structural Metals Quantification

Figure 11.5 is a spreadsheet that contains the completed structural metals' quantification for the roof framing (but does not include basement metals).

11.4 Metal Joists

Metal joists, also referred to as *open web steel joists,* are prefabricated lightweight trusses. Figure 11.6 shows various types of metal joists. The weight per metre is typically found in the manufacturer's catalogue. Figures 11.7 and 11.8 are excerpts from a bar joist catalogue that show weights and anchoring. Architects and engineers, to select the best depth and most economical joist, use the joist depth tables as a general guide. Engineering drawings usually identify joist mark, span, and loading and service requirements, and the truss manufacturers use this information to design and supply the joists.

To determine the quantity of bar joists, calculate the length (m) of joists required, and multiply this number by its weight per metre; the product represents the total weight. The cost per kilogram is considerably greater than that for some other types of steel (such as reinforcing bars or wide flange shapes) due to the sophisticated shaping and fabrication required.

Metal joists by themselves do not make an enclosed structure—they are one part of an assembly. The other materials included in this assembly are varied, and selection may be made on the basis of economy, appearance, sound control, fire-rating requirements, or any other criterion. When doing the takeoff of joists, the estimator should be aware of the other parts in the assembly; all parts must be included somewhere in the estimate. The mechanical and electrical requirements are generally quite compatible with steel joists; only large ducts have to be carefully planned for job installation.

Specifications.   Specifications will list the type of joists required, the type of attachment they will have to the rest of the structure, and the accessories and finish. Many

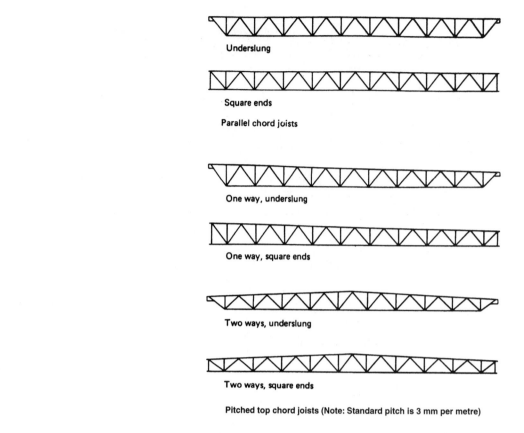

Underslung

Square ends

Parallel chord joists

One way, underslung

One way, square ends

Two ways, underslung

Two ways, square ends

Pitched top chord joists (Note: Standard pitch is 3 mm per metre)

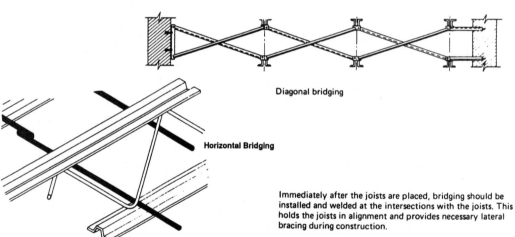

Diagonal bridging

Horizontal Bridging

Immediately after the joists are placed, bridging should be installed and welded at the intersections with the joists. This holds the joists in alignment and provides necessary lateral bracing during construction.

FIGURE 11.6 Common Steel Joist Shapes and Joist Bridging

specifications will enumerate industry standards for strength, the type of steel used in the joists, erection and attachment techniques, and finishes that must be met. Read the standards carefully. Accessories that may be specified are bridging, ceiling extensions, masonry wall anchors, bridging anchors, and header angles.

Estimating. Because structural metal is sold by the kilogram, the total number of kilograms required must be determined. First, take off the length (m) of each different type of joist required. Then, multiply the length of each type times the weight per metre to determine the total weight of each separate type. Also, estimate the accessories required—both their type and the number needed.

JOIST DEPTH SELECTION TABLE

METRIC

XXX	: **Mass of Joist (kg/m)**
XXX	: **% of service load to produce a deflection of L/360**

Span 6

Span (m)	Joist Depth (mm)	4.5 / 3.0	6.0 / 4.0	7.5 / 5.0	9.0 / 6.0	10.5 / 7.0	12.0 / 8.0	13.5 / 9.0	15.0 / 10.0	16.5 / 11.0	18.0 / 12.0	19.5 / 13.0	21.0 / 14.0	22.5 / 15.0
	300	9.1 / 88	9.7 / 69	11.2 / 64	12.8 / 64	14.6 / 65	16.8 / 65	18.9 / 65	20.8 / 65	22.7 / 64	24.7 / 64	26.3 / 63	30.8 / 64	30.8 / 64
	350	9.3 / 122	9.6 / 92	10.0 / 77	11.7 / 74	12.4 / 71	13.6 / 67	14.9 / 68	16.1 / 66	18.3 / 68	19.0 / 64	20.3 / 66	24.1 / 65	24.1 / 64
	400	9.4 / 162	9.9 / 121	9.9 / 97	10.6 / 85	12.0 / 84	13.1 / 82	13.6 / 76	15.1 / 74	15.9 / 74	16.9 / 73	19.4 / 76	21.6 / 74	21.6 / 71
6	450	9.9 / 200	10.1 / 155	10.1 / 124	10.5 / 108	11.0 / 97	12.6 / 94	13.5 / 93	14.8 / 91	15.5 / 86	16.4 / 87	17.4 / 83	20.1 / 83	20.7 / 88
	500	10.1 / 200	10.2 / 193	10.2 / 154	10.7 / 129	11.1 / 116	11.6 / 101	13.0 / 104	14.6 / 105	14.9 / 100	15.7 / 98	16.9 / 96	18.9 / 93	18.9 / 93
	550	10.7 / 200	10.8 / 200	11.1 / 188	11.1 / 157	11.6 / 134	11.9 / 123	13.5 / 120	14.6 / 121	15.4 / 110	15.8 / 106	16.2 / 103	16.9 / 103	18.4 / 107
	600	10.8 / 200	10.9 / 200	11.8 / 200	12.5 / 200	13.4 / 196	13.8 / 177	15.0 / 172	16.0 / 158	16.3 / 147	17.5 / 146	18.5 / 138	18.5 / 142	18.5 / 119

Span 8

Span (m)	Joist Depth (mm)	4.5 / 3.0	6.0 / 4.0	7.5 / 5.0	9.0 / 6.0	10.5 / 7.0	12.0 / 8.0	13.5 / 9.0	15.0 / 10.0	16.5 / 11.0	18.0 / 12.0	19.5 / 13.0	21.0 / 14.0	22.5 / 15.0
	400	9.2 / 67	11.5 / 65	14.0 / 66	16.0 / 65	20.5 / 65	20.6 / 63	24.0 / 65	26.5 / 65	28.3 / 64	31.6 / 64	34.4 / 64	36.5 / 65	38.0 / 66
	450	9.6 / 86	10.3 / 70	12.5 / 70	14.1 / 66	16.9 / 65	18.5 / 66	20.2 / 65	21.7 / 65	23.9 / 65	25.9 / 64	28.5 / 64	30.3 / 65	30.3 / 63
	500	9.7 / 107	10.3 / 84	11.9 / 78	13.4 / 73	15.8 / 70	16.0 / 68	17.0 / 74	17.3 / 69	19.1 / 69	20.5 / 68	22.9 / 65	24.9 / 67	26.0 / 68
8	550	10.4 / 131	10.6 / 98	11.6 / 86	13.3 / 84	14.5 / 80	15.6 / 80	16.0 / 76	17.1 / 81	17.9 / 77	19.5 / 77	22.7 / 74	24.8 / 74	24.8 / 73
	600	10.7 / 156	10.9 / 117	11.8 / 98	14.1 / 101	15.0 / 91	15.2 / 86	15.6 / 92	16.4 / 91	17.6 / 88	18.1 / 84	22.6 / 81	24.5 / 83	24.5 / 82
	650	12.2 / 200	13.6 / 200	13.7 / 176	14.3 / 151	15.2 / 126	15.3 / 110	15.4 / 98	15.6 / 98	16.6 / 104	17.9 / 96	20.0 / 92	22.4 / 98	22.8 / 94
	700	12.3 / 200	13.7 / 200	13.9 / 200	14.4 / 176	15.8 / 158	16.0 / 128	16.1 / 114	16.5 / 106	17.0 / 106	18.0 / 112	20.3 / 106	21.3 / 108	22.0 / 107

Span 10

Span (m)	Joist Depth (mm)	4.5 / 3.0	6.0 / 4.0	7.5 / 5.0	9.0 / 6.0	10.5 / 7.0	12.0 / 8.0	13.5 / 9.0	15.0 / 10.0	16.5 / 11.0	18.0 / 12.0	19.5 / 13.0	21.0 / 14.0	22.5 / 15.0
	500	11.6 / 66	13.5 / 65	16.8 / 64	18.2 / 65	21.8 / 65	24.7 / 64	31.5 / 64	33.1 / 64	33.6 / 64	37.0 / 65	42.0 / 68	45.5 / 69	45.5 / 64
	550	10.5 / 70	13.3 / 68	13.9 / 68	15.6 / 65	18.4 / 65	20.2 / 63	24.6 / 65	28.3 / 65	28.3 / 64	30.0 / 64	33.3 / 64	36.1 / 67	38.4 / 64
	600	11.1 / 83	13.2 / 77	13.6 / 76	14.4 / 70	17.2 / 71	18.8 / 69	21.8 / 67	23.9 / 65	24.8 / 67	26.4 / 65	28.6 / 64	31.7 / 65	35.2 / 68
10	650	11.8 / 132	13.4 / 112	13.7 / 89	14.2 / 83	16.0 / 78	17.8 / 76	20.7 / 74	22.7 / 72	23.2 / 72	25.3 / 73	27.0 / 69	28.9 / 70	31.8 / 72
	700	11.9 / 153	13.5 / 14	13.8 / 104	14.3 / 87	15.4 / 85	17.2 / 85	19.9 / 81	22.3 / 80	22.3 / 76	24.8 / 83	25.2 / 75	26.7 / 75	29.9 / 80
	750	12.1 / 177	13.6 / 133	14.0 / 120	14.4 / 100	15.7 / 95	16.8 / 98	18.3 / 90	19.9 / 90	21.6 / 87	23.1 / 88	25.0 / 88	26.5 / 89	28.3 / 87
	800	12.3 / 200	13.7 / 172	14.1 / 137	14.5 / 114	16.0 / 98	17.1 / 95	19.3 / 100	21.9 / 96	21.9 / 93	22.9 / 95	24.1 / 93	26.0 / 94	27.4 / 93

Span 12

Span (m)	Joist Depth (mm)	4.5 / 3.0	6.0 / 4.0	7.5 / 5.0	9.0 / 6.0	10.5 / 7.0	12.0 / 8.0	13.5 / 9.0	15.0 / 10.0	16.5 / 11.0	18.0 / 12.0	19.5 / 13.0	21.0 / 14.0	22.5 / 15.0
	600	13.9 / 65	15.0 / 65	18.4 / 66	21.4 / 64	26.6 / 64	32.8 / 64	32.8 / 64	36.7 / 66	42.4 / 68	46.1 / 68	50.8 / 68	50.8 / 64	54.6 / 65
	650	13.1 / 86	13.4 / 64	15.8 / 64	18.8 / 65	23.3 / 65	25.5 / 65	28.3 / 65	31.6 / 64	34.2 / 64	38.0 / 64	43.3 / 68	47.0 / 65	47.4 / 65
	700	13.5 / 100	13.5 / 75	14.4 / 67	17.6 / 68	20.5 / 64	21.9 / 64	24.9 / 66	27.5 / 66	29.5 / 65	31.8 / 64	36.1 / 66	37.5 / 65	41.5 / 67
12	750	13.5 / 115	13.6 / 87	14.6 / 74	16.5 / 75	18.2 / 70	21.1 / 70	23.4 / 70	25.3 / 68	27.9 / 69	31.1 / 68	32.9 / 68	36.0 / 71	40.9 / 74
	800	13.6 / 132	13.8 / 99	14.7 / 79	16.7 / 79	18.8 / 77	19.6 / 76	22.7 / 75	23.9 / 72	26.7 / 75	29.6 / 74	31.6 / 72	33.2 / 72	36.4 / 76
	900	13.8 / 168	14.0 / 126	14.9 / 101	16.8 / 93	19.0 / 94	19.8 / 88	21.4 / 87	23.6 / 89	25.2 / 88	27.4 / 85	28.9 / 85	30.9 / 84	33.5 / 82
	1 000	14.1 / 200	14.3 / 156	15.0 / 125	17.0 / 107	19.1 / 108	20.0 / 102	21.5 / 100	23.7 / 99	25.4 / 97	27.0 / 100	28.3 / 98	29.8 / 96	31.4 / 94

  Lightest joist: See margin for corresponding depth

FIGURE 11.7 Joist Loading Table

(Courtesy of Canam)

JOIST DEPTH SELECTION TABLE

METRIC | XXX | : Mass of Joist (kg/m)
| XXX | : % of service load to produce a deflection of L/360

Span 15 — Factored load (kN/m) / Service load (kN/m)

Joist Depth (mm)	4.5 / 3.0	6.0 / 4.0	7.5 / 5.0	9.0 / 6.0	10.5 / 7.0	12.0 / 8.0	13.5 / 9.0	15.0 / 10.0	16.5 / 11.0	18.0 / 12.0	19.5 / 13.0	21.0 / 14.0	22.5 / 15.0
750	14.8 / 68	18.5 / 64	22.3 / 64	26.5 / 65	31.1 / 65	35.4 / 64	42.2 / 68	45.7 / 67	50.0 / 66	53.5 / 66	58.8 / 65	63.7 / 63	67.5 / 64
800	13.7 / 67	16.9 / 64	20.3 / 65	24.0 / 65	27.4 / 64	31.6 / 65	35.6 / 64	40.1 / 64	43.1 / 64	46.8 / 64	50.3 / 64	54.2 / 65	59.8 / 64
900	13.8 / 86	14.8 / 71	18.1 / 70	21.2 / 69	23.8 / 67	26.9 / 67	29.8 / 65	32.8 / 66	36.8 / 68	40.0 / 65	43.7 / 69	51.7 / 76	51.7 / 71
1 000	14.0 / 106	14.9 / 80	17.1 / 84	19.4 / 77	22.5 / 76	25.3 / 76	27.4 / 74	31.0 / 74	34.5 / 75	39.7 / 83	41.8 / 79	43.3 / 77	44.8 / 75
1 100	14.3 / 129	15.1 / 97	17.3 / 94	19.6 / 93	21.5 / 86	24.1 / 86	27.4 / 88	29.3 / 83	32.3 / 84	35.8 / 83	38.3 / 85	42.3 / 90	43.7 / 86
1 200	15.6 / 154	15.6 / 116	17.5 / 103	19.8 / 101	21.7 / 99	24.6 / 98	27.6 / 95	29.6 / 97	31.1 / 95	33.7 / 92	37.5 / 96	39.2 / 94	43.1 / 97
1 300	15.9 / 182	15.9 / 140	17.6 / 122	19.9 / 110	21.8 / 108	24.8 / 112	27.7 / 111	29.9 / 108	31.5 / 106	34.0 / 105	35.1 / 102	38.0 / 105	42.9 / 112

Span 18 — Factored load (kN/m) / Service load (kN/m)

Joist Depth (mm)	4.5 / 3.0	5.4 / 3.6	6.3 / 4.2	7.2 / 4.8	8.1 / 5.4	9.0 / 6.0	9.9 / 6.6	10.8 / 7.2	11.7 / 7.8	12.6 / 8.4	13.5 / 9.0	14.4 / 9.6	15.3 / 10.2
900	17.0 / 65	21.7 / 71	26.7 / 76	31.9 / 80	35.4 / 85	39.6 / 86	42.1 / 94	44.6 / 94	46.5 / 91	47.4 / 98	49.1 / 99	51.2 / 98	52.9 / 97
1 000	15.0 / 68	18.9 / 72	22.9 / 78	27.0 / 80	31.1 / 84	36.2 / 87	38.1 / 87	40.9 / 90	41.6 / 91	43.1 / 92	46.4 / 93	46.7 / 100	47.8 / 101
1 100	14.2 / 75	18.1 / 81	20.8 / 83	25.5 / 89	28.6 / 91	30.7 / 92	31.4 / 96	36.5 / 100	37.8 / 103	38.4 / 104	39.0 / 104	41.3 / 98	46.2 / 115
1 200	14.6 / 89	17.2 / 97	20.5 / 95	23.9 / 98	25.4 / 101	27.1 / 101	29.0 / 105	31.6 / 106	33.4 / 111	35.1 / 108	36.2 / 125	38.6 / 117	41.9 / 110
1 300	15.0 / 105	17.9 / 106	19.1 / 105	20.0 / 109	23.0 / 112	25.2 / 113	28.1 / 114	30.7 / 121	32.3 / 128	33.8 / 125	34.1 / 119	36.1 / 138	38.7 / 130
1 400	16.3 / 122	18.1 / 108	20.3 / 117	21.9 / 117	23.9 / 126	26.0 / 127	26.4 / 126	28.4 / 128	30.9 / 130	31.7 / 138	33.0 / 136	35.4 / 130	38.0 / 152
1 600	16.9 / 160	19.0 / 149	21.3 / 143	22.9 / 142	24.3 / 152	27.0 / 156	27.4 / 150	29.2 / 153	31.1 / 172	31.5 / 165	32.5 / 165	34.8 / 159	37.3 / 169

Span 24 — Factored load (kN/m) / Service load (kN/m)

Joist Depth (mm)	4.5 / 3.0	5.4 / 3.6	6.3 / 4.2	7.2 / 4.8	8.1 / 5.4	9.0 / 6.0	9.9 / 6.6	10.8 / 7.2	11.7 / 7.8	12.6 / 8.4	13.5 / 9.0	14.4 / 9.6	15.3 / 10.2
1 200	22.2 / 65	25.5 / 64	30.9 / 65	33.7 / 64	42.0 / 71	42.8 / 64	47.4 / 66	52.1 / 66	55.3 / 66	60.4 / 65	69.1 / 70	70.7 / 66	75.4 / 67
1 300	20.4 / 66	23.3 / 65	27.5 / 64	30.5 / 64	33.6 / 64	37.9 / 64	42.1 / 65	44.7 / 65	49.1 / 66	52.9 / 66	57.2 / 68	66.2 / 72	66.6 / 68
1 400	21.0 / 74	23.0 / 68	27.0 / 70	29.2 / 69	32.6 / 68	34.7 / 66	38.7 / 68	42.7 / 69	44.7 / 67	49.8 / 71	53.5 / 72	58.1 / 68	64.2 / 80
1 600	21.3 / 91	23.2 / 83	25.8 / 84	28.5 / 83	30.2 / 80	32.4 / 78	35.8 / 77	42.1 / 86	44.0 / 82	45.5 / 80	50.3 / 77	54.2 / 89	54.8 / 84
1 800	22.9 / 107	24.4 / 101	26.4 / 96	29.3 / 98	31.3 / 96	32.8 / 91	35.6 / 92	39.3 / 93	43.8 / 98	44.9 / 94	50.0 / 100	50.3 / 90	51.6 / 93
2 000	23.2 / 126	24.6 / 117	27.2 / 117	30.0 / 113	31.7 / 112	33.5 / 111	36.1 / 107	41.5 / 119	43.0 / 114	44.7 / 110	45.9 / 109	50.0 / 113	51.5 / 109
2 200	25.2 / 200	27.6 / 142	30.9 / 135	32.4 / 131	33.3 / 127	34.3 / 122	36.5 / 118	42.3 / 134	43.6 / 129	44.9 / 128	45.7 / 124	46.4 / 120	51.3 / 125

Span 30 — Factored load (kN/m) / Service load (kN/m)

Joist Depth (mm)	4.5 / 3.0	5.4 / 3.6	6.3 / 4.2	7.2 / 4.8	8.1 / 5.4	9.0 / 6.0	9.9 / 6.6	10.8 / 7.2	11.7 / 7.8	12.6 / 8.4	13.5 / 9.0	14.4 / 9.6	15.3 / 10.2
1 600	29.2 / 69	32.8 / 65	35.9 / 64	40.9 / 66	43.5 / 64	50.1 / 67	53.2 / 66	59.8 / 64	63.4 / 65	72.5 / 70	75.8 / 66	79.8 / 66	84.8 / 66
1 800	28.1 / 87	31.1 / 74	35.1 / 72	39.2 / 74	43.3 / 75	45.6 / 73	51.1 / 76	55.7 / 77	59.3 / 76	65.9 / 77	68.8 / 72	72.6 / 72	81.3 / 78
2 000	27.6 / 87	30.7 / 88	34.0 / 84	36.7 / 82	43.0 / 87	44.8 / 84	47.0 / 80	52.8 / 83	56.7 / 89	63.5 / 87	68.2 / 87	68.8 / 84	77.7 / 89
2 200	27.9 / 101	31.0 / 98	34.7 / 99	36.8 / 94	43.3 / 102	45.6 / 98	46.1 / 92	52.4 / 96	53.3 / 93	60.0 / 106	61.6 / 99	63.8 / 93	64.0 / 88
2 400	29.7 / 115	33.0 / 114	35.7 / 108	37.7 / 105	44.3 / 113	45.9 / 109	48.6 / 107	52.9 / 108	54.7 / 106	60.2 / 109	62.0 / 119	65.4 / 111	70.2 / 111
2 600	31.3 / 131	36.5 / 168	38.2 / 146	38.9 / 118	45.0 / 128	46.4 / 120	48.8 / 117	53.2 / 123	55.2 / 117	60.7 / 121	62.5 / 117	66.9 / 139	70.4 / 123
2 800	37.4 / 200	37.7 / 195	39.3 / 170	39.8 / 151	46.3 / 149	46.9 / 134	49.0 / 131	53.8 / 132	55.9 / 132	61.0 / 132	62.9 / 132	68.3 / 134	71.0 / 130

Lightest joist: See margin for corresponding depth

FIGURE 11.7 continued

(Courtesy of Canam)

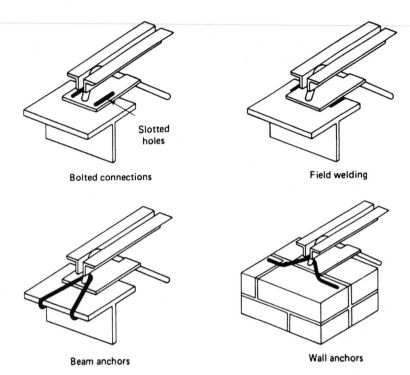

Slotted holes

Bolted connections

Field welding

Beam anchors

Wall anchors

FIGURE 11.8 End Anchorage

If the contractor is to erect the joists, an equipment list, costs, and the labour work hours are required. On small jobs, it is not unusual for the contractor to use its own forces to set the joists. Remember that all accessories must also be installed. If the joist erection is to be subcontracted, the general contractor's responsibilities must be defined, as well as who will install the accessories.

E X A M P L E   **11.2**   **Bar Joists**

Determine the quantity of bar joists required to frame the roof in drawing (see Figure 11.2). From this drawing, the bar joist quantity table, as shown in Figure 11.9, can be developed.

Mark	Quantity	Joist Depth (mm)	Span Base Length (m)	Factored Load (kN/m)	Spacing (m)
T1	1	1200	19	6.3	1.20
T2	1	1300	19	9.0	1.20

FIGURE 11.9 Bar Joist Takeoff

From Figure 11.7, the T1 weighs 19.3 kilograms per metre and the T2 weighs 26.0 kilograms per metre.

Length of T1 = 7.60 × 23 = 175 m

Length of T2 = 10.70 × 23 = 246 m

Kilograms of T1 = 175 m × 19.3 kilograms per metre = 3,778 kilograms

Kilograms of T2 = 246 m × 26.0 kilograms per metre = 6,396 kilograms

Total weight of bar joist = 10.2 tonnes

Joist Span	Work Hours per Tonne
Up to 10 metres	5.5–9.0
Over 10 metres	4.5–8.0

FIGURE 11.10 Bar Joist Installation Productivity Rates

Labour. The time required for the installation of steel joists may be taken from Figure 11.10. A crane is usually required for joists over 136 kilograms. The crane would reduce installation time on large projects and justify the cost of the crane, even if the joist weight were less than 136 kilograms.

11.5 Metal Decking

Metal decking is used for floor and roof applications. Depending on the particular requirements of the job, a wide selection of shapes, sizes, thicknesses, and accessories are available.

Roof applications range from simple decks over which insulation board and built-up roofing are applied, to forms and reinforcing over which concrete may be poured, and to decking that can receive recessed lighting and has acoustical properties. Depending on the type used and the design of the deck, allowable spans range from 2 m to about 12 m.

Decking for floor applications is equally varied from the simplest type used as a form and reinforcing for concrete, to elaborate systems that combine electrical and telephone outlets, electrical raceways, air ducts, acoustical finishes, and recessed lighting.

Decking is generally available unpainted, primed, painted, or galvanized. Accessories available include flexible rubber closures to seal the flutes, clips that fasten the decking to the purlins, lighting, and acoustical finishes.

Specifications. Determine the type of decking required, thickness, gauge of metal; finish required on the decking as received from the supplier, method of attachment, accessories required, and the specified manufacturers.

Items that are necessary for the completion of the decking and which must be included in other sections of the specifications include painting of the underside of the deck, acoustical treatment, openings, and insulation. The estimator must note these requirements and make specific exclusions for any work to be performed by other trades.

Estimating. Because metal decking is priced by the area, the first thing the estimator must determine is how many square metres are required. Again, a systematic plan should be used: Start on the floor on which the decking is first used and work up through the building, keeping the estimates for all floors separate.

Decking is usually installed by welding directly through the bottom of the rib, usually a maximum of 300 mm on centre, with side joints mechanically fastened not more than 910 mm on centre. The estimator will have to determine approximately how many weld washers will be required and how long it will take to install them. Otherwise, fastening is sometimes effected by clips, screws, and bolts.

The estimator should consult local dealers and suppliers for material prices to be used in preparing the estimate. For materials priced f.o.b., the dealer will require that the estimator add the cost of transporting the materials to the jobsite. Once at the jobsite, they must be unloaded, perhaps stored, and then placed on the appropriate floor for use. In most cases, one or two workers can quickly and easily position the decking and make preparations for the welder to make the connections.

The estimator must be especially careful on multifloor buildings to count the number of floors requiring the steel deck and keep the roof deck and any possible poured concrete for the first floor separate. Most estimators make a small sketch of the number of floors to help avoid errors. In checking the number of floors, be aware that often there may also be a lobby level, lower lobby level, and basement.

EXAMPLE 11.3 Decking

Determine decking for the sample project given in Figure 11.2.

$$\text{Roof area (m}^2\text{)} = 30.40 \times 18.30 = 556.32 \text{ m}^2$$
$$\text{Add 5\% for waste} - \text{Use 584 m}^2$$

From the labour productivity rates in Figure 11.11 and assuming a prevailing wage rate of \$33 per hour, the following labour cost calculations can be performed:

Steel Decking Thickness (mm)	Work Hours per m²
0.76	0.01 to 0.03
1.21	0.02 to 0.04
1.52	0.03 to 0.05

FIGURE 11.11 Steel Decking Productivity Rates

Using 0.76-mm-thick decking and 0.02 work hours per m²:
$$\text{Work hours} = 584 \text{ m}^2 \times 0.02 = 11.68 \text{ work hours}$$
$$\text{Basic labour cost (\$)} = 11.68 \text{ work hours} \times \$33 \text{ per work hour} = \$385.44$$

11.6 Miscellaneous Structural Metal

Other types of structural framing are sometimes used. They include structural aluminum or steel studs, joists, purlins, and various shapes of structural pipe and tubes available. The procedure for estimating each of these items is the same as that outlined previously for the rest of the structural steel.

1. Take off the various types and shapes.
2. Determine the weight in kilograms of each type required.
3. Figure the cost per tonne times the tonnage required equals the material cost.
4. Determine the work hours and equipment required and their respective costs.

11.7 Metal Erection Subcontractors

Most of the time these subcontractors can erect structural steel at considerable savings compared with the cost to the average general contractor. They have specialized, well-organized workers to complement the required equipment that includes cranes, air tools, rivet busters, welders, and impact wrenches. These factors, when combined with the experienced organization that specializes in one phase of construction, are hard to beat.

Using subcontractors never lets the estimator off the hook; an estimate of steel is still required for effective cost control.

11.8 Miscellaneous Metals

Metal Fabrications. The metal stairs, ladders, handrails, railings, gratings, floor plates, and any castings are all considered part of metal fabrications. These items are typically manufactured to conventional (standard) details.

The estimator should carefully review the drawings and specifications of each item. Next, possible suppliers should be called in to discuss pricing and installation. It may be possible to install with the general contractor's crews, or special installation or equipment may be needed, requiring the job to be subcontracted.

Ornamental Metal. Ornamental metals include ornamental stairs, prefabricated spiral stairs, ornamental handrails and railings, and ornamental sheet metal. These types of products are most commonly made of steel, aluminum, brass, and bronze and are often special orders.

The estimator should write a brief description of each item and must include the method of fabrication, the types of materials required, catalogue references (if from a specified supplier), and quantities required measured in metres or enumerated as appropriate. Attaching the relevant Division 5 sections to this scope of work and requesting quotations from fabricators and/or installers ensures that a good price and a good supplier can be found for the job.

Expansion Control Covers. Expansion control covers include the manufactured, cast or extruded metal expansion joint frames and covers, slide bearings, anchors, and related accessories. These materials are taken off by the metre, with the materials and any special requirements noted. Many times, a subcontractor doing related work installs them. For example, an expansion joint cover being used on an exterior block wall might be installed by the masonry subcontractor (if one is used). In other cases, the general contractor may assume the responsibility as it may also be specified in Division 10—Miscellaneous Specialties. The key is to be certain that both material and installation are included in the estimate somewhere.

11.9 Metal Checklist

Shapes:
  lengths
  sections
  quantities
  weights
  locations
  fasteners

Engineering:
  fabrication
  shop drawings shop painting
  testing
  inspection
  unloading, loading

erection
plumbing up

Fabrications:
metal stairs
ladders
handrails and railings
(pipe and tube)
gratings and floor plates
castings

Installation:
riveting
welding
bracing (cross and wind)
erection
bolts

Miscellaneous:
clips
ties
rods
painting
hangers
plates
anchor bolts

Ornamental:
stairs
prefab spiral stairs
handrails and railings
metal castings
sheet metal

Expansion control:
expansion joint cover
assemblies
slide bearings

Web Resources

www.cisc-icca.ca
www.cssbi.ca
www.sdi.org
www.steelplus.com
www.asce.org
www.aisc.org
www.canammanac.com

Review Questions

1. What two materials are most commonly used for structural framing metals and how are they priced?

2. Under what conditions might it be desirable for the contractor to use a structural subcontractor to erect the structural metal frame of the project?

3. Why should the estimator list each of the different shapes (such as columns and steel joists) separately?

4. What is the unit of measure for metal decks? What type of information needs to be noted?

5. How are fabricated metal and ornamental metal usually priced?

6. Determine the quantities for the structural steelwork required for the commercial building shown in Appendix D.

7. (Computer Exercise) Create an estimate for all the structural steelwork using Timberline Precision Estimating Basic by entering the data from Question 6. Change the prices to match prevailing market rates.

CHAPTER **12**

Wood

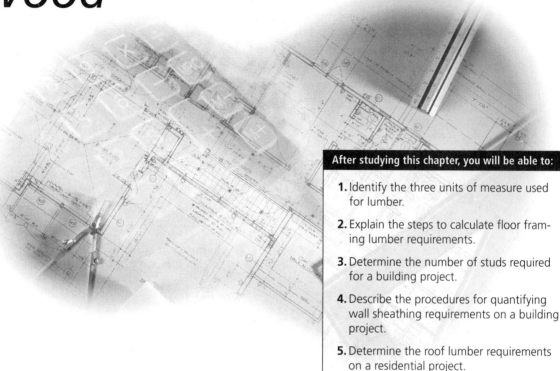

After studying this chapter, you will be able to:

1. Identify the three units of measure used for lumber.

2. Explain the steps to calculate floor framing lumber requirements.

3. Determine the number of studs required for a building project.

4. Describe the procedures for quantifying wall sheathing requirements on a building project.

5. Determine the roof lumber requirements on a residential project.

12.1 Frame Construction

SHEATHING
½" OSB
½ PLY - EXT WALLS
¾ PLY - FLOOR ⅝ T+G
⅜ OSB ROOF

DIMENTIONAL LUMBER
2×4 2×6 - walls
2×10 - 2×12 - Floors/ joists
2×10 - 2×12 - Lintels/ BRIDGING
2×4 2×6 Top + sill plates
2×12 stair stringers
4×4 4×6 6×6 POSTS

Wood frame construction is the most widely used system for the construction of residential buildings in Canada. The aspects of wood light framing construction discussed in this chapter primarily cover the rough carpentry work, which includes framing, sheathing, and subfloors. Flooring, roofing, drywall, and insulation are all included in their respective chapters, and so discussions of them are not repeated here.

Wood light frame construction makes use of dimension lumber and manufactured wood products of comparable size to build structural framing systems. These main structural members are used together with sheathing elements to provide rigidity for walls, floors, and roofs. Typically, light frame members are spaced no further apart than 600 mm. For some loading configurations, engineered wood products, such as light frame trusses, prefabricated wood I-joists, or other structural products, for example, laminated veneer lumber (LVL), parallel strand lumber (PSL) and Glulam, may be used as framing elements. Where large or clear spans are a requirement, light-framing members may be used in combination with heavy beams or columns to transfer loads directly to foundations.

Lumber is a general term, which includes boards, dimension lumber, and timber. It is a product that is manufactured by sawing logs into rough-sized lumber that are edged, resawn to final dimension, and cut to length. Canadian dimension lumber, timbers, and boards are manufactured according to the *Standard Grading Rules for Canadian Lumber* published by the National Lumber Grades Authority (NLGA). The NLGA rule is approved by the Canadian Lumber Standards Accreditation Board. Lumber manufactured and measured according to NLGA grading rules is considered Standard Lumber in Canada and meets the provisions of the Canadian Softwood Lumber CSA Standard CAN/CSA-0141M.

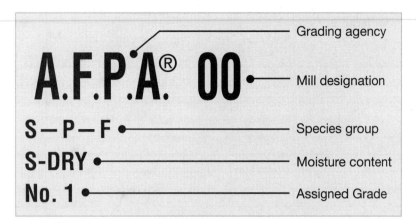

FIGURE 12.1 Lumber Grade Marks Used in Canada   *per 1000/ B.F.*

(Courtesy Canadian Wood Council)

MULTIPLES SOLD IN 2 FEET

LARGE QUANTITIES ARE SOLD BY VOLUME

Canadian dimension lumber (generally sawn lumber 38 mm wide and up to 286 mm deep) is manufactured to conform to grading rules meeting both Canadian and American requirements. Each piece of lumber is inspected to determine its grade, and a stamp (see Figure 12.1) is applied indicating the assigned grade, the mill of origin, a green or dry moisture content at time of manufacture, the species or species group, and the grading authority having jurisdiction over the mill of origin.

Dimension lumber is categorized into the following four groups: structural light framing, structural joists and planks, light framing, and studs. The table in Figure 12.2 shows the grades and uses for these groups.

Standard dimension lumber sizes produced in Canada are listed as surfaced dry sizes (19 percent moisture content or less) in the table in Figure 12.3. The availability of lumber sizes varies somewhat according to the area of the country from which the raw material originates. Lengths up to 6.1 m (20 feet) are commonly available from Western Canada and up to 4.9 m (16 feet) from Eastern Canada. Longer lengths are available on special order.

Grade Category	Size	Grades
Structural Light Framing (2" to 4" nom.)	38 to 89 mm thick and wide	Select Structural, No. 1, No. 2, No. 3
Structural Joists and Planks	38 to 89 mm (2" to 4" nom.) thick and 114 mm (5" nom.) or more wide	Select Structural, No. 1, No. 2, No. 3
Light Framing	38 to 89 mm (2" to 4" nom.) thick and wide	Construction, Standard, Utility
Studs	38 to 89 mm (2" to 4" nom.) thick and 38 to 140 mm (2" to 6" nom.) wide and 3 m (10') or less in length	Stud, Economy Stud

FIGURE 12.2 Dimensions Lumber—Grades and Uses

(Courtesy Canadian Wood Council)

Surfaced Dry (S-Dry), Size, mm	Surfaced Dry (S-Dry), Size, in. (actual)	Rough Sawn Size, in. (nom.)	Surface Green (S-Grn) Size, in. (actual)
38 x 38	1-1/2 x 1-1/2	2 x 2	1-9/16 x 1-9/16
x 64	x 2-1/2	2 x 3	x 2-9/16
x 89	x 3-1/2	2 x 4	x 3-9/16
x 140	x 5-1/2	2 x 6	x 5-5/8
x 184	x 7-1/4	2 x 8	x 7-1/2
x 235	x 9-1/4	2 x 10	x 9-1/2
x 286	x 11-1/4	2 x 12	x 11-1/2
64 x 64	2-1/2 x 2-1/2	3 x 3	2-9/16 x 2-9/16
x 89	x 3-1/2	2 x 4	x 3-9/16
x 140	x 5-1/2	2 x 6	x 5-5/8
x 184	x 7-1/4	2 x 8	x 7-1/2
x 235	x 9-1/4	2 x 10	x 9-1/2
x 286	x 11-1/4	2 x 12	x 11-1/2
89 x 89	3-1/2 x 3-1/2	4 x 4	3-9/16 x 3-9/16
x 140	x 5-1/2	2 x 6	x 5-5/8
x 184	x 7-1/4	2 x 8	x 7-1/2
x 235	x 9-1/4	2 x 10	x 9-1/2
x 286	x 11-1/4	2 x 12	x 11-1/2

Notes:
1. 38 mm (2" nominal) lumber is readily available as S-Dry.
2. S-Dry lumber is surfaced at a moisture content of 19 percent or less.
3. After drying, S-Green lumber sizes will be approximately the same as S-Dry lumber.
4. Tabulated metric sizes are equivalent to Imperial S-Dry sizes rounded to the nearest millimetre.
5. S-Dry is the final size for seasoned lumber in place and is the size used in design calculations.

FIGURE 12.3 Dimensions Lumber—Sizes

(Courtesy Canadian Wood Council)

The only safe way to estimate the quantity of lumber required for any particular job is to do a takeoff of each piece of lumber needed for the work. Since the time needed to do such an estimate is excessive, tables are included to provide as accurate material quantities as are necessary in as short a time as possible.

SPECS

-GRADE / LOCATION
- FASTNERS
- CODE
- CONNECTORS
- SILL GASKETS
- PRESSURE TREATED
- GLUE
- ENGINEERER LUMB
- STRINGERS

- TRUSSES

12.2 Buying Lumber

Dimension softwood lumber is sold by the local lumberyards or builder suppliers in standard lengths of 610 mm (2 foot) multiples ranging from 2,440 mm (8 feet) to 6,100 mm (20 feet). Lengths exceeding 6,100 mm (20 feet) have to be placed on a special/custom order and are priced at premium rates. Some special types, such as pre-cut studs, are cut to the exact desired length. Stock lengths are shown in Figure 12.4.

Hardwood and furniture grade lumber is ordered and sold by the *board foot* (imperial measure) or by the *cubic metre* (metric measure). When any quantity of lumber is purchased by the imperial measure, it is priced and sold by the thousand-feet-board measure, abbreviated *mfbm*. The estimator must calculate the number of board feet required on the job.

One board foot is equal to the volume of a piece of wood one inch thick and one foot square. By using Formula 12.1, the number of board feet can be quickly determined. The nominal dimensions of the lumber are used even when calculating board feet.

Metric Length (mm)	Nominal Length (feet)
2,440	8
3,050	10
3,660	12
4,270	14
4,880	16
5,490	18
6,100	20

FIGURE 12.4 Dimensions Lumber—Stock Lengths

Formula 12.1 Board Foot Measurement

$$N = P \times \frac{(T \times W)}{12} \times L$$

N = Number of feet (board measure)
P = Number of pieces of lumber
T = Thickness of the lumber (in inches)
W = Width of the lumber (in inches)
L = Length of the pieces (in feet)
12 = Inches (one foot), a constant—it does not change

EXAMPLE 12.1 Calculating Board Feet

Estimate the board feet of 10 pieces of lumber 2 × 6 inches and 16 feet long (it would be written 10—2 × 6s @ 16'-0" (see Figure 12.5b).

$$160 \text{ fbm} = 10 \times (2'' \times 6'') / 12'' \times 16'$$

This formula can be used for any size order and any number of pieces that are required.

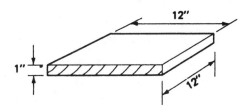

A. One foot of board measure

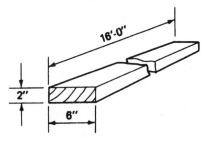

B. Sixteen feet of board measure

FIGURE 12.5 Board Measures, Example

continued

Metric lumber is sold by the cubic metre (m³). The volume is based on the actual size and calculated using Formula 12.2. Thickness and width are given in millimetres (mm) and the length is in metres (m).

Formula 12.2 Cubic Metre Measurement

$V = P \times T \times W \times L / 1{,}000{,}000$
V = Volume (cubic metre)
P = Number of pieces of lumber
T = Thickness of the lumber (in millimetres)
W = Width of the lumber (in millimetres)
L = Length of the pieces (in metres)
$1{,}000{,}000$ = Thickness (mm) × width (mm) converts to metres.

655.7 BF/m3

E X A M P L E 12.2 Calculating Cubic Metre

Estimate the cubic metre of one thousand pieces of lumber 38 × 140 mm and 4.88 metres (see Figure 12.6).

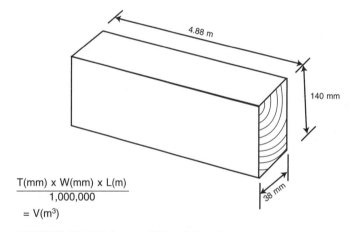

2×4
38 × 89.

$$\frac{T(mm) \times W(mm) \times L(m)}{1{,}000{,}000}$$
$= V(m^3)$

FIGURE 12.6 Volume of Metric Lumber

$$25.96 \text{ m}^3 = 1000 \times 38 \times 140 \times 4.88 / 1{,}000{,}000$$

This formula can be used for any size order and any number of pieces that are required. Figure 12.7 shows the conversion factors from board measure to metric measure. Multiply

Nominal Thickness (inches)	Surfaced Size (inches)	mm	DIMENSION LUMBER								
			Nominal Widths (inches)					Metric/Nomenclature (mm)			
			2(38)	3(64)	4(89)	6(140)	8(184)	10(235)	12(286)	14(337)	16(387)
2	2 1/2	38	1.321	1.484	1.547	1.623	1.599	1.634	1.657	1.674	1.682
2 1/2	2	51	—	1.593	1.661	1.742	1.717	1.755	1.779	1.797	1.806
3	2 1/2	64	—	1.66	1.737	1.822	1.796	1.835	1.861	1.879	1.889
3 1/2	3	76	—	—	1.768	1.854	1.828	1.868	1.894	1.913	1.922
4	3 1/2	89	—	—	1.812	1.900	1.873	1.914	1.941	1.960	1.970
4 1/2	4	102	—	—	—	1.936	1.908	1.950	1.977	1.997	2.007

Note: Multiply mfbm by the factors in the table to obtain m³.

FIGURE 12.7 Metric Lumber Conversions

12.3 Floor Framing

In beginning a wood framing quantity takeoff, the first portion estimated is the floor framing. As shown in Figure 12.8, the floor framing generally consists of a wood beam, sills, floor joists, joist headers, and subflooring. The first step in the estimate is to determine the grade (quality) of lumber required and to check the specifications for any special requirements. This information is then noted on the quantity takeoff sheets.

Wood Beams. When the building is of a width greater than that which the floor joists can span, a beam of some type is required. A builtup wood member or steel beam is often used. The sizes of the pieces used to build up the wood beam must be listed and the length of the beam noted. To accommodate large spans and loads, engineered wood beams that have shapes with webs and flanges are used.

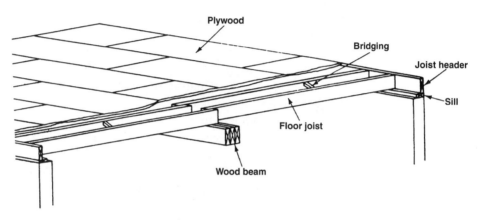

FIGURE 12.8 Floor Framing

E X A M P L E   **12.3**   **Wood Beam Takeoff**

From Figures 12.9 and 12.10, the inside foundation wall dimension and beam length can be determined.

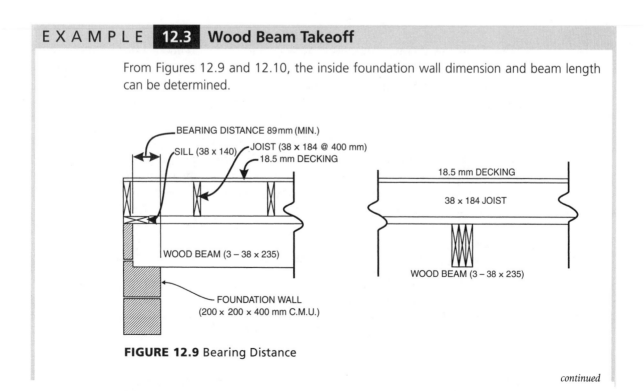

FIGURE 12.9 Bearing Distance

continued

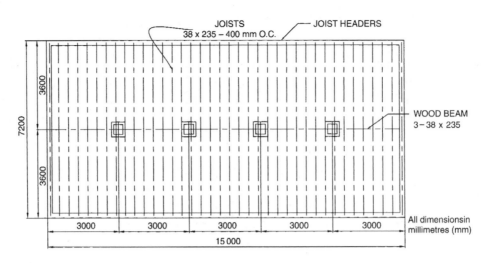

FIGURE 12.10 Floor Framing Plan

Inside dimension = Outside dimension − foundation wall thickness
Inside dimensions = 15.00 − 2 (0.20) m
Inside dimensions = 15.00 − 0.40
Inside dimensions = 14.60 m
Beam length = Inside dimension + bearing distance
Beam length = 14.60 + 2 (0.089) m
Beam length = 14.60 + 0.178
Beam length = 14.78 m
Because the beams consist of three 38 × 235, there are 44.34 m of 38 × 235.
Use 3,050 mm lengths to minimize waste.

Purchase Quantity		
Size	**Length (mm)**	**Pieces**
38 × 235	3,050	15

Sills. Sill plates are most commonly 38 × 89 and 38 × 140 and are placed on the foundation so that the length of sill plate required is the distance around the perimeter of the building. Lengths ordered will depend on the particular building. Not all frame buildings require a sill plate, so the details should be checked; but generally, where there are floor joists, there are sill plates.

The length of sill is often taken off as the distance around the building. An exact takeoff would require that the estimator allow for the overlapping of the sill pieces at the corners. This type of accuracy is required only when the planning of each piece of wood is involved on a series of projects, such as subdivision housing estates where the amount becomes significant.

E X A M P L E **12.4** **Sill Quantity**

The exact sill length is shown in the following calculation. However, the perimeter of 44.40 m is sufficient for this example.

$$\text{Sill length} = 2 (15.00 + 6.92) = 43.84 \text{ m}$$
Use 44 m.

continued

Using the building perimeter of 44 m, the following quantity of sill material would be purchased. Since most stock lumber comes in even increments, the purchase quantities have been rounded off to the nearest stock length. Furthermore, the sill lumber is typically treated and should be kept separate on the quantity takeoff, since treated lumber is more expensive.

Purchase Quantity		
Size	Length (mm)	Pieces
38 × 140	3,050	10
38 × 140	3,660	4

Wood Floor Joists. The wood joists should be taken off and separated into the various sizes and lengths required. The spacing most commonly used for joists is 400 mm on centre, but spacings of 300 mm and 600 mm are also found. The most commonly used sizes for floor joists are 38 × 140, 38 × 184, 38 × 235, and 38 × 286, although wider and deeper lumber is sometimes used.

To determine the number of joists required for any given area, the length of the floor is divided by the joist spacing, and then one joist is added for the extra joist that is required at the end of the span. Check for required blockings between adjacent joist under partitions; one extra joist should be added for each occurrence.

The length of the joist is taken as the inside dimension of its span plus 38 mm at each end for bearing on the wall or sill.

Joist Estimating Steps

1. From the foundation plan and wall section, determine the size of the floor joists required.
2. Determine the number of floor joists required by first finding the number of spaces, then adding one extra joist to enclose the last space.
3. Multiply by the number of bays.
4. Add one extra for partitions that run parallel to the joists.
5. Determine the required length of the floor joists.

EXAMPLE 12.5   Joist Quantity

Joist length 3.60 m (refer to Figures 12.11, 12.12, 12.13, and 12.14)
15,000 / 400 = 38 spaces (add 1 to convert from spaces to joists)
Use 39 joists
2 bays = 39 × 2 = 78 joists
Extra joists under walls (refer to Figure 12.11)
7 additional joists

Purchase Quantity		
Size	Length (mm)	Pieces
38 × 235	3,660	85

continued

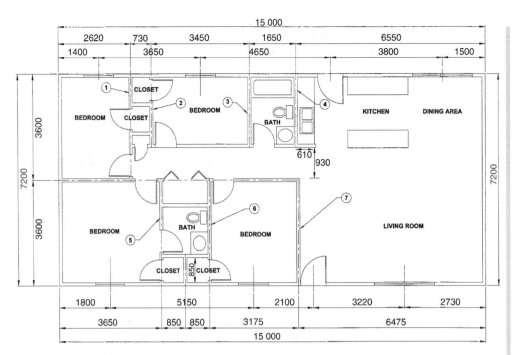

FIGURE 12.11 Floor Plan

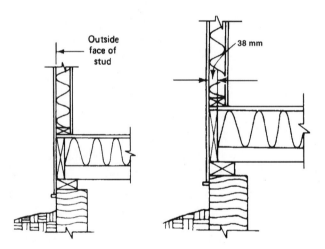

FIGURE 12.12 Joist Headers

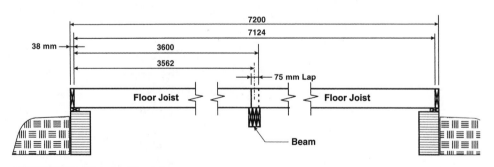

FIGURE 12.13 Joist Dimesions

continued

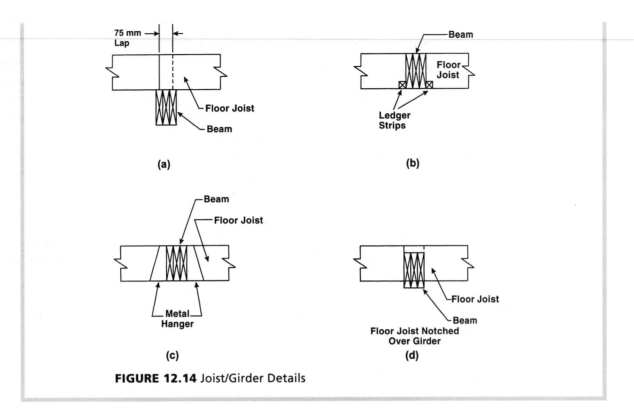

FIGURE 12.14 Joist/Girder Details

Trimmers and Headers. Openings in the floor, such as for stairs or fireplaces, are framed with trimmers running in the direction of the joists and headers that support the tail joists (Figure 12.15).

Unless the specifications say otherwise, when the header length (Figure 12.16) is 1.20 m or less most codes allow single headers to be used. For header lengths greater than 1.20 m, codes usually require double headers (Figure 12.17). For trimmer lengths less than 0.80 m, single trimmers are required and for lengths longer than 0.80 m double trimmers are required.

To determine the extra material required for openings, two situations are investigated:

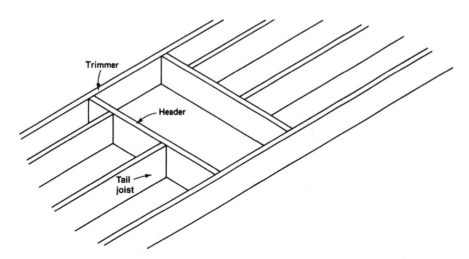

FIGURE 12.15 Framing for Floor Openings

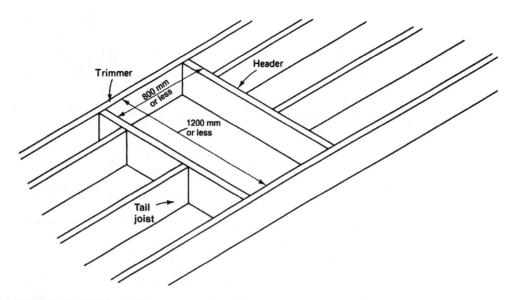

FIGURE 12.16 Single Trimmers and Headers

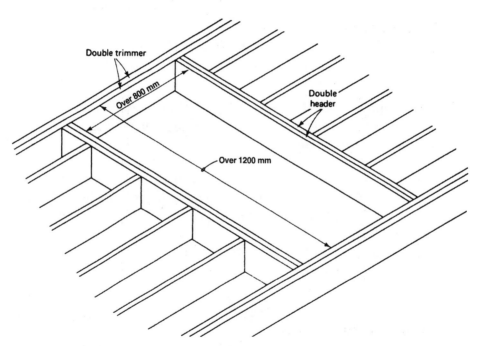

FIGURE 12.17 Double Trimmers and Headers

E X A M P L E **12.6** **Header and Trimmer Joists**

1. 900 mm × 1,200 mm opening as shown in Figure 12.18 (access to attic)
2. 900 mm × 2,400 mm opening as shown in Figure 12.18 (stairway opening)

continued

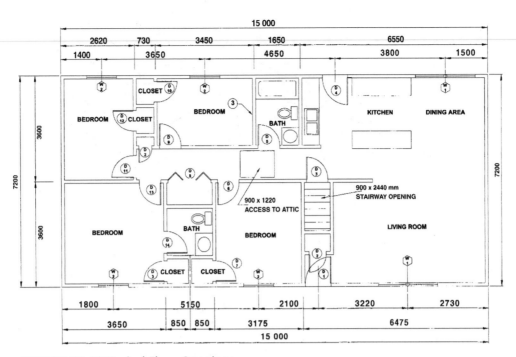

FIGURE 12.18 Typical Floor Openings

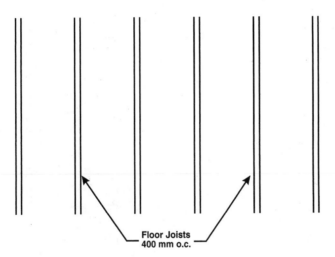

FIGURE 12.19 Joist Layout

The material is determined by:

1. Sketching the floor joists without an opening (Figure 12.19).
2. Locating the proposed opening on the floor joist sketch (Figure 12.20).
3. Sketching what header and trimmer pieces are required and how the cut pieces of the joist may be used as headers and trimmers (Figure 12.21).

continued

Purchase Quantity			
1.20 m Opening			
Size	**Length (mm)**	**Pieces**	**Comments**
38 × 235	3,660	3	Extra for trimmer joists
38 × 235	2,440	2	Extra for header joists
2.44 m Opening			
Size	**Length (mm)**	**Pieces**	**Comments**
38 × 235	3,660	3	Extra for trimmer joist Double headers cut from waste

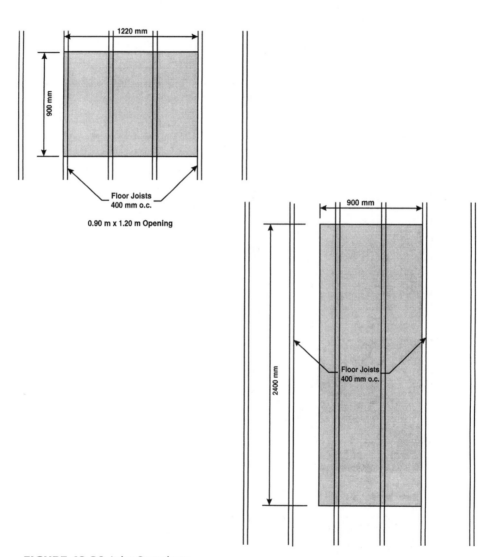

FIGURE 12.20 Joist Openings

continued

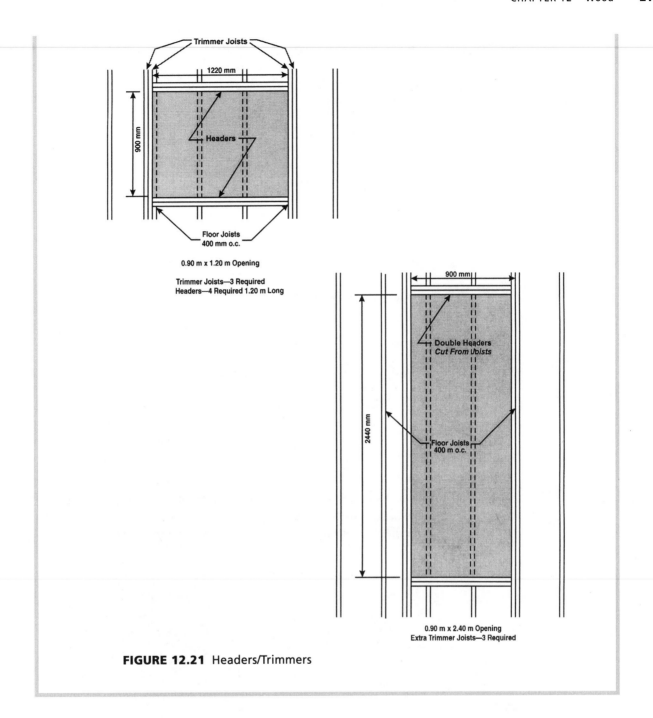

FIGURE 12.21 Headers/Trimmers

Joist headers are taken off next. The header runs along the ends of the joist to seal the exposed edges (Figure 12.22). The headers are the same size as the joists, and the length required will be two times the length of the building (see Figure 12.11).

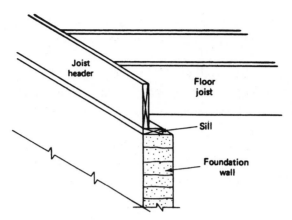

FIGURE 12.22 Joist Header Detail

E X A M P L E | **12.7** | **Joist Header Quantity**

From Figures 12.10 and 12.11, the joist header runs along the 15 m dimension of the building. Therefore, there are 30 m of joist headers.

Purchase Quantity		
Size	**Length (mm)**	**Pieces**
38 × 235	3,050	10

Bridging is customarily used with joists (except for glued nailed systems) and must be included in the costs. Codes and specifications vary on the amount of bridging required, but at least one row of cross-bridging is required between the joists. The bridging may be wood, 19 × 64 mm or 19 × 89 mm, or metal bridging (Figure 12.23).

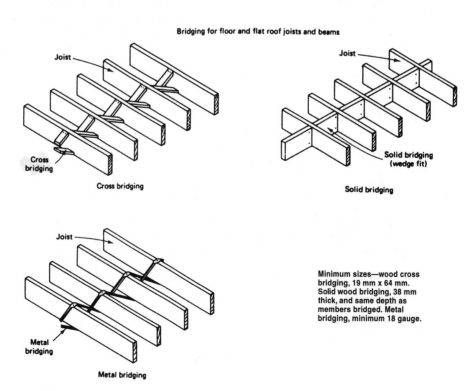

FIGURE 12.23 Bridging

The wood bridging must be cut, while the metal bridging is obtained ready for installation and requires only one nail at each end, half as many nails as would be needed with the wood bridging.

In this estimate, the specifications require metal bridging with a maximum spacing of 2.10 m between bridging or bridging and bearing. A check of the joists' lengths shows they are approximately 3.66 m long and that one row of bridging will be required for each row of joists.

Because each space requires two pieces of metal bridging, determine the amount required by multiplying the joist *spaces* (not the number of joists) times two (two pieces each space) times the number of rows of bridging required.

E X A M P L E   **12.8**   **Metal Bridging**

38 spaces between joists (refer to Example 12.5)
38 spaces × 2 pieces per space × 2 bays = 152 pieces

Sheathing/decking is taken off next. This sheathing is most commonly plywood or waferboard. First, a careful check of the specifications should provide the sheathing information required. The thickness of sheathing required may be given in the specifications or drawings (Figure 12.9). In addition, the specifications will spell out any special installation requirements, such as the glue-nailed system. The sheathing is most accurately estimated by doing a sketch of the area to be covered and planning the sheathing layout. Using 1.22 × 2.44 m sheets, a layout for the residence is shown in Figure 12.24.

Another commonly used method of estimating sheathing is to determine the area to be covered and divide the total area to be covered by the area of one sheet of sheathing (Example 12.9). Although both methods give almost the same answer, the use of square metres alone does not allow planning of sheet layout on irregular plans and does not allow properly for waste. List the quantity of sheathing required on the workup sheet, being certain to list all related information.

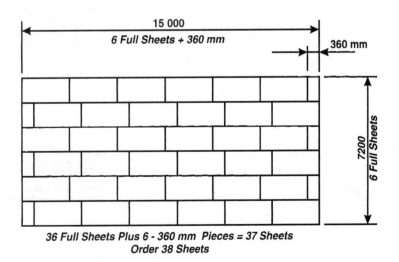

36 Full Sheets Plus 6 - 360 mm Pieces = 37 Sheets
Order 38 Sheets

FIGURE 12.24 Plywood Layout

E X A M P L E **12.9** **Plywood Quantity**

$$\text{Sheets of plywood} = \frac{\text{Area of coverage}}{2.98 \text{ m}^2 / \text{Sheet}}$$

$$\text{Sheets of plywood} = \frac{15.00 \times 7.20}{2.98 \text{ m}^2 / \text{Sheet}} = 36.24 \text{ sheets} - \text{Use 37}$$

Subflooring is sometimes used over the sheathing. Subflooring may be a pressed particle board (where the finished floor is carpet) or it may be plywood or waferboard (where the finished floor is resilient tile, ceramic tile, slate, and so on). A careful review of the specifications and drawings will show whether subflooring is required, as well as the type and thickness and location where it is required. It is taken off in sheets the same as sheathing.

E X A M P L E **12.10** **Flooring System Workup Sheet**

Figure 12.25 is the workup sheet for the flooring system.

Cost Code	Description		Pcs.	Length (m)	Width (mm)	Thickness (mm)	Quantity	Unit
	Wood Beam	Inside Dimensions = 14.60 m						
		Bearing Length = 89 mm						
		Actual Length of Wood Beam is 14.78 m						
		Wood Beam is 3-38 x 235 mm, nailed and glued						
		Order 15 - 38 x 235 @ 3050 (2" x 10" x 10')	15	3.05	235	38	0.41	m³
	Sill							
	38 x 140 Spr #2, treated	Foundation perimeter = 44 m						
		Order 10 - 38 x 140 @ 3050 (2" x 6" x 10')	10	3.05	140	38	0.16	m³
		Order 4 - 38 x 140 @ 3660 (2" x 10" x 12')	4	3.66	140	38	0.08	m³
	Floor Joists							
	38 x 235 DF #2	Use 3660 mm Joists						
		15000/400 = 38 spaces + 1 for end = 39 x 2 bays = 78 joists						
		Extra joists under partitions = 7						
		Extra joists at openings = 6						
		Order 91 - 38 x 235 @ 3660 (2" x 10" x 12')	91	3.66	235	38	2.97	m³
		Use 2440 mm Joists for 1.20 m opening - additional 2						
		Order 2 - 38 x 235 @ 2440 (2" x 10" x 8')	2	2.44	235	38	0.04	m³
	Joist Header	Both long dimensions - Use 15.00 m per side						
		Order 10 - 38 x 235 @ 3050 (2" x 10" x 10')	10	3.05	235	38	0.27	m³
	Bridging	38 spaces on each side = 76 spacees total						
	Use metal bridging	2 pcs per space = 152 pieces						152	Pcs.
	Plywood Decking	108 m2 of floor area/2.98 m2 per sheet = 37 sheets						
	15.5 mm T & G spruce sheathing	Order 37 sheets of 1220 x 2440 mm (4' x 8')						37	Sheets

FIGURE 12.25 Flooring Workup Sheet

12.4 Wall Framing

In this section, a quantity takeoff of the framing required for exterior and interior walls is done. Because the exterior and interior walls have different finish materials, they will be estimated separately. The exterior walls are taken off first, then the interior.

Exterior Walls

Basically, most of the wall framing consists of bottom plates, studs, double plates, headers, and finish materials (Figure 12.26).

Plates. The most commonly used assembly incorporates a double top plate and a single bottom plate, although other combinations may be used. The estimator first begins by reviewing the specifications and drawings for the thickness of materials (commonly 38×89 or 38×140), the grade of lumber to be used, and information on the number of plates required. The assembly in Figure 12.26 provides a 2,440 mm ceiling height, which is most commonly used and works economically with $1,220 \times 2,440$ mm sheets of plywood and gypsum board. When a 2,490 mm ceiling height is required, a double bottom plate is used (Figure 12.27).

Estimating Steps

1. Plates are required around the perimeter of the building. Since this is the same perimeter used to determine the sill material, the perimeter is already

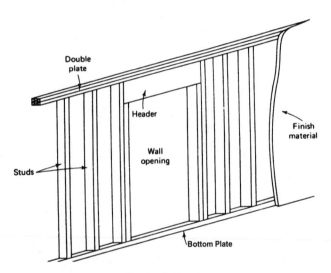

FIGURE 12.26 Wall Framing

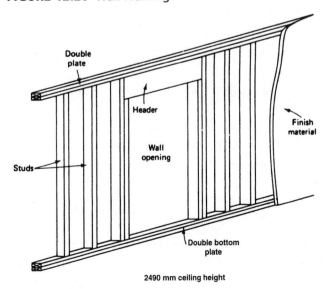

2490 mm ceiling height

FIGURE 12.27 Double Top and Bottom Plates

known. However, the sill is typically treated lumber and caution needs to be exercised to ensure that the treated lumber is not combined with the untreated lumber.

2. The total length of exterior plates is determined by multiplying the length of wall times the number of plates.

3. List this information on the workup sheet, and calculate the order quantities required.

E X A M P L E 12.11 Top and Bottom Plates

Perimeter = 44 m
Bottom plates 38 × 140, treated

Purchase Quantity		
Size	**Length (mm)**	**Pieces**
38 × 140 Treated	3,050	10 (Bottom plate)
38 × 140 Treated	3,660	4 (Bottom Plate)
38 × 140	3,050	30 (Double top plate)

Note: Interior plates will be done with the interior walls later in this portion of the takeoff, but the same basic procedure shown here will be used.

Studs. The stud takeoff should be separated into the various sizes and lengths required. Studs are most commonly 38 × 89 at 400 mm or 600 mm centres, or 38 × 140 at 600 mm on centre. The primary advantage of using 38 × 140 is that it allows for 140 mm of insulation compared with 89 mm of insulation with 38 × 89 studs.

Estimating Steps

1. Review the specifications and drawings for the thickness and spacing of studs and lumber grade.

2. The exterior studs will be required around the perimeter of the building; the perimeter used for the sills and plates may be used.

3. Divide the perimeter (length of wall) by the spacing of the studs to determine the number of spaces. Then, add one to close off the last space.

4. Add extra studs (check code requirements) for corners, wall intersections (where two walls join), and wall openings.

E X A M P L E 12.12 Exterior Wall Studs

Exterior walls 38 × 140 at 600 mm o.c.
Using the 44 m perimeter
44 / 0.60 = 74 spaces − Use 75 studs

Purchase Quantity		
Size	**Length (mm)**	**Pieces**
38 × 140	2,440	75

---

a. Corners using 38 × 89 studs: A corner is usually made up of three studs (Figure 12.28). This requires two extra studs at each corner. Estimate the extra material required by counting the number of corners and multiplying the number of corners by two.

b. Exterior corners using 2 × 6 studs: A corner is usually made up of three studs (Figure 12.29), requiring two extra studs. Reverse corner beads are used as backing to secure the drywall panel to the studs.

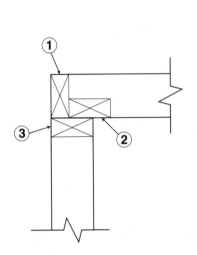

FIGURE 12.28 Corner Stud (38 × 89)

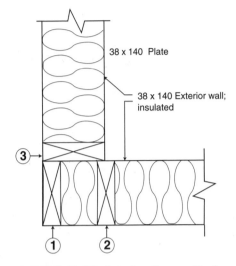

FIGURE 12.29 Exterior Corner Stud Detail (38× 140)

E X A M P L E **12.13** **Corner Studs**

Add 2 studs per corner.
4 corners × 2 studs per corner = 8 studs

Purchase Quantity		
Size	Length (mm)	Pieces
38 × 140	2,440	8

c. Wall intersections: Using 38 × 140 studs for the exterior wall and 38 × 89 for the interior partition, the wall intersection is made up of three studs (Figure 12.30). This requires two extra 38 × 140 studs and one extra 38 × 89 stud at each intersection. The extra material is estimated by counting the number of wall intersections (on the exterior wall) and multiplying that number of intersections by two.

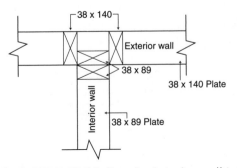

FIGURE 12.30 Exterior/Interior wall Intersection

E X A M P L E **12.14** **Interior Walls Intersecting Exterior Walls**

Refer to Figures 12.30 and 12.31.

9 interior walls (38 × 89 mm) intersect with the exterior (38 × 140 mm)

9 intersections × 2 studs per intersection = 18 studs (38 × 140)

9 intersections × 1 stud per intersection = 9 studs (38 × 89)

Purchase Quantity		
Size	Length (mm)	Pieces
38 × 89	2,440	9
38 × 140	2,440	18

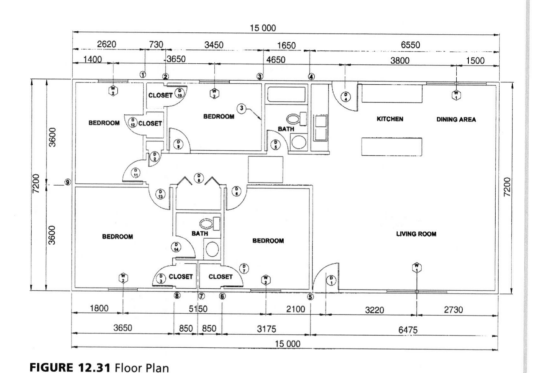

FIGURE 12.31 Floor Plan

d. Wall openings: Typically, extra studs are required at all openings in the wall (Figure 12.32). However, when the openings are planned to fit into the framing of the building, it requires less extra material. The illustrations in Figures 12.33a and 12.33b show the material required for a window, which has not been worked into the module of the framing. The layout in Figure 12.33a actually requires four extra studs; in 12.33b, three extra studs; and in 12.34a, two extra studs.

In this example, there is no indication that openings have been worked into the framing module. For this reason, an average of three extra studs were included for all windows and door openings.

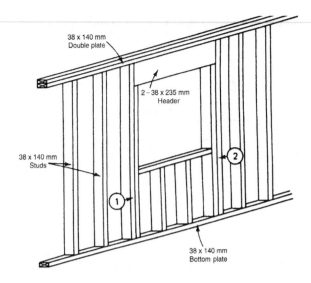

FIGURE 12.32 Wall Opening

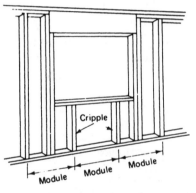

(a) Extra stud material
4 extra studs

Window not on module

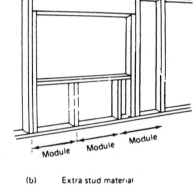

(b) Extra stud material
3 extra studs

Window partially on module

FIGURE 12.33 Wall Openings

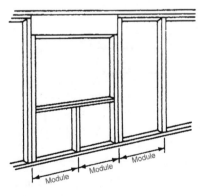

(a) Extra stud material
2 extra studs

(b) Framing without window

FIGURE 12.34 Wall Openings

EXAMPLE 12.15 **Cripple and Extra Studs for Wall Openings**

Refer to Figure 12.31.

8 openings × 3 studs per opening = 24 studs

Purchase Quantity		
Size	Length (mm)	Pieces
38 × 140	2,440	24

5. Additional studs are also required on the gable ends of the building (Figure 12.35), which will have to be framed with studs between the double plate and the rafter unless trusses are used (see Section 12.10). The specifications and elevations should be checked to determine whether the gable end is plain or has a louvre.

Estimating the studs for each gable end is accomplished by first drawing a sketch of the gable end and noting its size (Figure 12.36). Find the number of studs required by dividing the length by the stud spacing. Then, from the sketch find the approximate average height of the stud required and record the information on the workup sheet. Multiply the quantity by two since there are two gable ends on this building (Figure 12.37).

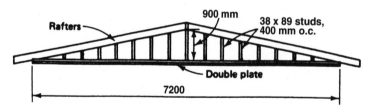

FIGURE 12.35 Gable End Frame

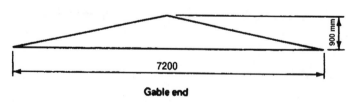

FIGURE 12.36 Gable End Sketch

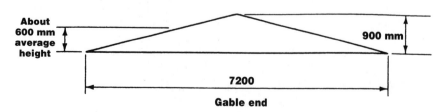

FIGURE 12.37 Gable Average Height

EXAMPLE **12.16** Gable End Studs

7.20 m gable end length / 0.40 centres = 18 spaces — Use 19 studs

Average height = 2/3 × h

Average height = 600 mm

Quantity must be doubled to compensate for the gable ends at both ends of the building.

Purchase Quantity		
Size	Length (mm)	Pieces
38 × 140	3,050	8

Headers.   Headers are required to support the weight of the building over the openings. A check of the specifications and drawings must be made to determine if the headers required are solid wood, headers and cripples, or plywood sheathing (Figure 12.38). For ease of construction, many carpenters and home builders feel that a solid header provides best results and they use two or three 38 × 235 as headers throughout the project, even in non-load-bearing walls. Shortages and higher costs of materials have increased the usage of plywood and smaller size headers.

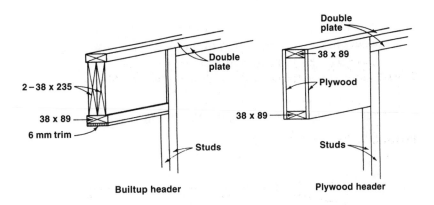

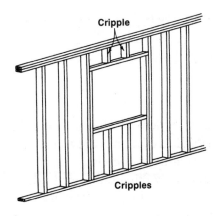

FIGURE 12.38 Headers

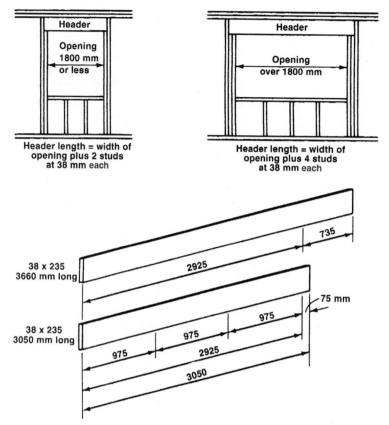

FIGURE 12.39 Header Example

The header length must also be considered. As shown in Figure 12.39, the header extends over the top of the studs and it is wider than the opening. Most specifications and building codes require that headers for openings up to 1,800 mm wide must extend over one stud at each end (Figure 12.39), and headers for openings 1,800 mm and wider must extend over two studs at each end.

List the number of openings, their width, the number of headers, their length, and the total length required.

E X A M P L E **12.17** **Header Quantity**

Number of Openings	Size	Header Length	Pieces
9	914 mm	990 mm	18
1	813 mm	889 mm	2

Purchase Quantity		
Size	Length (mm)	Pieces
38 × 235	3,050	6
38 × 235	2,440	1

Wall Sheathing.   Exterior wall sheathing may be a fiberboard material soaked with a bituminous material, insulation board (often urethane insulation covered with an aluminum reflective coating), waferboard, or plywood (Figure 12.40). Carefully check the specifications and working drawings to determine what is required (insulation requirements, thickness). Fiberboard and insulation board sheathing must be covered by another material (such as brick, wood, or aluminum siding), while the plywood may be covered or left exposed. All these sheathing materials are taken off first by determining the area required and then determining the number of sheets required. The most accurate takeoff is made by sketching a layout of the material required (as with the sheathing in floor framing). The estimator must check the height of sheathing carefully, as a building with a sloped soffit (Figure 12.41) may require a 2,740 mm length, while 2,440 mm may be sufficient when a boxed-in soffit is used (Figure 12.42).

Openings in the exterior wall are ignored, unless they are large and the sheathing that would be cut out can be used elsewhere. Otherwise, it is considered waste.

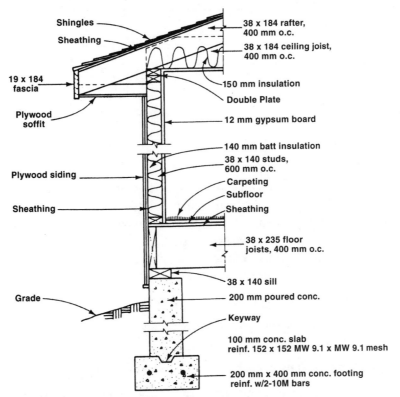

FIGURE 12.40 Typical Wall Section

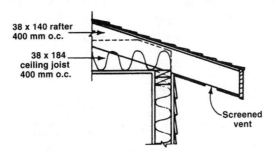

FIGURE 12.41 Sloped Soffit

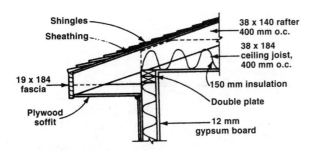

FIGURE 12.42 Boxed Soffit

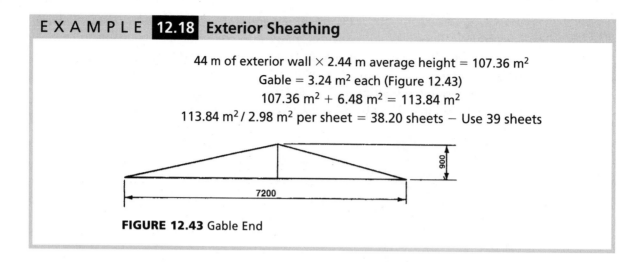

E X A M P L E 12.18 Exterior Sheathing

44 m of exterior wall × 2.44 m average height = 107.36 m²
Gable = 3.24 m² each (Figure 12.43)
107.36 m² + 6.48 m² = 113.84 m²
113.84 m² / 2.98 m² per sheet = 38.20 sheets − Use 39 sheets

FIGURE 12.43 Gable End

Interior Walls

Interior walls are framed with studs, top and bottom plates, and a finish on both sides of the wall. When estimating the material for interior walls, the first step is to determine from the specifications and drawings what thickness(es) of walls is (are) required. Most commonly, 38 × 89 studs are used, but 38 × 64 studs and metal studs (see Section 15.2) can be used. Also, the stud spacing may be different from the exterior walls.

Next, the length of interior walls must be determined. This length is taken from the plan by:

1. Using dimensions from the floor plan.
2. Scaling the lengths with a scale ruler.
3. Using a distance measurer or a digitizer over the interior walls.

On large projects, extreme care must be taken when using a scale since the drawing may not be done to the exact scale shown.

Any walls of different thicknesses (such as a 140-mm-thick wall, sometimes used where plumbing must be installed) and of special construction (such as a double or staggered wall) (Figure 12.44) may require larger stud or plate sizes.

Plates. Refer to the discussion under "Exterior Walls" earlier in this section.

Studs. Refer to the discussion under "Exterior Walls" earlier in this section. As in exterior walls, deduct only where there are large openings and take into account all corners, wall openings, and wall intersections.

Headers. Refer to the discussion under "Exterior Walls" earlier in this section.

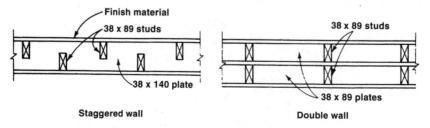

FIGURE 12.44 Special Wall Construction

E X A M P L E   **12.19**   **Interior and Exterior Wall Workup Sheets**

Following are the workup sheets (Figure 12.45) for the exterior and interior wall framing for the building found in Figure 12.11.

		ESTIMATE WORK SHEET						
Project H & C Cottage		**WALL FRAMING**		Estimate No.	1234			
Location Mountainville, BC				Sheet No.	1 of 2			
Architect C.K. Architects				Date	23/12/20xx			
Items Walls				By HYZ	**Checked** DCK			

Cost Code	Description		Pcs.	Length (m)	Width (mm)	Thickness (mm)	Quantity	Unit
	Bottom Plates	Exterior Perimeter = 44 m						
	38 x 140 Spr #2, treated	**Order 10 - 38 x 140 @ 3050 (2" x 6" x 10')**	10	3.05	140	38	**0.16**	m³
		Order 4 - 38 x 140 @ 3660 (2" x 10" x 12')	4	3.66	140	38	**0.08**	m³
		Interior walls - Total length = 41 m						
	38 x 89 Spr #2	**Order 14 - 38 x 89 @ 3050 (2" x 4" x 10')**	14	3.05	89	38	**0.14**	m³
	Top Double Plates	Exterior Perimeter = 44 m						
	38 x 140 Spr #2	**Order 30 - 38 x 140 @ 3050 (2" x 6" x 10')**	30	3.05	140	38	**0.49**	m³
		Interior walls - Total length = 41 m						
	38 x 89 Spr #2	**Order 28 - 38 x 89 @ 3050 (2" x 4" x 10')**	28	3.05	89	38	**0.29**	m³
	Studs	Exterior Perimeter = 44 m						
		44 m/0.60 m = 74 spaces Use 75						
		Exterior corners = 4 Use 8						
		Exterior/Interior Intersections = 9 Use 18						
		Exterior Openings = 8 Use 24						
	38 x 140 Spr #2	**Order 125 - 38 x 140 @ 2440 (2" x 6" x 8')**	125	2.44	140	38	**1.62**	m³
		Interior walls - Total length = 41 m						
		41 m/0.40 m = 103 spaces Use 104						
		Interior corners = 1 Use 2						
		Exterior/Interior Intersections = 9 Use 9						
		Interior/Interior Intersections = 16 Use 32						
		Interior Openings = 12 Use 36						
	38 x 89 Spr #2	**Order 183 - 38 x 89 @ 2440 (2" x 4" x 8')**	183	2.44	89	38	**1.51**	m³
	Misc. Studs	Gable Ends						
		38 pcs @ 0.60 m = 22.80 m						
	38 x 140 Spr #2	**Order 8 - 38 x 140 @ 3050 (2" x 6" x 10')**	8	3.05	140	38	**0.13**	m³
	Headers	Exterior count size m						
		9 openings @ 914 mm Use 38 x 235 18 990 mm 17.82						
		1 opening @ 813 mm Use 38 x 235 2 889 mm 1.78						
	38 x 235 Spr #2	**Order 6 - 38 x 235 @ 3050 (2" x 10" x 10')**	6	3.05	235	38	**0.16**	m³
		Order 1 - 38 x 235 @ 2440 (2" x10" x 8')	1	2.44	235	38	**0.02**	m³
		Interior						
		1 opening @ 533 mm Use 38 x 235 2 609 mm 1.22						
		2 openings @ 787 mm Use 38 x 235 4 863 mm 3.45						
		8 openings @ 838 mm Use 38 x 235 16 914 mm 14.62						
		1 opening @ 1600 mm Use 38 x 235 2 1676 mm 3.35						
		Total **22.64**						
	38 x 235 Spr #2	**Order 7 - 38 x 235 @ 3660 (2" x10" x12')**	7	3.66	235	38	**0.23**	m³
	Exterior Sheathing	Walls 44 m x 2.44 m high = 107.36 m2						
		Gables = 2 x 3.24 m2 = 6.48 m2						
		113.84 m2 / 2.98 m2 per sheet = USE 39 Sheets						
	7.5 mm sheathing							**39**	Sheets

FIGURE 12.45 Stud Workup Sheet

12.5 Ceiling Assembly

In this section, a quantity takeoff of the framing required for the ceiling assembly of a wood frame building is done. The ceiling assembly will require a takeoff of ceiling joists, headers, and trimmers that is quite similar to the takeoff done for the floor assembly.

First, a careful check of the specifications and drawings must be made to determine if the ceiling and roof are made up of joists and rafters (Figure 12.46), often called "stick construction," or prefabricated wood trusses that are discussed in Section 12.10.

Ceiling joists, from the contract documents, determine:

1. Size, spacing, and grade of framing required.

2. The number of ceiling joists required by dividing the spacing into the length and adding one (the same as done for floor joists).

3. The length of each ceiling joist. (Don't forget to add one-half of any required lap.)

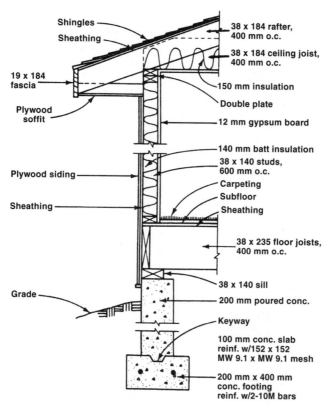

FIGURE 12.46 Typical Wall Section

E X A M P L E **12.20** **Ceiling Joists**

15.00 m / 0.40 m = 38 spaces − Use 39 joists

Purchase Quantity		
Size	Length (mm)	Pieces
38 × 184	3,660	78

The headers and trimmers required for any openings (such as stairways, fireplaces, and attic access openings) are considered next. This is taken off the same as for floor framing (see Section 12.3).

E X A M P L E **12.21** **Attic Opening**

A 1,125 × 900 mm attic opening is required. Refer to Figure 12.47.

continued

Purchase Quantity Trimmer Joists		
Size	Length (mm)	Pieces
38 × 184	3,660	2

Purchase Quantity Header Joists		
Size	Length (mm)	Pieces
38 × 184	2,440	2

The ceiling finish material (drywall) is estimated in Chapter 15.

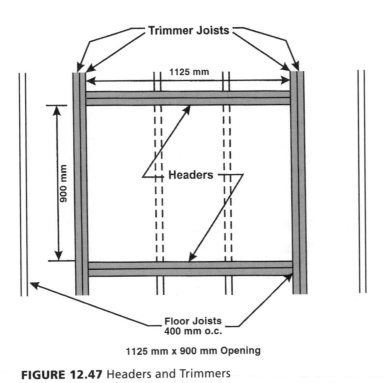

FIGURE 12.47 Headers and Trimmers

12.6 Roof Assembly

In this section, a quantity takeoff of the framing required for the roof assembly of a wood frame building is done. The roof assembly will require a takeoff of rafters, headers and trimmers, collar ties (or supports), ridge, lookouts, sheathing, and the felt that covers the sheathing. If trusses are specified, a separate takeoff for roof and ceiling assembly will not be made, and it would be estimated as discussed in Section 12.10.

Rafters. Roof rafters should be taken off and separated into the sizes and lengths required. The spacing most commonly used for rafters are 300 mm, 400 mm, and 600 mm on centre. Rafter sizes of 38 × 140, 38 × 184, 38 × 235, and 38 × 286 are most common. The lengths of rafters should be carefully taken from the drawings or worked out by the estimator if the drawings are not to scale. Be certain to add any required overhangs to the lengths of the rafters. The number of rafters for a pitched roof can be determined in the same manner as the number of joists. The principle of pitch versus slope should be understood to reduce mistakes. Figure 12.48 shows the difference between pitch and slope, and Figure 12.49 shows the length of rafters required for varying pitches and slopes.

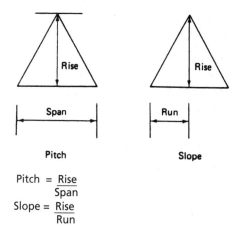

Pitch = $\dfrac{\text{Rise}}{\text{Span}}$

Slope = $\dfrac{\text{Rise}}{\text{Run}}$

Example: Span 12 m, Rise 3m

Pitch $= \dfrac{3}{12} = \dfrac{1}{4}$ Pitch

Slope $= \dfrac{3}{6} = \dfrac{6}{12} = 500$ mm/m

FIGURE 12.48 Roof Pitch and Slope Calculations

Pitch of roof	Slope of Elevation	For Length of Rafter Multiply Length of Run By
1/12	1 in 6	1.014
1/8	1 in 4	1.03
1/6	1 in 3	1.054
5/24	1 in 2.4	1.083
1/4	1 in 2	1.118

FIGURE 12.49 Pitch/Slope Rafter Length Factors

Rafters, from the contract documents, determine:

1. Size, spacing, and grade of framing required.

2. The number of rafters required (divide spacing into length and add one). If the spacing is the same as the ceiling and floor joists, the same amount will be required.

3. The length of each rafter (Figure 12.50). (Don't forget to allow for slope or pitch, and overhang.)

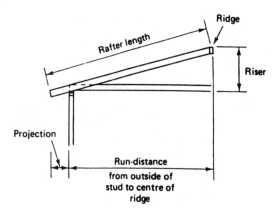

FIGURE 12.50 Rafter Length

EXAMPLE **12.22** **Rafter Quantity**

250 mm/m slope length factor is 1.031 (see Figure 12.49)

3.60 m run to centre line + 0.46 m projection = 4.06 m total run length

4.06 m total run length × 1.03 = 4.18 m actual rafter length

Use 4,270 mm rafters

15 m building length + 0.60 m overhang = 15.60 m ridge length

15.60 m / 0.40 m = 39 spaces

Use 40 rafters per side

Purchase Quantity		
Size	Length (mm)	Pieces
38 × 184	4,270	80

Headers and Trimmers. Headers and trimmers are required for any openings (such as chimneys) just as in floor and ceiling joists. For a complete discussion of headers and trimmers, refer to Section 12.3, "Floor Framing."

Collar Ties. Collar ties are used to keep the rafters from spreading (Figure 12.51). Most codes and specifications require them to be a maximum of 1,500 mm apart or every third rafter, whichever is less. A check of the contract documents will determine:

a. Required size (usually 19 × 140 or 38 × 89)

b. Spacing

c. Location (how high up)

Determine the number of collar ties required by dividing the total length of the building (used to determine the number of rafters) by the spacing and add one to close up the last space.

The length required can be a little harder to determine. Most specifications do not spell out the exact location of the collar ties (up high near ridge, low closer to joists, halfway in between), but the typical installation has the collar ties about one-third down from the ridge with a length of 1.50 m to 2.40 m, depending on the slope and span of the particular installation.

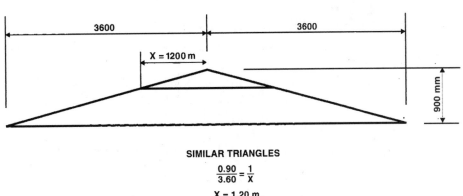

SIMILAR TRIANGLES

$$\frac{0.90}{3.60} = \frac{1}{X}$$

X = 1.20 m

Collar length = 2.40 m

FIGURE 12.51 Collar Length

E X A M P L E **12.23** **Collar Ties**

78 rafter pairs without gable end overhang
39 rafter pairs
39 / 3 = 13 spaces
Use 14 collars

From Figure 12.51, collars are 2.40 m long

Purchase Quantity		
Size	Length (mm)	Pieces
38 × 89	2,440	14

Ridge. The *ridge board* (Figure 12.52) is taken off next. The contract documents must be checked for the size of the ridge board required. Quite often, no mention of ridge board size is made anywhere on the contract documents. This will have to be checked with the consultant (or whoever has authority for the work). Generally, the ridge is 19 mm thick and one size wider than the rafter. In this case, a 38 × 184 rafter would be used with a 19 × 235 ridge board. At other times a 38 mm thickness may be required.

The length of ridge board required will be the length of the building plus any side overhang.

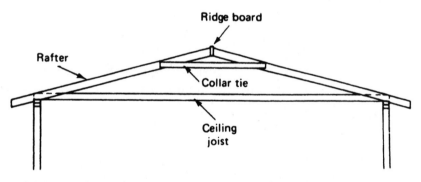

FIGURE 12.52 Ridge Board

E X A M P L E **12.24** **Ridge**

There are 15.60 m of ridge from the edge of the overhang to the far overhang.

Purchase Quantity		
Size	Length (mm)	Pieces
19 × 235	2,440	7

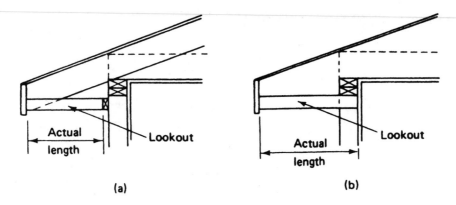

FIGURE 12.53 Lookouts

Lookouts. Lookouts are often required when a soffit is boxed in as shown in Figure 12.53. The number of lookouts is found by dividing the total length of boxed-in soffit by the spacing and adding one extra lookout for each side of the house on which it is needed. (If required, front and back add two extra lookouts.) The length of the lookout is found by reviewing the sections in details. The detail in Figure 12.53a indicates that the lookout would be the width of the overhang minus the actual thickness of the fascia board and the supporting pieces against the wall. In Figure 12.53b the lookout is supported by being nailed to the stud wall; its length is equal to the overhang plus stud wall minus the actual thickness of the fascia board.

Sheathing Materials. Sheathing for the roof is taken off next. A review of the contract documents should provide the sheathing information needed:

1. Thickness or identification index (maximum spacing of supports)
2. Veneer grades
3. Species grade
4. Any special installation requirements

The sheathing is most accurately estimated by making a sketch of the area to be covered and planning the sheathing sheet layout. For roof sheathing, be certain that the rafter length that is required is used and that it takes into consideration any roof overhang and the slope (or pitch) of the roof. The roof sheathing for the residence is then compiled on the workup sheets.

Because the carpenters usually install the roofing felt (which protects the plywood until the roofers come), the roofing felt is taken off here or under roofing depending on the estimator's preference. Because felt, shingles, and builtup roofing materials are estimated in Chapter 13, that is where it will be estimated for this residence.

EXAMPLE **12.25** Roof Decking

Refer to Figure 12.54.
Actual rafter length = 4.18 m
Ridge length = 15.60 m
Area of roof = 2 sides × 4.18 m rafter length × 15.60 m ridge length = 130.42 m^2
130.42 m^2 of deck / 2.98 m^2 per sheet = 43.77 sheets − Use 44 sheets

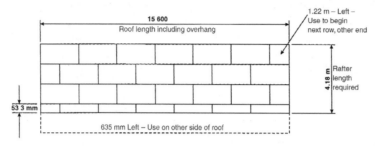

FIGURE 12.54 Plywood Sheet Layout

12.7 Trim

The trim may be exterior or interior. Exterior items include mouldings, fascias, cornices, and corner boards. Interior trim may include base, moldings, and chair rails. Other trim items may be shown on the sections and details or may be included in the specifications. Trim is taken off by the length required and usually requires a finish (paint, varnish, and so on), although some types are available prefinished (particularly for use with prefinished plywood panels).

EXAMPLE **12.26** Exterior Trim

The exterior trim for the residence is limited to the fascia board and soffit.

Fascia board. 19 × 184, Fir, 48 m required − Order 50 m.
2(15.60 m) + 4 (4.18 m) = 47.92 or 48 m

Soffit, plywood. 12.5 mm thick, A-C, exterior
0.46 mm wide, 31.20 m = 14.35 m^2 = 4.81; order five − 1,220 × 2,440 sheets

Baseboard. Exterior wall = 44 m; interior wall = 41 m × 2
(both sides) minus doors (8m), 89 mm ranch mould,
118 m required − order 125 m

EXAMPLE **12.27** Roof Estimate Workup Sheet

The following workup sheet (Figure 12.55) is for the roofing system for the residence in Figure 12.11.

		ESTIMATE WORK SHEET						
Project H & C Cottage		**ROOF FRAMING**				Estimate No. 1234		
Location Mountainville, BC						Sheet No. 1 of 1		
Architect C.K. Architects						Date 23/12/20xx		
Items Roof Framing						By HYZ Checked DCK		

Cost Code	Description		Pcs.	Length (m)	Width (mm)	Thickness (mm)	Quantity	Unit
	Ceiling Joists	15.00 m / 0.40 m = 38 spaces - USE 39 Joists / side						
		Order 78 - 38 x 184 @ 3660 (2" x 8" x 12')	78	3.66	184	38	2.00	m³
		Attic Opening 1125 mm x 900 mm						
		Order 2 - 38 x 184 @ 3660 (2" x 8" x 12')	2	3.66	184	38	0.05	m³
		Order 2 - 38 x 184 @ 2440 (2" x 8" x 8')	2	2.44	184	38	0.03	m³
	Rafters							
		Actual Rafter Length = 4.18 m - USE 4270 mm rafters						
		Building Length + Overhang = 15.60 m						
		15.60 m / 0.40 m = 39 spaces - Use 40 Rafters / side						
		Order 80 - 38 x 184 @ 4270 (2" x 8" x 14')	80	4.27	184	38	2.39	m³
	Collar Ties	39 Spaces w/ 1 collar tie per 3rd rafter						
		39 Spaces / 3 = 13 Spaces - USE 14 collar ties						
		Collar Ties at 1/3 from Ridge = 2.40 m						
		Order 14 - 38 x 89 @ 2440 (2" x 4" x 8')	14	2.44	89	38	0.12	m³
	Ridge	Ridge - 15.60 m						
		Order 7 - 19 x 235 @ 2440 (1" x10" x 8')	7	2.44	235	19	0.08	m³
	Lookouts	Same Number of 38 x 89 for Lookouts as Rafters - USE 80						
		Each Lookout is 400 mm - USE 2440 mm - 38 x 89						
		Order 14 - 38 x 89 @ 2440 (2" x 4" x 8')	14	2.44	89	38	0.12	m³
	Roof Decking	4.18 m x 15.60 m = m² of decking per side = 65.21 m²/side						
		130.42 m² of deck on roof/ 2.98 m² per sheet = 43.77 USE 44						44	Sheets
	Exterior Trim	Fascia Board : 2(15.60) + 4 (4.18) = 47.92 m						
		Order 12 - 19 x184 @ 4270 (1" x 10" x14')	12	4.27	184	19	0.18	m³
		Soffit Plywood: 0.46 m x 31.20 m = 14.35 m²						
		Order 5 Sheets						5	Sheets
		Baseboard: 44 m + 2(41 m) - 8 m = 118 m						
		Order 28 pcs. @ 4270 (14')						28.00	Pcs.

FIGURE 12.55 Roof Framing Workup Sheet

12.8 Labour

Labour may be calculated at the end of each portion of the rough framing or it may be done for all the rough framing at once. Many estimators will use a square metre/foot figure for the rough framing based on the cost of past work and taking into consideration the difficulty of work involved. Such a job as the small residence being estimated would be considered very simple to frame and would receive the lowest square metre/foot cost. The cost would increase as the building became more involved. Many builders use framing subcontractors for this type of work. The subcontractor may price the job by the square metre/foot or as a lump sum. All these methods provide the easiest approaches to the estimator.

When estimators use their own workforce and want to estimate the time involved, they usually use records from past jobs, depending on how organized they are. The labour would be estimated for the framing by using the appropriate portion of the table from Figure 12.56 for each portion of the work to be done.

Light framing	Unit	Work Hours
Sills	30 m	2.0 to 4.5
Joists, floor and ceiling	MBFM	16.0 to 24.0
Walls, interior, exterior (including plates)	MBFM	18.0 to 30.0
Rafters,		
gable roof	MBFM	18.5 to 30.0
hip roof	MBFM	22.0 to 35.0
Cross bridging, wood	100 sets	4.0 to 6.0
metal	100 sets	3.0 to 5.0
Plywood, floor	10 m²	1.0 to 2.0
wall	10 m²	1.2 to 2.5
roof	10 m²	1.4 to 2.8
Trim,		
fascia	30 m	3.5 to 5.0
soffit	30 m	2.0 to 3.5
baseboard	30 m	1.5 to 2.5
moulding	30 m	2.0 to 4.0

FIGURE 12.56 Work Hours Required for Framing

E X A M P L E **12.28** **Priced Framing Estimate**

The following summary sheet (Figure 12.57) is for all of the framing in the residence in Figure 12.11.

Project	H & C Cottage			ESTIMATE SUMMARY SHEET							Estimate No.	1234			
Location	Mountainville, BC										Sheet No.	1 of 1			
Architect	C.K. Architects										Date	23/12/20xx			
Items	Framing										By HYZ	Checked DCK			

Cost Code	Description	Qty.	Waste Factor	Purch. Quan.	Unit	Crew	Prod. Rate	Wage Rate	Work Hours	Unit Cost Lab.	Unit Cost Mat.	Unit Cost Equip.	Lab.	Mat.	Equip.	Total
	FOUNDATION															
	Wood Beam; 38 x 235	0.41	0%	0.41	m³			22.80	5.00	278.05	636.47		114.00	260.95		374.95
	Sill; 38 x 140 treated	0.24		0.24	m³			22.80	3.00	285.00	975.97		68.40	234.23		302.63
	Floor Joists; 38 x 235	3.01		3.01	m³			22.80	34.00	257.54	636.47		775.20	1,915.77		2,690.97
	Joist Headers; 38 x 235	0.27		0.27	m³			22.80	3.30	278.67	636.47		75.24	171.85		247.09
	Metal Bridging	152		152	Pcs.			22.80	4.60	0.69	3.20		104.88	486.40		591.28
	Plywood Decking; 15.5 mm T & G	37		37	Sheets			22.80	18.20	11.22	33.60		414.96	1,243.20		1,658.16
	WALL FRAMING															
	Bottom Plate; 38 x 140 treated	0.24		0.24	m³			22.80	2.10	199.50	975.97		47.88	234.23		282.11
	Bottom Plate; 38 x 89	0.14		0.14	m³			22.80	2.00	325.71	600.00		45.60	84.00		129.60
	Top Double Plate; 38 x 140	0.49		0.49	m³			22.80	6.60	307.10	936.54		150.48	458.90		609.38
	Top Double Plate; 38 x 89	0.29		0.29	m³			22.80	4.40	345.93	600.00		100.32	174.00		274.32
	Studs; 38 x 140	1.62		1.62	m³			22.80	45.11	634.88	936.54		1,028.51	1,517.19		2,545.70
	Studs; 38 x 89	1.51		1.51	m³			22.80	20.25	305.76	600.00		461.70	906.00		1,367.70
	Misc. Studs - Gable Ends; 38 x 140	0.13		0.13	m³			22.80	1.80	315.69	936.54		41.04	121.75		162.79
	Headers; 38 x 235	0.41		0.41	m³			22.80	6.20	344.78	783.35		141.36	321.17		462.53
	Exterior Sheathing	39		39	Sheets			22.80	25.80	15.08	17.50		588.24	682.50		1,270.74
	ROOF CEILING															
	Ceiling Joists; 38 x 184	2.08		2.08	m³			22.80	26.20	287.19	620.39		597.36	1,290.41		1,887.77
	Rafters; 38 x 184 x 4270	2.39		2.39	m³			22.80	37.30	355.83	620.39		850.44	1,482.73		2,333.17
	Collar Ties; 38 x 89 x 2440	0.12		0.12	m³			22.80	1.50	285.00	600.00		34.20	72.00		106.20
	Ridge; 19 x 235 x 2440	0.08		0.08	m³			22.80	2.88	820.80	347.61		65.66	27.81		93.47
	Lookouts; 38 x 89 x 2440	0.12		0.12	m³			22.80	1.50	285.00	600.00		34.20	72.00		106.20
	Roof Decking	44		44	Sheets			22.80	33.10	17.15	20.00		754.68	880.00		1,634.68
	TRIM															
	Fascia Board; 19 x 184 x 4270	0.18		0.18	m³			22.80	8.00	1 013.33	191.32		182.40	34.44		216.84
	Soffit Plywood	55			Sheets			22.80	3.50	15.96	35.20		79.80	176.00		255.80
	TOTAL								296.34				$6,757	$12,848		$19,604

FIGURE 12.57 Framing Estimate

12.9 Wood Systems

Wood is used as a component in several structural systems, among them wood trusses, laminated beams, wood decking, and box beams. Wood is used for a variety of reasons.

The wood trusses are economical, whereas the laminated beams and box beams are both economical and used for appearance.

12.10 Wood Trusses

Various wood truss manufacturers can supply wood joists with spans in excess of 45 m. The cost savings of these types of trusses, compared with steel, range as high as 25 percent; less weight is involved and the trusses are not as susceptible to transportation and erection damage. Decking is quickly and easily nailed directly to the chords. Trusses of almost any shape (design) are possible. Ducts, piping, and conduit may be easily incorporated into the trusses. The trusses are only part of the system and must be used in conjunction with a deck of some type. Typical truss shapes are shown in Figure 12.58.

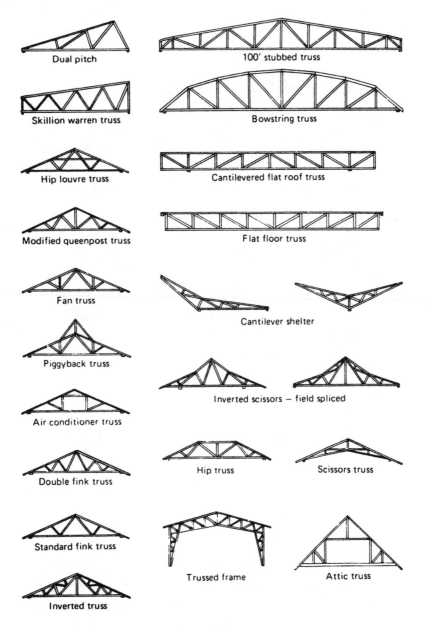

FIGURE 12.58 Typical Wood Truss Shapes

Specifications. Check the type of truss required. If any particular manufacturer is specified, note how the members are to be attached (to each other and to the building) and what stress-grade-marked lumber is required. Note also any requirements regarding the erection of the trusses and any finish requirements.

Estimating.   The estimator needs to determine the number, type, and size required. If different sizes are required, they are kept separate. Any special requirements for special shapes are noted, with a written proposal from the manufacturer or supplier. If the estimator is not familiar with the project, arrangements should be made to see the contract documents for the complete price. A check is made to see how the truss is attached to the wall or column supporting it.

If the trusses are to be installed by the general contractor, an allowance for the required equipment (booms, cranes) and workers must be made. The type of equipment and number of workers will depend on the size and shape of the truss and the height of erection. Trusses 120 m long may be placed at the rate of 100 m^2 to 150 m^2 of coverage per hour, using two mobile cranes mounted on trucks and seven workers.

12.11 Laminated Beams and Arches

Laminated structural members are pieces of lumber glued under controlled temperature and pressure conditions. The glue used may be either interior or exterior. They may be rectangular beams or curved arches, such as parabolic, bowstring, V, or A (Figure 12.59). The wood used includes Douglas fir, southern pine, birch, maple, and redwood. Spans just over 100 m have been built by using laminated arches in a cross

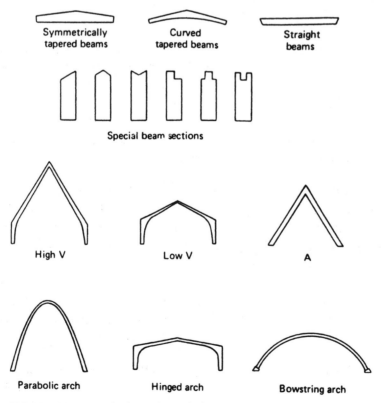

FIGURE 12.59 Typical Laminated Shapes

vault. They are generally available in three grades: industrial appearance, architectural appearance, and premium appearance grade. The specifications should be carefully checked so that the proper grade is used. They are also available prefinished.

Specifications. Check for quality control requirements, types of adhesives, hardware, appearance grade, finish, protection, preservative, and erection requirements.

Estimating. The cost of materials will have to come from the manufacturer or supplier, but the estimator performs the takeoff. The takeoff should list the length required of each type and the size or style of beam or arch. Note also the type of wood, appearance grade, and finishing requirements. If the laminated shapes are not prefinished, be certain that finishing is covered somewhere in the specifications.

If the laminated beams and arches are to be installed by the general contractor, consider who will deliver the material to the site, how it will be unloaded, where it will be stored, and how much equipment and how many workers will be required to erect the shapes. Erection time varies with the complexity of the project. In a simple erection job, beams may be set at the rate of six to eight lengths per hour, while large, more complex jobs may require four hours or more for a single arch. Carefully check the fastening details required—the total length of each piece, how pieces will be braced during construction, and whether an arch is one piece or segmental.

12.12 Wood Decking

Wood decking is available as a solid timber decking, plank decking, and laminated decking. Solid timber decking is available either in natural finish or prefinished; the most common sizes are 38 × 140, 64 × 140, and 89 × 140. The most commonly used woods for decking are southern pine, western red cedar, inland white fir, western white spruce, and redwood.

Plank decking is tongue-and-groove decking fabricated into panels. The most common panel size is 533 mm wide; lengths are up to 7.32 m. Installation costs are reduced by using this type of deck. Note wood species, finish, size required, and appearance grade.

Laminated decks are available in a variety of thicknesses and widths. Installation costs are reduced substantially. Note the wood species, finish, and size required. Appearance grades are also available.

Estimating. No matter what type of wood deck is required, the takeoff is the area to be covered. Take particular notice of fastening details and size of decking required. Consider also if the decking is priced delivered to the jobsite or must be picked up, where the decking will be stored, how it will be moved onto the building, and the number of workers required.

Solid decking requires between 12 and 16 work hours for an mfbm (thousand board feet of measure), while plank decking may save 5 to 15 percent of installation cost on a typical project. Laminated decking saves 15 to 25 percent of the installation cost, since it is attached by nailing instead of with heavy spikes.

Specifications. Check for wood species, adhesives, finishes, appearance grade, fastening, size requirements, and construction of the decking. Size of decking should also be noted.

12.13 Plywood Systems

The strength and versatility of plywood has caused it to be recommended for use in structural systems. The systems presently in use include box beams, rigid frames, folded plates, and space planes for long-span systems. For short-span systems, stressed skin panels and curved panels are used. The availability, size, and ease of shaping of plywood make it an economical material to work with. Fire ratings of one hour can be obtained also. Plywood structural shapes may be supplied by a local manufacturer, shipped in, or built by the contractor's workforce. The decision will depend on local conditions, such as suppliers available, workers, and space requirements.

Specifications. Note the thickness and grade of plywood, whether interior or exterior glue is required, fastening requirements, and finish that will be applied. All glue used in assembling the construction should be noted, and any special treatments, such as fire or pressure preservatives, must also be taken off.

Estimating. Take off the size and shape of each member required. Total the length of each different size and shape. If a manufacturer is supplying the material, it will probably be priced by the metre of each size or as a lump sum. If the members are to be built by the contractor's workforce, either on or off the jobsite, a complete takeoff of lumber and sheets of plywood will be required.

12.14 Wood Checklist

Wood:
- species
- finish required
- solid or hollow
- laminated
- glues
- appearance grade
- special shapes
- primers

Fastening:
- bolts
- nails
- spikes

- glue
- dowels

Erection:
- cranes
- studs
- wood joists
- rafters
- sheathing
- insulation
- trim
- bracing
- bridging
- plates
- sills

Web Resources

www.cmhc.ca
www.cwc.ca
www.canply.org
www.chpva.ca
www.lumber.org
www.southernpine.com
www.wrcla.org

Review Questions

1. What unit of measure is used for lumber?

2. Determine the fbm and m³ for the following order:

   190 − 2 × 4 − 8'-0"   38 × 89   × 2.4
   1,120 − 2 × 4 − 12'-0"   38 × 89   × 3.6
   475 − 2 × 6 − 16'-0"   38 × 140   × 4.8
   475 − 2 × 12 − 16'-0"   38 × 286   × 4.8
   18 − 2 × 8 − 18'-0"   38 × 184   × 5.4

3. How would you determine the number of studs required for a project?

4. How do you determine the number of joists required?

5. What is the difference between pitch and slope, as the terms pertain to roofing?

6. Determine the length of rafter required for each of the following conditions if the run is 4.88 m.

   1/12 pitch
   1/6 pitch
   3 in 12 slope
   4 in 12 slope

7. What unit of measure is used for plywood when it is used for sheathing? How is plywood waste kept to a minimum?

8. What unit of measure is most likely to be used for laminated beams?

9. What unit of measure is most likely to be used for wood decking? How is it determined?

10. What unit of measure is used for wood trim? If it requires a finish, where is this information noted?

11. Determine the quantities for the floor framing required for the residential building shown in Appendix C.

12. Determine the quantities for the wall framing required for the residential building shown in Appendix C.

13. Determine the quantities for the roof framing required for the residential building shown in Appendix C.

14. (Computer Exercise) Create an estimate for all the wood framing using Timberline Precision Estimating Basic by entering the data from Questions 11 to 13. Change the prices to match prevailing market rates.

242

CHAPTER **13**

Thermal and Moisture Protection

After studying this chapter, you will be able to:

1. Describe the difference between waterproofing and dampproofing.

2. Compile quantities of membrane waterproofing for pricing.

3. Identify the unit of measure for asphalt shingles.

4. Determine the amount of insulation for walls and ceilings on a residential project.

5. Identify the types of roof covering materials used on building projects.

13.1 Waterproofing

Waterproofing refers to materials which seal concrete surfaces to prevent the passage of water and resist the hydrostatic pressures to which a wall or floor might be subjected. Dampproofing consists of treating the concrete surfaces to retard the passage or absorption of water or water vapour. Waterproofing can be effected by the admixture mixed with the concrete, the *integral method;* by placing layers of waterproofing materials on the surface, *membrane waterproofing;* and by the *metallic method.*

13.2 Membrane Waterproofing

Membrane waterproofing (Figure 13.1) consists of a buildup of tar or asphalt and membranes (plies) into a strong impermeable blanket. This is the only method of waterproofing that is dependable against hydrostatic head. In floors, the floor waterproofing must be protected against any expected upward thrust from hydrostatic pressure.

The actual waterproofing is provided by an amount of tar or asphalt applied between the plies of reinforcement. The purpose of the plies of reinforcement is to build up the amount of tar or asphalt that meets the waterproofing requirements and to provide strength and flexibility to the membrane.

Reinforcement plies are of several types, including a woven glass fabric with an open mesh, saturated cotton fibre, and tarred felts.

Newly applied waterproofing protection membranes should always be protected against rupture and puncture during construction and backfilling. Materials applied over the membrane include building board and rigid insulation for protection on exterior walls against damage during construction and backfilling.

1 ft³ = 62.4 lbs

1.0 GALS = 10 lbs

1 l = kg

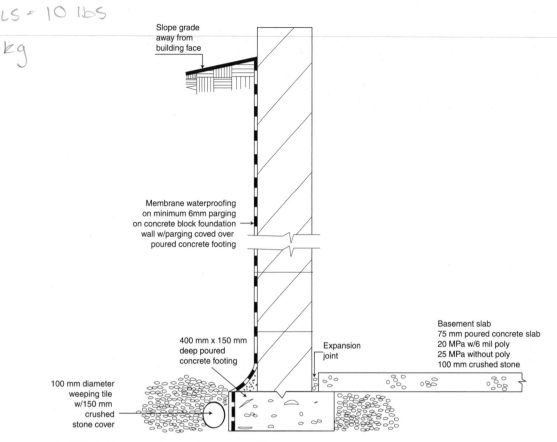

FIGURE 13.1 Membrane Waterproofing

Another type of waterproofing is the use of sprayed-on asphalt with chopped fibreglass sprayed on simultaneously. The fibreglass reinforcement helps the resultant film bridge hairline cracks that may occur in the wall. The sprayed-on waterproofing requires that one to two days of good weather pass before backfilling can begin. Take care during backfilling so that the film is not damaged.

Estimating Membrane Waterproofing. The unit of measurement is the square metre or the square (1 square = 100 s.f.). To estimate the quantity of reinforcement felt required, the estimator determines the area of the walls and floors that require membrane waterproofing, keeping walls and floors separate.

The specifications must be checked to determine the number of membrane (plies) required, the kind and weight of reinforcement ply, the number of coatings, type of coating (tar or asphalt), and the kilograms of coating material required to complete each square metre of waterproofing. The estimator also checks to determine what industry requirements must be met and whether the specifications limit the manufacturer's materials that may be used.

If one layer of reinforcement is specified, the amount must allow for laps over the footings and at the top of the wall, as well as the lap required over each strip of reinforcement. A 100 mm side lap of layers plus the top and bottom laps will require an additional 20 percent. A 150 mm lap will require an additional 25 percent to the area being waterproofed. Lap is generally listed in the specifications. The number of plies used usually ranges from two to five, and the amount to be added for laps and extra material drops to about 10 percent of the actual wall or floor area.

Material	Plies					
	1 Ply 100 mm Lap	1 Ply 150 mm Lap	2 Ply	3 Ply	4 Ply	5 Ply
Reinforcement m^2	11	12	21	31	41	51
Mopping of Bituminous Coatings	2	2	3	4	5	6
Kilograms of Bituminous Coatings	27–32	27–32	41–45	54–61	68–77	82–93

FIGURE 13.2 Materials Required for Membrane Waterproofing per Square

Most felt and cotton fabric reinforcement is available in rolls of 40 m^2. Glass yarn reinforcement is generally available in rolls of 42 m^2. Different manufacturers may have various size rolls and sheets available of certain types of reinforcements.

The amount of tar required varies depending on the number of plies; the amount used per square metre should be in the specifications. Figure 13.2 shows the amount of reinforcing and coating required. Membrane waterproofing is similar to applying builtup roofing, but the cost is greater for vertical surfaces than for horizontal surfaces and there is generally less working space on the vertical.

The area of protective coating must also be determined; this is the same as the area to be waterproofed. The type of protective coating, thickness, and method of installation are included in the specifications.

The sprayed-on asphalt and chopped fibreglass require special equipment to apply, and most manufacturers specify that only if these are applied by approved applicators will they be responsible for the results. Materials required per square metre are 3 to 4 litres of asphalt and 0.3 to 0.5 kilograms of fibreglass. The surface that is being sprayed must be clean and dry.

Labour. A specialist subcontractor approved by the manufacturers of the product almost always installs membrane waterproofing. A typical subcontractor's bid (to the general contractor) is shown in Figure 13.3. Typical labour productivity rates are shown in Figure 13.4.

> *Agrees to furnish and install 3-ply builtup membrane waterproofing in accordance with the contract documents for the sum of --------------------- $8,225.00*

FIGURE 13.3 Subcontractor's Bid for Membrane Waterproofing

Type of Work	Work Hours per m^2
Cleaning	0.03–0.35
Painting Brush, per Coat	0.05–0.20
Parging, per Coat	0.1–0.03
Metallic	0.15–0.30
Waterproofing	
Asphalt or Tar, per Coat	0.07–0.20
Felt or Reinforced, per Layer	0.12–0.3

FIGURE 13.4 Work Hours Required for Applying Waterproofing and Dampproofing

13.3 Integral Method

The admixtures added to the cement, sand, and gravel generally are based on oil/water-repellent preparations. Calcium chloride solutions and other chemical mixtures are used for liquid admixtures and stearic acid for powdered admixtures. These admixtures are added in conformance with manufacturers' recommendations. Most of these admixtures enhance the water resistance of the concrete but offer no additional resistance to water pressure.

Estimating. The amount to be added to the mix varies considerably, depending on the manufacturer and composition of the admixture. The powdered admixtures should be mixed with water before they are introduced into the mix. Liquid admixtures may be measured and added directly, which is a little less trouble. The costs involved derive from delivering the admixture to the jobsite, moving it into a convenient place for use, and adding it to the mix. The containers may have to be protected from the weather, particularly the dry admixture containers.

13.4 Metallic Method

Metallic waterproofing employs a compound that consists of graded fine-iron aggregate combined with oxidizing agents. The principle involved is that when the compound is properly applied, it provides the surface with a coating of iron that fills the capillary pores of the concrete or masonry and, as the iron particles oxidize, the compound creates an expanding action that fills the voids in the surface and becomes an integral part of the mass. Always apply in accordance with the manufacturer's specifications and always apply the compound to the inside of the wall.

Estimating. The estimator needs to check the specification for the type of compound to be used and the thickness and method of application specified. From the drawings he or she can determine the area to be covered and, using the manufacturer's information regarding square metres coverage per litre, determine the number of litres required.
    Labour will depend on the consistency of the compound, the thickness, and method of application required as well as the convenience afforded by the working space. Equipment includes scaffolds, planks, ladders, mixing pails, and trowels.

13.5 Dampproofing

Dampproofing is designed to resist dampness and is used on foundation walls below grade and exposed exterior walls above grade. It is not intended to resist water pressure. The methods used to dampproof include painting the wall with bituminous materials below grade and transparent coatings above grade. *Parging* with a rich cement base mixture is also used below grade, often with a bituminous coating over it.

13.6 Painting Method

Among the most popular paints or compounds applied by brush or spray are tar, asphalt, cement washes, and silicone-based products. This type of dampproofing may be applied with a brush or mop or may be sprayed on. The type of application and the number and thickness of coats required will be determined by the specifications and manufacturer's recommendations.

Below grade on masonry or concrete, the dampproofing material is often black mastic coating that is applied to the exterior of the foundation. It is applied with a brush, mop, or spray. The walls must be thoroughly coated with the mastic, filling all voids or holes. The various types available require from one to four coats to do the job, and square metres per litre vary from 1.25 to 2.50. The more porous the surface, the more material it will require.

Transparent dampproofing of exterior masonry walls is used to make them water repellent. The colourless liquids must be applied in a quantity sufficient to completely seal the surface. Generally, manufacturers recommend one or two coats, with the square metres per litre ranging from 1.30 to 5.

Estimating Painting. To estimate these types of dampproofing, it is necessary first to determine the types of dampproofing required, the number of coats specified (or recommended by the manufacturer), and the approximate number of square metres that one litre of material will cover so that the amount of material required may be determined. If two coats are required, twice the material is indicated.

The amount of labour required to do the work varies with the type of compound being used, the number of coats, the height of work, and the method of application. When the surface to be dampproofed is foundation walls below grade and the work is in close quarters, it will probably take a little longer. Sprayed-on applications require much less labour time but more expensive equipment. High buildings require scaffolds that will represent an added cost factor.

Equipment that may be required includes ladders, planks, mixing cans, brushes, mops, spraying equipment, and scaffolding.

E X A M P L E   **13.1** **Foundation Walls**

The below-grade foundation wall on the building in Figure 13.5 must be brushed with a bituminous product on the 2.54-m-high foundation walls to within 50 mm of the finished grade. Figure 13.5 shows the building layout and Figure 13.6 is a tabulation of the length (m) of foundation wall. The coverage rate is 1.50 m² per litre.

Assume the height to be sprayed is 2.49 m.
Area to be dampproofed = 32.92 m × 2.49 m
Area to be dampproofed = 82 m²
Litres of asphalt = 82 m² / 1.50 m² per litre = 54.67 litres
Use 55 litres.
Using the labour productivity rates from Figure 13.4 and a local prevailing wage rate of $21.20 per hour, the labour costs can be determined.
Work hours = 0.13 work hours per m² × 82 m² = 10.66 work hours
Basic labour cost = Work hours × wage rate
Basic labour cost = 10.66 work hours × $21.20 per work hour = $226

continued

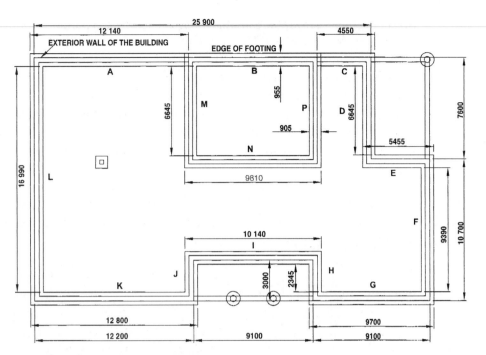

FIGURE 13.5 Foundation Walls

Side	Length of 2.54-m-High Foundation Wall (m)
B	9.21
M	7.25
N	9.21
P	7.25
Total	32.92

FIGURE 13.6 Foundation Walls Requiring Dampproofing

13.7 Parging (Plastering)

The material used to parge exterior portions of foundation walls below grades is a mixture of Portland cement, fine aggregates (sand), and water. A water-repellent admixture is often added to the mix and is trowelled or sprayed in place; the minimum thickness is about 6 mm. Compounds with a cement base are also available from various manufacturers.

On some projects the specifications require parging and the application of a coating of bituminous material in a liquid form. The bituminous type of coating is discussed in Section 13.6.

Estimating Parging. To determine the amount of material required for parging, the estimator first checks the specifications to determine exactly the type of materials required. The materials may be blended and mixed on the job or they may be pre-blended compounds. Information concerning the specially formulated compounds must be obtained from the manufacturer. The job-mixed parging requires that the mix proportions be determined. The next step is to determine the area to be covered and the thickness. With this information the volume (m^3) required can be calculated, or the number of litres determined if it is preblended.

The amount of labour will depend on the amount of working space available and the application technique. Labour costs will be higher when the materials are mixed on the job, but the material cost will be considerably lower.

Equipment required includes a mixer, trowel or spray accessories, scaffolding, planks, shovels, and pails. Water must always be available for mixing purposes.

13.8   Insulation

Ridgide
Batt
Loose

Insulation in light frame construction may be placed between the framing members (studs or joists) or nailed to the rough sheathing. It is used in the exterior walls and the ceiling of most buildings.

Insulation placed between the framing members may be pumped in or laid in rolls or in sheets, while loose insulation is sometimes placed in the ceiling. Roll insulation is available in approximate widths of 275, 375, 475, and 575 mm to fit snugly between the spacing of the framing materials. Sheets of the same widths and shorter lengths are also available. Rolls and sheets are available unfaced, faced on one side, faced on both sides, and foil faced. The insulating materials may be of glass fibre or mineral fibre. Nailing flanges, which project about 50 mm on each side, lap over the framing members and allow easy nailing or stapling. To determine the length (m) required, the area of wall to be insulated is easily determined by multiplying the distance around the exterior of the building by the height to which the insulation must be carried (often the gable ends of a building may not be insulated). Add any insulation required on interior walls for the gross area. This gross area should be divided by the factor given in Figure 13.7. If the studs are spaced 400 mm on centre, the 375-mm-wide insulation plus the width of the stud equals 400 mm, so a batt 1 metre long will cover an area of 0.40 m^2.

Ceiling insulation may be placed in the joists or in the rafters. The area to be covered should be calculated, and if the insulation is placed in the joists, the length of the building is multiplied by the width. For rafters (gabled roofs) the methods shown in Chapter 12 (for pitched roofs) should be used: the length of the building times the rafter length from the ridge to plate times two (for both sides). Divide the area to be covered by the factor given in Figure 13.7 to calculate the length (m) of a given width roll.

Ceiling insulation may also be poured between the joists; a material such as vermiculite is most commonly used. Such materials are available in bags and may be levelled easily to any desired thickness. The volume of material required must be calculated. Loose mineral or fibreglass may also be "blown in" between the ceiling joists. Because this material compresses easily it is typically not done until all construction foot traffic is completed.

Stud Spacing (mm)	Insulation Width (mm)	To Determine Length (m) of Insulation Required Multiply Area (m^2) by	OR	Length (m) of Insulation Required per m^2 of Wall Area
300	275	3.33		3.33
400	375	2.50		2.50
500	475	2.00		2.00
600	575	1.67		1.67

FIGURE 13.7 Insulation Requirements

Insulation board that is nailed to the sheathing or framing is estimated by the square metre. Sheets in various thicknesses and sizes are available in wood and mineral fibres. The area to be covered must be calculated and the number of sheets determined.

For all insulation add 5 percent waste when net areas are used. Net areas will give the most accurate takeoff. Check the contract documents to determine the type of insulation, thickness or R-value required, and any required methods of fastening. When compiling the quantities, estimators keep each different thickness or type of insulation separate. Figure 13.8 shows typical R-values for the more common stud wall application.

R-Value		Stud Wall	o.c. (mm)	Coverage m²	Pieces	Dimensions (mm)		
Metric	Imperial					Width	Length	Thickness
RSI 2.4	R 13.5	38 x 89	600	5.58	8	584	1,194	89
RSI 2.4	R 13.5	38 x 89	400	5.55	12	387	1,194	89
RSI 3.8	R 21.5	38 x 140	600	3.48	5	584	1,194	140
RSI 3.8	R 21.5	38 x 140	400	3.70	8	387	1,194	140

FIGURE 13.8 Typical Insulation R-Values

E X A M P L E   **13.2**   **Residential Building Insulation**

Determine the insulation required for the residential building introduced in Chapter 12 and shown in Figure 13.9.

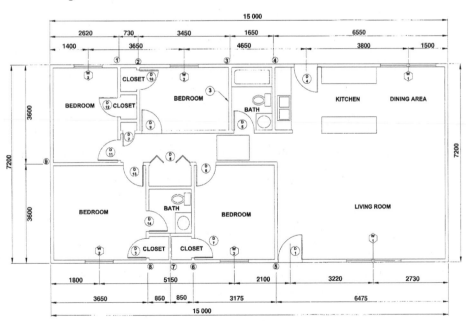

FIGURE 13.9 Residential Floor Plan

Floor: 140 mm thick RSI 3.5 (R20) roll insulation
Area of floor = 15.00 m × 7.20 m = 108 m²
1 roll = 0.38 m × 17.07 m
Effective coverage per roll = 0.40 m × 17.07 m = 6.83 m²
Rolls of insulation = 108 m² of floor / 6.83 m² of coverage per roll
Use 16 rolls

continued

Ceiling: 190 mm RSI 7.0 (R40) roll insulation
Area of ceiling = Area of floor
Area of ceiling = 108 m²
1 roll = 0.38 m × 9.75 m
Effective coverage per roll = 0.40 m × 9.75 m = 3.90 m²
Rolls of insulation = 108 m² of ceiling / 3.90 m² of coverage per roll
Use 28 rolls

Walls: 89 mm RSI 2.4 (R13.5) roll insulation
Area of wall = 2.34 m high × 44.40 m of exterior walls = 103.90 m²
1 roll = 0.38 m × 17.07 m
Effective coverage per roll = 0.40 m × 17.07 m = 6.83 m²
Rolls of insulation = 103.90 m² of wall / 6.83 m² of coverage per roll
Use 16 rolls

13.9 Roofing

Roofing is considered to include all the material that actually covers the roof deck. It includes any felt (or papers) as well as bituminous materials that may be placed over the deck. These are commonly installed by the roofing subcontractor and included in that portion of the work. Flashing required on the roof, at wall intersections, joints around protrusions through the roof, and the like is also included in the roofing take-off. Miscellaneous items, such as fibre blocking, cant strips, curbs, and expansion joints, should also be included. The sections and details, as well as the specifications, must be studied carefully to determine what is required and how it must be installed.

13.10 Roof Areas

In estimating the area to be roofed, estimators must consider the shape of the roof (Figure 13.10). When measuring a flat roof with no overhang or with parapet walls, they take the measurements of the building from the outside of the walls. This method is used for parapet walls because it allows for the turning up of the roofing on the sides of the walls. If the roof projects beyond the building walls, the dimensions used must be the overall outside dimensions of the roof and must include the overhang. The drawings should be carefully checked to determine the roofline around the building, as well as where and how much overhang there may be. The floor plans, wall sections, and details should be checked to determine the amount of overhang and type of finishing required at the overhang. Openings of less than 1 m² should not be deducted from the area being roofed.

To determine the area of a gable roof, multiply the length of the ridge (A to C) by the length of the rafter (A to B) by two (for the total roof surface). The area of a shed roof is the length of the ridge (A to C) times the length of rafter (A to B). The area of a regular hip roof is equal to the area of a gable roof that has the same span, pitch, and length. The area of a hip roof may be estimated the same as the area of a gable roof (the length of roof times the rafter length times two).

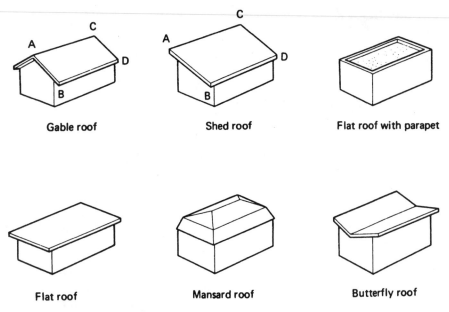

Gable roof   Shed roof   Flat roof with parapet

Flat roof   Mansard roof   Butterfly roof

FIGURE 13.10 Typical Roof Shapes

The length of rafter is easily determined from the span of the roof, the overhang, and slope. Refer to Figure 12.50 and Example 12.22 for information required to determine the lengths of rafters for varying pitches and slopes.

13.11 Shingles

Asphalt Shingles

Available in a variety of colours, styles, and exposures, strip asphalt shingles are 305 and 381 mm wide and 914 mm long. They are packed in bundles that contain enough shingles to cover 2.35 or 4.65 m² of roof area. The *exposure* (amount of shingle exposed to the weather) generally is 102, 114, or 127 mm. Individual shingles 305 to 406 mm long is sometimes also used. Asphalt shingles may be specified by the weight per square metre, which may vary from 8 to 16 kilograms. Shingles may be fire rated and wind resistant, depending on the type specified.

In determining the area to be covered, always allow one extra course of shingles at the eaves, as the first course must always be doubled, and any additional for eaves protection. Hips and ridges are taken off by the metre and considered 305 mm wide to determine the area of shingles required. Waste averages 5 to 8 percent. Galvanized, large-headed nails, 12.7 to 44.5 mm long, are used on asphalt shingles. From 0.07 to 0.14 kilograms of nails are required per square metre.

Asphalt shingles are generally placed over an underlayment of building paper or roofing felt. The felt is specified by the type of material and weight per square metre. The felt should have a minimum top lap of 50 mm and end lap of 100 mm; to determine the area of felt required, multiply the roof area by a lap and waste factor of 5 to 8 percent. To estimate the number of bundles of shingles, add the number of sheets of roof plywood and add an extra bundle for every 7.50 m of ridge or hip.

E X A M P L E   **13.3**   **Asphalt Shingle Roofing**

Using the residence that was introduced in Chapter 12, determine the felt and shingle requirements. In Example 12.22, the rafter length of 4.18 m and ridge length of 15.60 m was determined. The first step is to determine the number of 6.8 kilogram roofing felt that is required.

Roof area = 4.18 m rafter length × 15.60 m ridge length × 2 sides = 130.42 m²

Slope: 3 on 12

Felt requirements: 2 layers – lapped 483 mm

Area of felt in a roll = 914 mm wide × 43.89 m long = 40.12 m²

Formula 13.1 Felt Coverage

$$\text{Roll coverage} = \text{Square metre in roll} \times \frac{\text{Exposure}}{\text{Width}}$$

Exposure = Roll width − lap

Exposure = 914 mm − 483 mm = 431 mm

Roll coverage = 40.12 m² × (431 mm / 914 mm) = 18.92 m²

Rolls of felt required = 130.42 m² / 18.92 m² per roll = 6.89 rolls

Use 7 rolls.

Use 107 kilogram shingles

Starter course = 15.60 m of eaves per side × 2 sides = 31.20 m

(Assume 305 mm wide)

Starter course = 9.52 m²

Squares of roofing = 130.42 m² + 9.52 m² / 9.3 m² per square = 15.05 squares

Order 16 squares

Ridge shingles = 127 mm exposure

15.60 m of ridge = 15,600 mm of ridge

Pieces of ridge shingles = 15,600 mm of ridge / 127 mm per shingle = 123 pieces

Labour. Subcontractors who specialize in roofing will do this work and they may price it on a unit basis (per square) or a lump sum. The time required to install shingles is shown in Figure 13.11.

Roofing Material	Work Hours per Square (9.3 m²)
Shingles	
Asphalt (strip)	.8 to 4.0
Asphalt (single)	1.5 to 6
Wood (single)	3.0 to 5.0
Metal (single)	3.5 to 6.0
Heavyweight Asphalt	Add 50%
Tile	
Clay	3.0 to 5.0
Metal	3.5 to 6.0
Builtup	
2 Ply	.8 to 1.4
3 Ply	1.0 to 1.6
4 Ply	1.2 to 2.0
Aggregate Surface	.3 to .5

FIGURE 13.11 Roofing Labour Productivity Rates

Wood Shingles

Available in various woods-the best of which are cypress, cedar, and redwood—wood shingles come hand-split (rough texture) or sawed. The hand-split variety is commonly referred to as *shakes*. Lengths are 406, 457, and 610 to as long as 813 mm, but random widths of 102 to 305 mm are common. Wood shingles are usually sold by the square (9.3 m^2) based on a 254 mm weather exposure. A double-starter course is usually required; in some installations, roofing felt is also needed. Valleys will require some type of flashing, and hips and ridges require extra material to cover the joint; thus extra-long nails become requisite. Nails should be corrosion resistant, and 1 to 2 kg of 44.5- to 50.8-mm-long nails are required per square.

13.12   Builtup Roofing

Builtup roofing consists of layers of overlapping roof felt with each layer set in a mopping of hot tar or asphalt. Such roofing is usually designated by the number of plies of felt that are used: For example, a three-ply roof has four coats of bituminous material (tar or asphalt) and three layers of felt. Although builtup roofing is used primarily for flat and near-flat roofs, it can be used on inclines of as great as 750 mm/m, providing certain special bituminous products are used.

The specifications must be read carefully because no one type of system exists for all situations. There may be vapour barrier requirements and varying amounts of lapping of felt. Different weights and types of felts and bituminous materials may be used. The deck that is to receive the roofing must be considered, as must the service it is intended to sustain.

Many specifications require a bond on the roof. This bond is furnished by the material manufacturer and supplier through the roofing contractor. It generally guarantees the water tightness of the installation for a period of 10, 15, or 20 years. During this period of time, the manufacturer will make repairs on the roof that become necessary due to the normal wear and tear of the elements to maintain the roof in a watertight condition. Each manufacturer has its own specifications or limitations in regard to these bonds, and the roofing manufacturer's representative should be called in for consultation. Special requirements may pertain to approved roofers, approved flashing, and deck inspection; photographs may also be necessary. If it appears that the installation may not be approved for bonding, the consultant should be so informed during the bid period. The manufacturer's representative should call on the consultant and explain the situation so it can be worked out before the proposals are due.

The takeoff follows the same general procedure as any other roofing:

1. The number of squares to be covered must be noted.
2. Base sheets—a base sheet or vapour barrier is used, when required; note the number of plies, lap (usually 100 mm), mass per ply per square.
3. Felt—determine the number of plies, mass of felt per square, and type of bituminous impregnation (tar or asphalt).
4. Bituminous material—note the type, number of coats, and the kilograms per coat per square (some specifications call for an extra—heavy topcoat or pour).
5. Aggregate surface (if required)—the type and size of slag, gravel, or other aggregate, and the kilograms required per square should all be noted.
6. Insulation (if required)—note type, thickness, and any special requirements (refer also to Sec. 13.8).

7. Flashing—note type of material, thickness, width, and length required.

8. Trim—determine the type of material, size, shape and colour, method of attachment, and the length required.

9. Miscellaneous materials—blocking, cant strips, curbs, roofing cements, nails and fasteners, caulking, and taping are some examples.

Bituminous materials are used to cement the layers of felt into a continuous skin over the entire roof deck. Types of bituminous material used are coal tar pitch and asphalts. The specifications should indicate the amount of bituminous material to be used for each mopping so that each ply of felt is fully cemented to the next. In no instance should felt be allowed to touch felt. The mopping between felts averages 11 to 14 kg, while the top pour (poured, not mopped), which is often used, may be from 30 to 34 kg per square. It is this last pour into which any required aggregate material will become embedded.

Felts are available in 6.8 and 13.6 kg weights, 914 mm widths, and rolls 40 or 20 m². With a 50 mm lap, a 40 m² roll will cover 37.16 m², while the 20 m² roll will cover 18.6 m². In builtup roofing, a starter strip 305 to 457 mm wide is applied; then over that one, a strip 914 mm wide is placed. The felts that are subsequently laid overlap the preceding felts by 483, 627, or 838 mm, depending on the number of plies required. Special applications sometimes require other layouts of felts. The specifications should be carefully checked to determine exactly what is required as to weight, starter courses, lapping, and plies. Waste averages about 8 to 10 percent of the required felt, which allows for the material used for starter courses.

Aggregate surfaces, such as slag or gravel, are often embedded in the extra-heavy top pouring on a builtup roof. This aggregate acts to protect the membrane against the elements, such as hail, sleet, snow, and driving rain. It also provides weight against wind uplift. To ensure embedment in the bituminous material, the aggregate should be applied while the bituminous material is hot. The amount of aggregate used varies from 110 to 220 kg per square. The amount required should be listed in the specifications. The amount of aggregate required is estimated by the tonne.

Since various types of materials may be used as aggregates, the specifications must be checked for the size, type, and gradation requirements. It is not unusual for aggregates, such as marble chips, to be specified. This type of aggregate will result in much higher material costs than gravel or slag. Always read the specifications thoroughly.

The joints between certain types of roof deck materials—such as precast concrete, gypsum, and wood fibre—may be required to have caulking and taping of all joints. This application of flashing cement and a 150-mm-wide felt strip will minimize the deleterious effects of uneven joints on the roofing membrane, movement of the units, and moisture transmission through the joint; it also reduces the possibility of bitumen seepage into the building. The length of joints to be covered must be determined for pricing purposes.

13.13 EPDM

This type of roof is completely in a single layer, thereby the term "single ply." The rubber used to make the large sheets is ethylene propylene diene terpolymer (EPDM)—a highly weather-resistant, synthetic polymer designed for thermal stability and blended with carbon black, oils, and other additives. EPDM membranes are anchored to the building using one of three techniques.

Ballasted is the most common and uses stone or concrete pavers that will resist the potential for the roof to blow off. The membrane is laid over nonanchored insulation boards. The membrane is seamed together and covered with a minimum of 0.42 kg/m² of nominal 38 mm, rounded river-bottom stone.

Mechanically attached systems incorporate either the use of polymer or metal attachment bars and fasteners or metal attachment plates and fasteners to hold the membrane to the roof deck. Wind and thermal forces are transferred from the membrane to the fastener and ultimately to the deck, which allows the system to pass high wind uplift tests.

Fully adhered uses either contact adhesive or spray foam to attach the membrane to the top of an approved substrate. Membranes backed with fleece can be adhered using hot asphalt. Great care should be taken during this application in that asphalt-based products should not come in direct contact with the EPDM membrane. As with asphalt, the adhesives used have a distinct odour, which may be offensive to some occupants of the building where the roof system is being installed. The use of water-based adhesives versus solvent-based adhesives greatly reduces odour.

Estimate EPDM systems by the square metre; note the type of system and method fastening required. Consult the local sales representative for the required information. The specifications and manufacturers' recommendations should be checked to determine any special requirements; only then can these systems be adequately estimated.

13.14 Thermoplastic or Weldable

Thermoplastic sheets are heat welded with hot air used to create a seam by fusing the separate pieces of sheets together. These sheets are manufactured from thermoplastic resins, the most common being PVC and TPO/EP. They are reinforced with either a fibreglass or polyester scrim fabric.

Thermoplastic systems are most often mechanically attached to the deck through the lap area. However, they can be ballasted or fully adhered, like EPDM roof systems. When properly welded, the seams are as strong as the sheet itself.

Estimate PVC systems by the square metre; note the method of jointing and fastening required. Consult the local sales representative for the required information. The specifications and manufacturers' recommendations should be checked to determine any special requirements prior to the estimate completion.

13.15 Metal Roofing

The two basic types of metal roofing are architectural and structural. The difference is that architectural systems need the metal to be supported on a continuous deck, while structural systems can be placed directly on metal or wood supports. Structural systems are generally watertight systems with some manufacturers allowing their systems to be installed on slopes, down to 20 mm/m. Architectural systems generally require quick drainage to function and are sloped no less than 150 mm/m, but most applications call for 250 mm/m slope.

Galvanized-coated steel is the most popular metal for both applications. Galvalume and acrylume are other coatings that are commonly used. Other metals used for architectural applications include aluminum, zinc, and copper.

Basically, in this portion, we are considering flat and sheet roofing. The basic approach to estimating this type is the same as that used for other types of roofing: First the number of squares to be covered must be determined; then consideration must be given to available sheet sizes (often 356 × 508 mm and 508 × 711 mm for tin) and quantity in a box. Items that require special attention are the fastening methods and the type of ridge and seam treatment.

Copper roofing comes in various sizes. The weights range from 40 to 60 g per m², and the copper selected is designated by its weight (454 g copper). The copper may have various types of joining methods; among the most popular are standing seam and flat seam. The estimator must study the drawings and specifications carefully, as the type of seam will affect the coverage of a copper sheet. When using a standing seam, 64 mm in width are lost from the sheet size (in actual coverage) to make the seam. This means that when standing seams are used, 12 percent must be added to the area being covered to allow for the forming of the seam; if end seams are required, an allowance must also be added for them. Each different roofing condition must be planned to ensure a proper allowance for seams, laps, and waste.

13.16 Slate

Slate is available in widths of 150 to 400 mm and lengths of 300 to 660 mm. Not all sizes are readily available, and it may be necessary to check the manufacturer's current inventory. The basic colours are blue-black, grey, and green. Slate may be purchased smooth or rough textured. Slate shingles are usually priced by the amount required to cover a square (9.3 m²) if the manufacturer's recommended exposure is used. If the exposure is different from the manufacturer's, Formulas 13.2 and 13.3 (Section 13.17) may be used. To the squares required for the roof area, add 0.3 m² per metre of length for hips and rafters. Slate thicknesses and corresponding weights are shown in Figure 13.12, assuming a 75 mm lap. Each shingle is fastened with two large-head, solid copper nails, 32 mm or 38 mm long. The felt required under the slate shingles must also be included in the estimate. Waste for slate shingles varies from 8 to 20 percent, depending on the shape, number of irregularities, and number of intersections on the roof.

13.17 Tile

The materials used for roofing tiles are cement, metals, and clay. To estimate the quantities required, the roof areas are obtained in the manner described earlier. The number, or length, of all special pieces and shapes required for ridges, hips, hip starters,

Thickness of slate (mm)	4.76	6.35	9.53	12.70	19.05	25.40
Mass per square (75 mm) lap	317–362	408–454	590–635	771–816	1,134–1,270	1,542–1,633

FIGURE 13.12 Approximate Mass (kg) of Slate per Square (9.3 m²)

terminals, and any other special pieces must be carefully taken off. If felt is used under the tile, its cost must also be included in the estimate of the job, and furring strips must be installed for certain applications.

As unfamiliar installations arise, the estimators may call the manufacturer, the manufacturer's representative, or a local dealer. These people can review the project with the estimators and help them arrive at material takeoffs, and perhaps even suggest a local subcontractor with experience in installing the tile.

13.18 Sheets, Tiles, and Shingles Formula

To determine the number of sheets, tiles, or shingles required to cover a square (9.3 m^2) of roof area for any required lap or seam, the following formula may be used.

Formula 13.2 Number of Sheets, Tiles, or Shingles

$$N = \frac{(\text{One square}) \times (\text{Square millimetres})}{(W - S) \times (L - E)}$$

N = Number of sheets (tiles or shingles)
W = Width of sheet (millimetres)
L = Length of sheet (millimetres)
S = Side lap or seam lap requirement
  (for some types there may be none)
E = End lap or seam requirements

To determine the square metre of any given roofing sheet, tile, or shingles required to cover one square, Formula 13.3 may be used:

Formula 13.3 Square Metre of Material per Square

$$A = \frac{9.3 \times W \times L}{(W - S) \times (L - E)}$$

A = Square metre of material per square

13.19 Liquid Roofing

Liquid roofing materials were developed primarily for free-form roofs. Composition varies among the manufacturers, but the application sequences are similar. First, primer is applied liberally over the entire surface. Then, all major imperfections are caulked, joints taped, and flashing is applied at all intersecting surfaces. Next, three coats of liquid roofing are applied; depending on the deck and slope, they may be either base and finish coats or only the finish coat. Finish coats are available in a variety of colours. Equipment is simple, consisting only of rollers, hand tools, brushes, and a joint-tape dispenser. Primarily, the shape of the roof, ease of moving about, and so forth will affect labour. On some buildings, it is necessary to erect scaffolding to apply the roofing; in such cases, the cost obviously will increase accordingly. Before bidding this type of roofing, estimators will discuss the project with the manufacturer's representative so that the latest technical advice may be incorporated in the bid.

13.20 Flashing

Flashing is used to help keep water from getting under the roof covering and from entering the building wherever the roof surface meets a vertical wall. It usually consists of strips of metal or fabric shaped and bent to a particular form. Depending on the type of flashing required, it is estimated by the piece, length (m) with the width noted, or square metre. Materials commonly used include copper, asphalt, plastic, rubber, composition, and combinations of these materials. Bid the gauge of thickness and width specified. Expansion joints are estimated by the metre, and it may be necessary that curbs be built up and the joint cover either prefabricated or job assembled. Particular attention to the details on the drawings is required so that the installation is understood.

13.21 Insulation

Insulation included as a part of roofing is of the type that is installed on top of the deck material. This type of insulation is rigid and is in a sheet (or panel). Rigid insulation may be made of urethane, fibreboards, or perlite and is generally available in lengths of 900 mm to 3,600 mm in 300 mm increments and widths of 300, 400, 600, and 1,200 mm. Thicknesses of 12.7, 19, 25, 38, 50, 64, 75, and 100 mm are available. Insulation is estimated by the square with a waste allowance of 5 percent, provided there is proper planning and utilization of the various sizes.

When including the insulation, keep in mind that its installation will require extra materials, either in the line of additional sheathing paper, moisture barriers, mopping, or a combination of these. Also, the specifications often require two layers of insulation, usually with staggered joints, which requires twice the area of insulation (to make up two layers), an extra mopping of bituminous material, and extra labour. Estimators must read the specifications carefully and never bid on a project they do not fully understand.

NOTE: Some roofing manufacturers will not bond the performance of the roof unless the insulation meets their specifications. This item should be checked and if the manufacturer or the representative sees any problem, the consultant should be notified so that the problem may be cleared up during the bid period.

13.22 Roofing Trim

Trim—such as gravel stops, fascia, coping, ridge strips, gutters, downspouts, and soffits—is taken off by the length (m). All special pieces used in conjunction with the trim (e.g., elbows, shoes, ridge ends, cutoffs, corners, and brackets) are estimated by the number of pieces required.

The usual wide variety of materials and finishes is available in trim. Not all trim is standard stock; much of it must be specially formed and fabricated, which adds considerably to the cost of the materials and to the delivery time. Estimators must be thorough and never assume that the trim needed is standard stock.

13.23 Labour

The labour cost of roofing will depend on the hourly output and hourly wages of the workers. The output will be governed by the incline of the roof, size, irregularities in the plan, openings (skylights and so on), and the elevation of the roof above the ground since the higher the roof, the higher the materials have to be hoisted to the work area.

Costs can be controlled by the use of crews familiar with the type of work and experienced in working together. For this reason, a specialty contractor with equipment and trained personnel handles much of the roofing done on projects.

13.24 Equipment

Equipment requirements vary considerably, depending on the type of roofing used and the particular job being estimated. Most jobs will require hand tools for the workers, ladders, some scaffolding, and some type of hoist, regardless of the type of roofing being applied. Specialty equipment for each particular installation may also be required. Builtup roofing may demand that mops, buggies, and heaters (to heat the bituminous materials) be used; some firms have either rotary or stationary felt layers. Metal roofing requires shears, bending tools, and soldering outfits.

Equipment costs are estimated either by the square or by the job, with the cost including such items as transportation, setup, depreciation, and replacement of miscellaneous items.

13.25 Checklist

Paper
Felt
Composition (roll)
Composition (builtup)
Tile (clay, metal, concrete)
Shingles (wood, asphalt, slate)
Metal (copper, aluminum, steel)
Insulation
Base
Solder:
  paints
  plaster
  foundation walls
  slabs
  sump pits
  protective materials
  exterior
  interior
  admixtures
  drains
  pumps

Flashing
Ridges
Valleys
Fasteners
Trim
Battens
Blocking (curbs)
Cant strips
Waterproofing:
  integral
  membrane

Dampproofing:
  integral
  parge
  vapour barriers
  bituminous materials
  drains
  foundation walls
  slabs
  pumps

Web Resources

www.nrca.net
www.asphaltroofing.org
www.roofing.ca
www.naima.ca

Review Questions

1. What is the difference between waterproofing and dampproofing?

2. What is membrane waterproofing, and how is it estimated?

3. What is parging, and what unit of measure is used?

4. What is the unit of measure for roll batt insulation? What type of information should be noted on the estimate?

5. What is the unit of measure for shingle roofing? What type of information should be noted on the estimate?

6. What is the unit of measure for builtup roofing? What type of information should be noted on the estimate?

7. Determine the quantities of the insulation for the walls and ceiling required for the residential building shown in Appendix C.

8. Determine the quantities of the insulation for the walls and roof required for the commercial building shown in Appendix D.

9. Determine the quantities for the builtup roofing and associated works for the commercial building shown in Appendix D.

CHAPTER 14

Doors and Windows

After studying this chapter, you will be able to:

1. Determine the window and/or curtain wall and door materials required for a building project.

2. Describe the accessories required for windows and curtain wall.

3. Identify the factors to be considered in pricing a door.

4. Describe the ways hardware can be handled on a project.

5. Identify the four types of glass used on building projects.

14.1 Window and Curtain Wall Frames

Window and curtain wall frames may be made of wood, steel, aluminum, bronze, stainless steel, or plastic. Each material has its particular types of installation and finishes, but from the estimator's viewpoint, there are two basic types of window: stock and custom made.

Stock windows are more readily available and, to the estimator, more easily priced as to the cost per unit. Mockup samples, brochures, and current price lists are important reference information in a contractor's library. The estimator can count the number of units required and list the accessories to work up the material cost.

Custom-made frames cannot be accurately estimated. Approximate figures can be worked up on the basis of the area and type of window, but exact figures can be obtained only from the manufacturer. In this case, the estimator will call either the local supplier or manufacturer's representative to be certain that they are bidding the job. Often, copies of the drawings and specifications are sent to them, which they may use to prepare a quotation.

When checking proposals for the windows on a project, the estimator needs to note whether the glass or other glazing is included, where delivery will be made, and whether installation is done by the supplier, a subcontractor, or the general contractor. The proposal must include all the accessories that may be required, including mullions, screens, and sills, and the material being bid must conform to the specifications.

If the contractor is going to install the windows, the estimator needs to check if they will be delivered preassembled or if they must be assembled on the job. The job may be priced by the square metre or length (m) of frame, but the most common method is to bid in a lump sum.

Shop drawings should always be required for windows and curtain wall frames because even stock sizes vary slightly in terms of masonry and rough openings required for their proper installation. Custom-made windows always require shop drawings so that the manufactured sizes will be coordinated with the actual job conditions.

This type of work is generally done by a specialized manufacturer and is usually subject to a subtrade price. If the estimator decides to do a complete takeoff of materials required, he or she should (1) determine the length of each different shape required of each type of material, (2) determine the type and thickness of glazing required, and (3) calculate the sheet sizes required.

Wood Frames. More than 60 percent of all stock window frames are made from Ponderosa pine. Other commonly used species are southern pine and Douglas fir. Custom-made frames may be of any of these species or of some of the more exotic woods, such as redwood and walnut. Finishes may be plain, primed, preservative treated, or even of wood that is shielded with vinyl.

The same general estimating procedure is followed here as in other frames since these frames may be stock or custom made. Shop drawings should be made and carefully checked so that all items required are covered by the proposal. If painting is required, it must be covered in that portion of the specifications.

E X A M P L E **14.1** **Residential Windows**

Determine the quantity of windows for a residential building. The window schedule (Figure 14.1) typically denotes all window sizes and types. The estimator simply must crosscheck the drawings to verify the list and count the number of each size and type (Figure 14.2).

Windows			
No	**Type**	**Size**	**Manufacturer**
1	Casement	2,300 x 1,625	Maple Leaf - C616/C1016/C616
2	Slider	1,225 x 1,225	" - S1210
3	Slider	1,225 x 1,025	" - S120
4	Casement	1,225 x 1,025	" - C1210
5	Basement	1,200 x 400	" - B124
6	Casement	1,225 x 1,225	" - C1212
7	Casement	1,225 x 1,025	" - C1210
8	Casement	1,825 x 1,525	" - C1815
9	Casement	1,825 x 1,525	" - C1815

FIGURE 14.1 Sample Window Schedule

continued

Size	Type/Description	Quantity
2,300 x 1,625	Wood Casement	1
1,225 x 1,025	Wood Casement	2
1,225 x 1,225	Wood Casement	1
1,825 x 1,525	Wood Casement	2
TOTAL		**6**

FIGURE 14.2 Window Takeoff Schedule

Installation:

6 wood windows in masonry
Productivity rate (Figure 14.3) 3 work hours / window
Work hours = 6 windows × 3 work hours per window = 18 work hours
Basic labour cost $ = 18 work hours × $19.60 per work hour = $352.80

Windows	Work Hours per Window
Wood:	
In Wood Frames	2.0 to 3.5
In Masonry	2.5 to 5.0
Metal:	
In Wood Frames	2.0 to 3.5
In Masonry	2.0 to 4.0
Over 1.20 m^2	Add 20 %
Bow and Bay Windows	2.5 to 4.0

FIGURE 14.3 Work Hour Productivity Rates

Aluminum Windows and Curtain Wall. The shapes for aluminum mullions used in curtain walls or storefronts are made by an extrusion process and roll forming. Finishes include mill finish (natural silvery sheen), anodizing (protective film; it may be clear or a variety of colour tones), paint, lacquer, and coloured or opaque coatings. Obviously the finish required must be noted because it will affect the price of the materials. The specifications will require certain thicknesses of metal throughout the construction of the frames.

Care on the job is important because exposed aluminum surfaces may be subject to splashing with plaster, mortar, or concrete masonry cleaning solutions that often cause corrosion of the aluminum. Exposed aluminum should be protected during construction with either a clear lacquer or strippable coating. Concealed aluminum that is in contact with concrete, masonry, or absorbent material (such as wood or insulation) and may become intermittently wet must be protected permanently by coating either the aluminum or adjacent materials. Coatings commonly used include bituminous paint and zinc-chromate primer.

The costs of such protection must be included in, and are part of, the cost of the project. If the frames are to be installed by a subcontractor, the estimator must check to see if that entity has included the required protection in the quotation; if not, the estimator will have to make an allowance for it. Also, who will apply the protection? If a strippable coating is used on exposed aluminum the estimator must note who will remove the coating, if the subcontractor has included this in the price, or if the general contractor will handle it. Any item of work that must be done costs money, so the estimator must ensure that all items are included.

Before submitting a bid, the estimator should check the details showing how the windows are to be installed. They may have to be installed as the construction progresses (where the window is built into the wall), or they may be slipped into openings after the building has been constructed. The method is important, as there is always a time lag for approvals, and manufacture (of custom units) and coordination will be important.

E X A M P L E 14.2 Commercial Building

Determine the quantity of windows in a small commercial building. Figure 14.4 is an excerpt of the window schedule for this project.

No.	Type	Size	Material	Description	Remarks
\multicolumn{6}{c}{**Window Schedule**}					
1	W1	2,000 x 1,800	Aluminum	Double Casement	
2	W2L	2,000 x 2,000	Aluminum	Double Hung	
3	W2R	2,000 x 2,000	Aluminum	Double Hung	
4	W3R	2,000 x 1,800	Aluminum	Double Hung	
5	W3L	2,000 x 1,800	Aluminum	Double Hung	
6	W4	1,200 x 975	Aluminum	Fixed Panels	
7	W5	975 x 1,800	Aluminum	Fixed Panels	
8	W3R	2,000 x 1,800	Aluminum	Double Hung	
9	W3L	2,000 x 1,800	Aluminum	Double Hung	
10	W4	1,200 x 975	Aluminum	Fixed Panels	
11	W1	2,000 x 1,800	Aluminum	Double Casement	
12	W5	975 x 1,800	Aluminum	Fixed Panels	
13	W2L	2,000 x 2,000	Aluminum	Double Hung	
14	W2R	2,000 x 2,000	Aluminum	Double Hung	
15	W1	2,000 x 1,800	Aluminum	Double Casement	
16	W4	1,200 x 975	Aluminum	Fixed Panels	
17	W3R	2,000 x 1,800	Aluminum	Double Hung	
18	W3L	2,000 x 1,800	Aluminum	Double Hung	
19	W1	2,000 x 1,800	Aluminum	Double Casement	

FIGURE 14.4 Window Schedule

continued

Double casement	2,000 mm × 1,800 mm Count	4
Double hung	2,000 mm × 1,800 mm Count	6
Double hung	2,000 mm × 2,000 mm Count	4
Fixed panels	975 mm × 1,800 mm Count	2
Fixed panels	1,200 mm × 975 mm Count	3

Productivity rate (see Figure 14.3) 3 work hours / window

Work hours = 19 windows × 3 work hours per window = 57 work hours

Basic labour cost $ = 57 work hours × $19.60 per work hour = $1,117.20

E X A M P L E 14.3 Storefront Glass

Perform a takeoff of the materials for the storefront glass required for the building entrance as per Figures 14.5, 14.6, and 14.7.

A subcontractor, usually the curtain wall supplier, almost always installs aluminum curtain walls. The time required for curtain wall installation is shown in Figure 14.8.

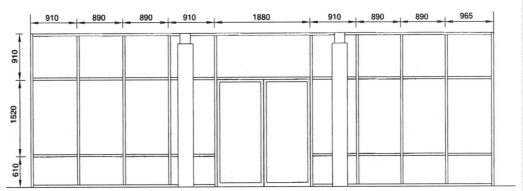

All dimensions in millimetres (mm)

FIGURE 14.5 Storefront Glass

Aluminum Tubing

Length (mm)	Pcs.	Length (m)
3,040	10	30.40
910	12	10.92
890	16	14.24
965	4	3.86
1,880	2	3.76
Total (m)		**63.18**

FIGURE 14.6 Aluminum Tubing Takeoff

continued

Insulating Glass

Size (mm)	Pieces	m² Each	Total m²
910 x 910	3	0.83	2.49
910 x 1,520	3	1.38	4.14
910 x 610	3	0.56	1.68
890 x 910	4	0.81	3.24
890 x 1,520	4	1.35	5.40
890 x 610	4	0.54	2.16
965 x 910	1	0.88	0.88
965 x 1,520	1	1.47	1.47
965 x 610	1	0.59	0.59
1,880 x 910	1	1.71	1.71

FIGURE 14.7 Insulating Takeoff

Metal Curtain Walls	m² per Work Hour
Up to 1 Storey High	0.75–1.66
1 to 3 Stories	0.55–1.10
Over 3 Stories	0.55–1.00

Includes glazing. The larger the opening for the glass or panels the more work can be done per work hour.

FIGURE 14.8 Curtain Wall Productivity Rates

14.2 Accessories

The items that may be required for a complete job include glass and glazing, screening, weather-stripping, hardware, grilles, mullions, sills, and stools. The specifications and details of the project must be checked to find out what accessories must be included so that the estimator can make a list of each accessory and what restrictions apply for each.

Glass. *Glass* is discussed in Section 14.10. At this point, estimators note whether it is required, what thickness, type, and quality, and if it is to be a part of the unit or installed at the jobsite. The area of each type of glass must be known as well.

Glazing. *Glazing* is the setting of whatever material is installed in the frames. Often, the material is glass, but porcelain, other metal panels, exposed aggregate plywood, plastic laminates, and even stone veneer and precast concrete have been set in the frames. Many stock frames are glazed at the factory, but it is not uncommon for stock windows, especially steel frames, to be glazed on the job. Custom-made frames are almost always glazed on the job. The subcontractor and supplier will help determine who glazes them. Glazing costs will depend on the total amount involved, size and quality of the material to be glazed, and the type of glazing method used. If the frame is designed so that the material can be installed from inside the building, no scaffolding will be required and work will proceed faster. On wood frames, glazing compounds are usually employed, whereas on metal frames, either glazing compounds or neoprene gaskets are used.

Screens. Screening mesh may be painted or galvanized steel, plastic-coated glass fibre, aluminum, or bronze. If specified, ensure that the screens are included in the quotations received. If the selected subcontractor or supplier for the frames does not include the screens, a source for those items must be obtained elsewhere. If so, be certain that the screens and frames are compatible and that a method of attaching the screen to the frame is determined. The various sizes required will have to be noted so an accurate price for the screening can be determined.

Hardware. Most types of frames require *hardware*, which may consist only of locking and operating hardware with the material of which the frame is made. Because the hardware is almost always sent unattached, it must be applied at the jobsite. Various types of locking devices, handles, hinges, and cylinder locks on sliding doors are often needed. Materials used may be stainless steel, zinc-plated steel, aluminum, or bronze, depending on the type of hardware and where it is being used.

Weatherstripping. Most specifications require some type of *weatherstripping*, and many stock frames come with weatherstripping factory installed—in that case, the factory-installed weatherstripping must be the type specified. Some of the more common types of weatherstripping are vinyl, polypropylene woven pile, neoprene, metal flanges and clips, polyvinyl, and adhesive-backed foam. It is usually sold by the metre for the rigid type and by the roll for the flexible type. The estimator will do the takeoff in metres, and work up a cost.

Mullions. *Mullions* are the vertical bars that connect adjoining sections of frames and sashes. The mullions may be of the same material as the frame, or a different material, colour, or finish. Mullions may be small T-shaped sections that are barely noticeable or large elaborately designed shapes used to accent and decorate. The mullions should be taken off by the metre with a note as to thicknesses, finish, and colour.

Sills. The sill is the bottom member of the frame. The member on the exterior of the building, just below the bottom member of the frame, is also a sill. This exterior sill serves to direct water away from the window itself. These exterior sills may be made of stone, brick, precast concrete, tile, metal, or wood. The details of the frame and its installation must be studied to determine the type of material, its size, and how it fits in. Basically, there are two types of sills: the slip sill and the engaged sill. The slip sill is slightly smaller than the opening for the frame and can be slipped in place after the construction of the walls is complete or either just before or just after the frame has been installed (depending on the exact design). The engaged sill must be installed as the walls go up, since it is wider than the opening and extends into the wall construction. Sills are taken off by the length (m); notes and sketches that show exactly what is required should accompany the takeoff.

Stools. The interior member at the bottom of the frame (sill) is called the *stool*. The stool may be made of stone, brick, precast concrete, terrazzo, tile, metal, or wood. It may be of the slip or engaged variety. The quantity should be taken off in metres with notes and sketches showing material size and installation requirements.

Flashing. Flashing may be required at the head and sill of the frame. The specifications and details will state if it is required and the type. Usually, the flashing is installed when the building is being constructed, but the installation details diagram how it is to be installed. Also, the estimator checks on who buys the flashing and who is supposed to install it. These seemingly small items should not be neglected because they amount to a good deal of time and expense.

Lintels. The horizontal supporting member of the opening is called a *lintel.* It may be wood, concrete, steel, or block with steel reinforcing. The lintels are not installed as part of the window but, rather, are installed as the building progresses. At this time, however, the estimator double-checks that the lintels have been included, both as material and installation costs in the relevant subtrade quotations.

14.3 Doors

Doors are generally classified as interior or exterior, although exterior doors are often used in interior spaces. The list of materials of which doors are made includes wood, aluminum, steel, glass, stainless steel, bronze, copper, plastics, and hardboard. Doors are also grouped according to the mode of their operation. Some different types of operation are illustrated in Figure 14.9. Accessories required include glazing, grilles and louvres, weatherstripping (for sound, light, and weather), moulding, trim, mullions, transoms, and more. The frames and hardware must also be included.

Many specially constructed doors serve a particular need: Some examples are fire-rated, sound-reduction, and lead-lined doors. The frames may be of as many different materials as the doors themselves, and the doors sometimes come prehung in the frame. The doors may be prefinished at the factory or site finished.

Wood Doors. Wood doors are basically available in two types: solid and hollow core. Solid-core doors may have a core of wood block or low-density fibres and are generally available in thicknesses of 29 mm and 32 mm for fibre core and 32 mm through 64 mm in the wood-block core. Widths of 1,500 mm are available in both types with a height of 3,000 mm as maximum for wood-block core. The fibre-core door is available in 2,030 mm, 2,134 mm, and 2,438 mm heights, but different manufacturers offer different sizes. In checking the specifications, the estimator notes the type of core required and any other special requirements, such as the number of plies of construction and the type of face veneer.

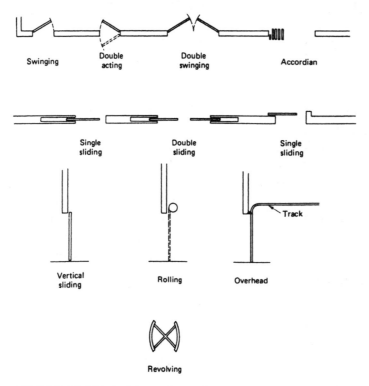

FIGURE 14.9 Typical Door Operations

Hollow-core doors are any doors in which the cores are not solid. The core may have interlocking wood grid strips, ribs, struts, or corrugated honeycomb. The specifications should spell out exactly what type is required. Thicknesses usually range from 29 mm to 32 mm, with widths of 300 mm to 1,200 mm and heights to 2,400 mm.

The face veneer required in the specifications should be checked carefully, as there is a considerable difference in prices. Refer to Figure 14.10 for a quick reference cost guide to various face veneers. From the specifications, the estimator determines whether the doors are to be prefinished at the factory or site finished. He or she decides who will finish the doors if they must be site finished and what type of finish is required. Doors that are prefinished at the factory may require some touchup on the job, and damaged doors will have to be replaced. No matter what type of door is required, the items to be checked are the same. The specifications must be read carefully.

Metal Doors. The materials used in a metal door may be aluminum, steel, stainless steel, bronze, or copper. The doors may be constructed of metal frames with large glazed areas, hollow metal, tin clad, or a variety of other designs. The important thing is to read the specifications, study the drawings, and bid what is required. Often, special doors, frames, and hardware are specified and the material custom made. Finishes may range from lacquered and anodized aluminum and primed or prefinished steel to natural and satin finish bronze. Bid only that which is specified.

Door Swings. The swing of the door is important to the proper coordination of door, frame, and hardware. The names are often confused, so they are shown in Figure 14.11 to act as a reminder and learning aid. When checking door, frame, and hardware schedules, estimators must check the swing as well.

Veneer	Index Value
Rotary sound ph. mahogany	72
Rotary sound natural birch	92
Rotary good birch	98
Rotary premium grade birch	100
Rotary premium red oak	100
Rotary premium select red birch	116
Plain sliced premium red oak	133
Plain sliced premium African mahogany	150
Plain sliced premium natural birch	165
Plain sliced premium cherry	170
Plain sliced premium walnut	180
Sliced quartered premium walnut	250
Plain sliced premium teak	250
Laminated plastic faces (premachined)	250

Each veneer has an index value comparing it with the price of Rotary cut premium grade birch. Fluctuation in veneer costs will cause variations in the index values.

FIGURE 14.10 Veneer Cost Comparison

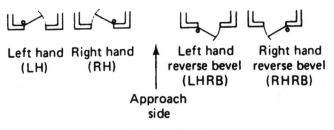

Left hand (LH)   Right hand (RH)   Left hand reverse bevel (LHRB)   Right hand reverse bevel (RHRB)

Approach side

FIGURE 14.11 Door Hand and Swing

Underwriters Label Classification	Heat Transmission Rating	Most Commonly Recommended Locations in Building
A 3 hr. situations	Temperature rise 30 minutes 120°C max.	Opening in fire walls. Areas of high hazard contents. Curtain and division walls.
A* 3 hr. situations		Openings in fire walls. Areas of high hazard contents. Curtain and division walls.
B 1 ½ hr. situations	Temperature rise 30 minutes 120°C max.	Openings in stairwells, elevator shafts, vertical shafts.
B* 1 ½ hr. situations		Openings in stairwells, elevator shafts, vertical shafts.
C* ¾ hr. situations		Openings in room partitions and corridors.
D* 1 ½ hr. situations		Openings in exterior walls where exposure to fire is severe.
E* ¾ hr. situations		Openings in exterior walls where exposure to fire is moderate.

*Not rated for heat transmission

FIGURE 14.12 Fire Door Requirements

Fire-Rated Doors. *Fire-rated doors* are produced under the factory inspection and labelling service program of Underwriters Laboratories, Inc. (UL) and can be identified by the labels on the door. The label states the rating given to that particular door. The ratings given are also related to the temperatures at which the doors are rated. Doors labelled in this manner are commonly called "labelled doors." The letters used, hourly ratings, temperatures, and common locations are given in Figure 14.12. Fire doors may be made of hollow metal, sheet metal, composite wood with incombustible cores, and other composite constructions. Special hardware will be required and the frame may also have to be labelled.

Fire-rated doors may be conventionally operated or be horizontal or vertical slide doors that close when the fusible link releases.

Acoustical Doors. *Acoustical doors* are doors specially designed and constructed for use in situations when sound control is desired. The doors are generally metal or wood, with a variety of cores used. Because the doors available offer a wide range of sound control, the estimator should use caution in selecting the right door to meet the specifications. When high sound control is required, double doors may be desirable. Acoustical doors alone will not solve the problem of noise transmission around the door. An automatic drop seal is used for the bottom of the door, and adjustable doorstops (or other types of sound control devices) are used at the doorjamb and head where the door meets the frame. Often, the seals and devices are sold separately from the door. In these cases, estimators determine the length (m) of the seal needed and include its cost and labour charges in the estimate.

Overhead Doors. *Overhead doors* are available in all sizes. They are most commonly made of wood and steel, although aluminum and stainless steel are sometimes used.

Overhead doors are first designated by the type of operation of the door: rolling, sectional, canopy, and others. The estimator must determine the size of door, size of opening, type and style of door, finishes required, door hanger type, installation details, and whether it is hand, electric, or chain operated. The overhead door is priced as a unit with all required hardware. Often, the supplier will act as a subcontractor and install the doors.

Miscellaneous Doors. There are a wide variety of specialized doors available to fill a particular need. Among these are rolling grilles, which roll horizontally or vertically to provide protection or control traffic. Other specialty doors are revolving, dumbwaiter, rubber shock-absorbing doors, cooler and freezer doors as well as blast- and bulletproof doors. The estimating procedures are basically the same, regardless of the type of door required.

Folding Doors and Partitions. *Folding (accordion) doors* and *partitions* offer the advantage of increased flexibility and more efficient use of the floor space in a building. They are available in fibreglass, vinyl, and wood, or in combinations of these materials. They can be made to form a radius and have concealed pockets and overhead track; a variety of hardware is also available. Depending on the type of construction, the maximum opening height ranges from 2,400 mm to 6,400 mm and the width from 2,400 mm to 9,100 mm. The doors may be of steel construction with a covering, rigid fibreboard panels, laminated wood, or solid wood panels. Specially constructed dividers are also available with higher sound-control ratings. The exact type required will be found in the specifications, as will the hardware requirements. Both the specifications and drawings must be checked for installation details and the opening sizes required.

14.4   Prefitting and Machining (Premachining) Doors

The doors may be machined at the factory to receive various types of hardware. Factory machining can prepare the door to receive cylindrical, tubular, mortise, unit, and rim locks. Other hardware, such as finger pulls, door closers, flush catchers, and hinges (butts), are also provided for. Bevels are put on the doors, and any special requirements are handled.

Premachining is popular, as it cuts site labour costs to the minimum, but coordination is important because the work is done at the factory from the hardware manufacturer's templates. This means that approved shop drawings, hardware and door schedules, and the hardware manufacturer's templates must be supplied to the door manufacturer.

From the estimator's point of view, premachining takes an item that is difficult to estimate and simplifies it considerably. Except where skilled carpenters are available, premachined doors offer cost control with maximum results. For this reason, they are being used more and more frequently.

14.5   Prefinishing Doors

Prefinishing of doors is the process of applying the desired finish on the door at the factory instead of finishing the work on the job. Doors that are premachined are often prefinished as well. Various coatings are available, including varnishes, lacquer, vinyl, and polyester films (for wood doors). Pigments and tints are sometimes added to achieve the desired visual effect. Metal doors may be prefinished with baked-on enamel or vinyl-clad finishes. Prefinishing can save considerable finishing time and generally yields a better result than site finishing. Doors that are prefinished should also be premachined so that they will not have to be "worked on" on the job. The prefinished door must be handled carefully and installed at the end of the job so that the finish will not be damaged; it is often difficult to repair a damaged finish. Care during handling and storage is a requisite.

EXAMPLE 14.4 Residential Doors

Determine the doors required for a residence as per the door schedule (Figure 14.13). Just as with the window schedule, this table should list all the unique types of doors. The task for the estimator is to find and count the different types of doors.

No.	Type	Size	Remarks
1	Exterior metal	44 × 864 × 2,030	Steel w/sidelites
2	Aluminum	44 × 1,525 × 2,030	Glass
3	Hollow core	35 × 762 × 2,030	Wood—Colonial
4	Solid core	35 × 710 × 2,030	Wood—Colonial
5	Hollow core	35 × 610 × 2,030	Wood—Colonial
6	Overhead	44 × 2,030 × 2,030	Aluminum
7	Slider	35 × 1,420 × 2,030	Vinyl Clad

FIGURE 14.13 Residence Door Schedule

1 – 864 × 2,030 steel door with glass sidelites
2 – 1,525 × 2,030 aluminum and glass sliding doors
5 – 762 × 2,030 hollow core (h.c.), wood, prefinished
2 – 710 × 2,030 solid core, wood, prefinished
3 – 610 × 2,030 hollow core, wood, prefinished
1 – 2,030 × 2,330 overhead garage door, 4 panel, aluminum
4 – 1,420 × 2,030 vinyl clad, bi-pass, prefinished

The labour costs are typically expressed as so many work hours per door. Figure 14.14 shows productivity rates for installing doors.

Doors and Frames	Work Hours Per Unit
Residential, Wood	
Prehung	2.0 to 3.5
Pocket	1.0 to 2.5
Not prehung	3.0 to 5.0
Overhead	4.0 to 6.0
Heavy duty	Add 20%
Commercial	
Aluminum Entrance, per door	4.0 to 6.0
Wood	3.0 to 5.0
Metal, prefitted	1.0 to 2.5

FIGURE 14.14 Door Installation Productivity Rates

continued

10 prehung hollow core at 2 work hours per door = 20 work hours
2 prehung solid core at 2.5 work hours per door = 5 work hours
1 overhead door at 5 work hours per door = 5 work hours
4 bi-pass doors at 3 work hours per door = 12 work hours
1 exterior door with sidelites at 6 work hours = 6 work hours
48 work hours × $21.60 per work hour = $1036.80

E X A M P L E **14.5** **Commercial Building**

The door schedule for a commercial building not only shows all the doors but their location, fire rating, and required hardware. Figure 14.15 is a typical door schedule. If the drawings are that detailed, the takeoff simply becomes an accuracy check.

	Door							Frame					Hardware			Remarks
No.	Width	Height	Thick	Type	Mat.	Finish	Glass	Mat.	Finish	Type	Jamb	Head	Panic Hardware	ULC Closer	ULC Rating Door & Frame	
100	900	2,150	45	D1	WD.	Nat. Clear	-	HM	P	F1	A	D				
101	900	2,150	45	D1	WD.	Nat. Clear	-	HM	P	F1	A	D				
102	900	2,150	45	D1	WD.	Nat. Clear	-	HM	P	F1	A	D				
103	900	2,150	45	D1	WD.	Nat. Clear	-	HM	P	F1	B	E				
104	900	2,150	45	D1	WD.	Nat. Clear	-	HM	P	F1	B	E				
105A	900	2,150	45	D3	WD.	Nat. Clear	TEMP.	HM	P	F19	6/A5	G		•	45 MIN	
105B	4-400	2,150	45	D7	WD.	Nat. Clear	-	HM	P	F25	B	E				BI-FOLD
106	900	2,150	45	D1	WD.	Nat. Clear	-	HM	P	F1	B	E				
107	900	2,150	45	D1	WD.	Nat. Clear	-	HM	P	F3	B	E				
108A	900	2,150	45	D2	WD.	Nat. Clear	TEMP.	HM	P	F3	B	E		•		
108B	900	2,150	45	D1	WD.	Nat. Clear	-	HM	P	F1	B	E		•	45 MIN.	
109	2-900	2,150	45	D1	HM.	P		HM	P	F15	C	F	•	•		
110	2-900	2,150	45	D3	HM.	P	GWG	HM	P	F14	15/A5	6/A6	•	•		

FIGURE 14.15 Commercial Building Door Schedule

14.6 Door Frames

The door frames are made of the same type of materials as the doors. In commercial work, the two most common types are steel and aluminum. The steel frames (also called *door bucks)* are available in 1 mm, 0.8 mm, or 0.6 mm cold rolled primed steel construction. Steel frames are available knocked down (KD), set up and spot welded (SUS), or set up and arc welded (SUA). Many different styles and shapes are available, and the installation of the frame in the wall varies considerably. Usually, steel frames are installed during the building of the surrounding construction. For this reason, it is important that the door frames are ordered quickly; slow delivery will hinder the progress of the job. Also, the frames must be anchored to the surrounding construction.

The sides of the door frame are called the *jambs;* the horizontal pieces at the top are called the *heads.* Features available on the frames include a head 100 mm wide, lead lining for X-ray frames, anchors for existing walls, base anchors, a sound-retardant strip placed against the stop, weatherstripping, and various anchors. Fire-rated frames are usually required with fire-rated doors.

When frames are ordered, the size and type of the door, the hardware to be used, and the swing of the door all must be known. Standard frames may be acceptable on some jobs, but often special frames must be made.

Factory finishes for steel frames are prime coated and must be finished on the job. Aluminum, bronze, brass, and stainless steel have factory-applied finishes but have to be protected from damage on the jobsite. Wood frames are available primed and prefinished.

When steel frames are used in masonry construction, it is critical that the frames be installed as the masonry is laid, which means that the frames will be required very early in the construction process. Because door frames for most commercial buildings must be specially made from approved shop drawings and coordinated with the doors and hardware used, the estimator must be keenly aware of this item and how it may affect the flow of progress on the project. Generally, basic door frames are installed by the general contractor's own forces.

14.7   Hardware

The hardware required on a project is divided into two categories: rough and finished. *Rough hardware* comprises the bolts, screws, nails, small anchors, and any other miscellaneous fasteners. This type of hardware is not included in the hardware schedule, but it is often required for installation of the doors and frames. *Finished hardware* is the hardware that is exposed in the finished building and includes such items as hinges (butts), hinge pins, door-closing devices, locks and latches, locking bolts, kickplates, and other miscellaneous articles. Special hardware is required on exit doors.

Finished hardware for doors is either completely scheduled in the specifications, or a cash allowance is made for the purchase of the hardware. If a cash allowance is made, this amount is included in the estimate (plus sales taxes and so on) only for the purchase of materials. This allowance is for the net value of the hardware purchase. Installation costs, overheads, and delivery charges are all extra. When hardware is completely scheduled in the specification or on the drawings, this schedule should be sent to a hardware supplier for a price. Only on small projects will the estimator figure a price on hardware, unless the firm is experienced in this type of estimating. The cost for installation of finished hardware will vary depending on what type and how much hardware is required on each door and whether the door has been premachined.

14.8   Accessories

Items that may be required to complete the job include weatherstripping, sound control, light control, and saddles. The specifications and details will spell out what is required. A list containing each item must be made. The takeoff for accessories should be made in metres or the number of each size piece; for example, "five saddles, 910 mm long."

Weatherstripping for the jambs and head may be metal springs, interlocking shapes, felt or sponge, neoprene in a metal frame, and woven pile. At the bottom of the door (sill), the weatherstripping may be part of the saddle, attached to the door, or both. It is available in the same basic types as used for jambs and heads, but the attachment may be different. Metals used for weatherstripping may be aluminum, bronze, or stainless steel.

Saddles are most commonly wood, aluminum, or bronze. Various shapes, heights, and widths may be specified. Sound control and light control usually employ felt or sponge neoprene in aluminum, stainless steel, or bronze housing. The sill protection is usually automatic, closing at the sill, but at the jambs and head it is usually adjustable.

14.9   Checklist for Doors and Frames

1. Sizes and number required
2. Frame and core types specified
3. Face veneer specified (wood and veneer doors)
4. Prefinished or job finished (if so, specify the finish)
5. Prehung or job-hung (if so, specific installer)
6. Special requirements
   a. louvres
   b. windows
   b. fire rating
   b. lead lining
   b. sound control
7. Type, size, style of frame, and the number of each required
8. Method of attachment of the frame to the surrounding construction
9. Finish required on the frame and who will apply it
10. Hardware—types required and installer
11. Accessories—types required and time to install them

Everything takes time and costs money. At the construction site or in the factory, someone must do every job, so all requirements, materials, and work hours will be included in the estimate.

14.10   Glass

Glass is the most common material to be glazed into the frames for windows, curtain walls, storefronts, and doors. The most commonly used types are plate glass, clear window glass, wire glass, and patterned glass. Clear window glass is available in thicknesses of 2.2 mm to 58.4 mm; the maximum size varies with the thickness and type. Generally available as single and double strength, heavy sheet and picture glass with various qualities are available in each classification. Clear window glass has a characteristic surface wave that is more apparent in the larger sizes.

Plate glass is available in thicknesses of 4.8 mm to 31.8 mm and as heavy-duty polished plate glass, rough, or polished plate glass. The more common types are regular, grey, bronze, heat absorbing, and tempered.

Wire glass is available with patterned and polished finishes, and with various designs of the wire itself. The most common thickness is 6.4 mm and it is also available in colours. This type of glass is used when fire-retardant and safety characteristics (breakage) may be required.

Patterned glass, used primarily for decoration, is available primarily in 5.6 mm and 3.2 mm thicknesses. Pattern glass provides a degree of privacy and yet allows diffused light into the space.

Other types of glass available include a structural-strength glass shaped like a channel, and tempered, sound control, laminated, insulating, heat- and glare-reducing, coloured, and bullet-resisting glass.

The frame may be single glazed (one sheet of glass) or double glazed (two sheets of glass) for increased sound and heat insulation. If the specifications call for double glazed, twice as many square metres of glass will be required.

Glass is estimated by the square metre with sizes taken from the working drawings. Because different frames may require various types of glass throughout the project, special care must be taken to keep each type separate. Also to be carefully checked is which frames need glazing, since many windows and doors come with the glazing work already completed. The types of setting blocks and glazing compound required should be noted as well.

Review Questions

1. What accessories should be checked for when taking off windows and curtain wall?

2. Define glazing. Why must the estimator determine who will perform the required glazing?

3. What information is required to price a door?

4. Describe the advantages in prefitting and prefinishing doors?

5. Why should the type of finish required on the door and door frames be noted on the workup sheet?

6. Describe briefly the ways hardware may be handled on a project.

7. What precautions must an estimator make when using an allowance, from the specifications, in the estimate?

8. What is the unit of measure for glass, and why should the various types and sizes required be listed separately?

9. Determine the window and door materials required for the residential building shown in Appendix C.

10. Determine the window and door materials required for the commercial building shown in Appendix D.

CHAPTER 15
Finishes

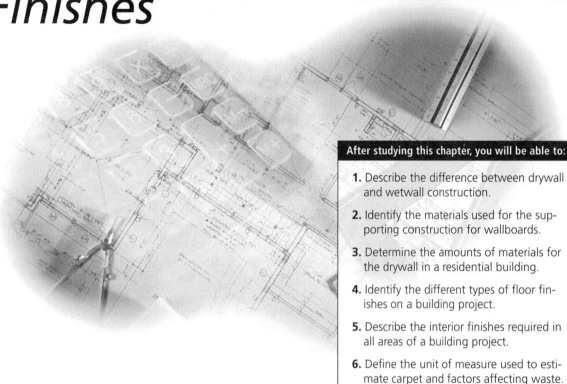

After studying this chapter, you will be able to:

1. Describe the difference between drywall and wetwall construction.

2. Identify the materials used for the supporting construction for wallboards.

3. Determine the amounts of materials for the drywall in a residential building.

4. Identify the different types of floor finishes on a building project.

5. Describe the interior finishes required in all areas of a building project.

6. Define the unit of measure used to estimate carpet and factors affecting waste.

15.1 Drywall and Wetwall Construction

Although drywall construction utilizes wallboard and wetwall utilizes plaster and stucco, many components of the two systems of construction are interchangeable. Both require that supporting construction be applied under them, and the same types are used for both. Many of the fasteners, attachments, and accessories are the same or very similar.

All supporting systems and furring should be installed in accordance with the specifications, and the manufacturer's recommendations for spacing, accessories, and installation should be consulted. If the specifications by the consultant are stricter than the manufacturer's, it is those specifications that must be followed. If the project specifications are less stringent than those of the manufacturer, it is advisable to call the consultant's office to be certain of what must be included in the bid.

Many manufacturers will not guarantee the performance of their materials on the job unless those materials are installed in accordance with their recommendations.

15.2 Supporting Construction

Wallboard can be applied directly to wood, metal, concrete, or masonry that is capable of supporting the design loads and provides a firm, level, plumb, and true base.

Wood and metal supporting construction often consists of self-supporting framing members, including wall studs, ceiling joists, and roof trusses. Wood and metal furring members, such as wood strips and metal channels, are used over the supporting construction to plumb and align the framing, concrete, or masonry.

Concrete and Masonry. Concrete and masonry often have wallboard applied to them. When used, either exterior or below grade, furring should be applied over the concrete or masonry to protect the wallboard from damage due to moisture in the wall; this is not required for interior walls. Furring may also be required to plumb and align the walls. The actual thickness of the wall should be checked so that the mechanical and electrical equipment will fit within the wall thickness allowed. Any recessed items, such as fire extinguishers and medicine cabinets, should be carefully considered.

Wood Studs. The most common sizes used are 38 × 89 and 38 × 64, but larger sizes may be required on any particular project: Spacing may vary from 300 to 600 mm on centre, again depending on job requirements. Openings must be framed around and backup members should be provided at all corners. The most common method of attachment of the wallboard to the wood studs is by nailing, but screws and adhesives are also used. Many estimators take off the wood studs under rough carpentry, particularly if they are load bearing. The number of wood studs required can be determined as illustrated in Section 12.4. Special care should be taken with staggered and double walls so that the proper amount of material is estimated.

Wood Joists. The joists themselves are estimated under rough carpentry (see Chapter 12). When the plans require the wallboard to be applied directly to the joists, the bottom faces of the joists should be aligned in a level plane. Joists with a slight crown should be installed with the crown up, and if slightly crooked or bowed joists are used, it may be necessary to straighten and level the surface with the use of nailing stringers or furring strips. The wallboard may be applied by nailing or screwing.

Wood Trusses. When used for the direct application of wallboard, trusses often require cross-furring to provide a level surface for attachment. Stringers attached at third points will also help align the bottom chord of wood trusses, and a built-in chamber is suggested to compensate for further deflection. Because the trusses are made up of relatively small members spanning large distances, they have a tendency to be more difficult to align and level for the application of wallboard.

Metal Studs. The metal studs (Figure 15.1) most commonly used are made of 25-gauge (.0179) or 20-gauge (.0312), cold-formed steel, electrogalvanized to resist corrosion. Most metal studs have notches at each end and knockouts located about 600 mm on centre to facilitate pipe and conduit installation. The size of the knockout, not the size of the stud, will determine the maximum size of pipe, or other material, that can be passed through. Often, when large pipes, ducts, or other items must pass vertically or horizontally in the wall, double stud walls are used, spaced the required distance apart. Studs are generally available in thicknesses of 41 mm, 64 mm, 92 mm, and 152 mm. The metal runners used are also 25-gauge (.0179 mm) steel and sized to complement the studs.

A variety of systems have been developed by the manufacturers to meet various requirements of attachment, sound control, and fire resistance. Many of the systems have been designed for ease in erection and yet are still demountable for revising room arrangements. The estimator must carefully determine exactly what is required for a particular project before beginning the takeoff.

The wallboard is typically attached with screws and, in certain applications, with adhesives or nails. Different shapes of studs are available to accommodate either the screws or the nails. Metal studs and runners are sold by the metre. Once the length of

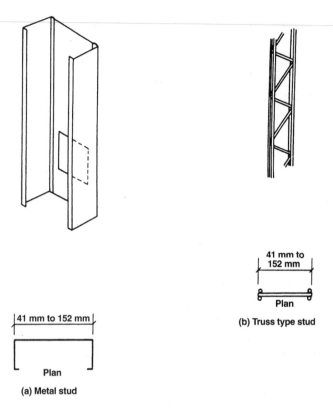

41 mm to 152 mm

Plan

(b) Truss type stud

41 mm to 152 mm

Plan

(a) Metal stud

FIGURE 15.1 Metal Studs

studs has been determined, it is easy to calculate the quantity required. The length of each different type of wall must be determined. The walls must be separated according to thickness or type of stud, backing board and wallboard, as well as according to any variations in ceiling height, application techniques, and stud spacing.

Once a listing of wall lengths and heights has been made, the number of pieces may be determined by the principle illustrated in Section 12.4. Once again, special care should be taken with staggered and double walls so that the proper amount of material may be estimated.

Open-Web Joists. The joists themselves are estimated under metals (see Chapter 11). A joist, however, may be used as a base for wallboard. Because the bottom chords of the joists are seldom well aligned and the spacing between joists is often excessive, the most common methods of attachment are with the use of furring and with a suspension system. Each of these methods is discussed in this chapter.

Metal Furring. Metal furring (Figure 15.2) is used with all types of supporting construction. It is particularly advantageous where sound control or noncombustible assemblies are required. Various types of channels are available. Cold-rolled channels, used for drywall or wetwall construction, are made of 16-gauge (1.4 mm) steel, 6.4 mm to 51 mm wide and available in lengths of up to 6.10 metres. These channels must be wire tied to supporting construction. They are used primarily as a supporting grid for the lighter drywall channels to which the wallboard may be screw attached.

Drywall channels are 25-gauge (.0179 mm) electrogalvanized steel and are designed for screw attachment of wallboard; nailable channels are also available. The channels may be used in conjunction with the cold-rolled channels or installed over the wood,

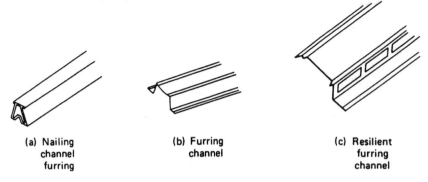

(a) Nailing
channel
furring

(b) Furring
channel

(c) Resilient
furring
channel

FIGURE 15.2 Metal Furring

steel, masonry, or concrete supporting the construction. These drywall channels may be plain or resilient. The resilient channels are often used over wood and metal framing to improve sound isolation and to help isolate the wallboard from structural movement.

Metal furring is sold by the length (m). The estimator needs to determine the size and types of furring required, the area to be covered, and then the length of each type. Also note the type and spacing of fasteners. Labour and equipment will depend on the type of supporting construction, height and length of walls, shape of walls (straight or irregular), and fastening.

Wood Furring. Strips are often used with wood frame, masonry, and concrete to provide a suitably plumb, true, or properly spaced supporting construction. These furring strips may be 19 × 38 mm spaced 400 mm on centre, or 38 × 38 mm spaced 600 mm on centre. Occasionally, larger strips are used to meet special requirements. They may be attached to masonry and concrete with cut nails, threaded concrete nails, and powder or air-actuated fasteners.

When the spacing of the framing is too great for the intended wallboard thickness, cross-furring is applied perpendicular to the framing members. If the wallboard is to be nailed to the cross-furring, the furring should be a minimum of 38 × 38 mm in order to provide sufficient stiffness to eliminate excessive hammer rebound. The furring (19 × 38 or 19 × 64 mm) is often used for screw- and adhesive-attached wallboard.

The furring is attached by nailing with the spacing of the nails 400 or 600 mm on centre. The estimator will have to determine the length of furring required, the nailing requirements, equipment, and work hours. The labour will vary depending on the height of the wall or ceiling, whether straight or irregular walls are present, and the type of framing to which it is being attached.

E X A M P L E   **15.1**   **Steel Studs**

Using the small commercial building floor plan shown in Figure 15.3 for an example, all the interior walls are framed with 92-mm-wide metal studs, 400 mm on centre. The wall between the men's and women's washrooms is framed with 152-mm-wide studs. The quantification of steel studs is performed in virtually the same manner, as are wood studs. First, identify the length of wall by stud thickness. In this example (Figure 15.4), the following lengths are found:

continued

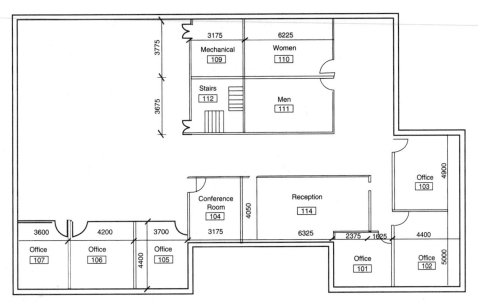

FIGURE 15.3 Floor Plan

92-mm-wide walls − 108.26 m

152-mm-wide wall − 6.23 m (Dividing wall between washrooms)

Number of stud spaces = Length of wall / spacing

Number of stud spaces = 108.26 m / 0.40 m = 271 spaces

Add 1 to get the number of studs

Use 272 studs − 3.66 m long (92 mm wide)

Number of stud spaces = 6.23 m / 0.40 m = 16 spaces

Use 17 studs − 3.66 m long (152 mm wide)

92 mm Wide Studs × 3.660 m Long			
Room	Wall Length (m)	Room	Wall Length (m)
107	3.60	103	4.90
	4.40	109	3.78
106	4.20		3.18
	4.40		3.78
105	3.70	112	3.68
104	3.18		3.18
	4.05		3.68
114	4.05		3.18
	6.33	111	6.23
	4.05		3.68
	2.38		3.68
101	2.38	110	3.78
	1.63		3.78
102	4.40		
	5.00		
Sub-Total	**57.75**		**50.51**
Total			**108.26**

FIGURE 15.4 Length of Studs

continued

Just as with the wood studs, the number of openings, intersections, and corners need to be counted.

13 openings — Add 3 studs per opening — 39 studs

10 interior intersections — Add 2 studs per intersection — 20 studs

5 interior corners — Add 2 studs per corner — 10 studs

Total studs (92 mm) = 272 + 39 + 20 + 10 = 341 studs (92 mm)

Add 5 percent for waste — Order 358 (92 mm × 3.66 m) studs

Add 5 percent for waste — Order 18 (152 mm × 3.66 m) studs

Top and bottom runner track

92 mm — 210.16 m

152 mm = 12.46 m

Since runner track comes in 6.10 m pieces, the following would be required (8 percent waste factor):

Order 38 pieces of 92 mm × 6.10 m runner track

Order 3 pieces of 152 mm × 6.10 m runner track

15.3 Suspended Ceiling Systems

When the plaster, wallboard, or tiles cannot be placed directly on the supporting construction, the wallboard is suspended below by a structural system. This may be required if the supporting construction is not properly aligned and true, or if lower ceiling heights are required.

A large variety of systems are available for use in drywall construction, but basically, they can be divided into two classes (Figure 15.5): exposed grid systems and concealed grid systems. Within each group, many different shapes of pieces are used to secure the plaster wallboard or tile, but basically, the systems consist of hangers, main tees (runners), cross tees (hangers), and furring channels. No matter which type is used, such accessories as wall mouldings, splines, and angles must be considered. For wetwall construction, a lath of some type is required.

The suspension system and wallboard may also be used to provide recessed lighting, acoustical control (by varying the type of wallboard and panel), fire ratings, and air distribution (special tile and suspension system).

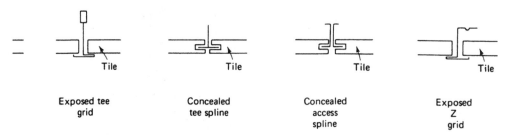

Exposed tee grid   Concealed tee spline   Concealed access spline   Exposed Z grid

FIGURE 15.5 Typical Grid Systems

The supporting construction for ceilings generally utilizes wood joists and trusses, steel joists and suspended ceilings, and concrete and masonry, sometimes in conjunction with the various types of furring materials available.

Drywall construction is generally estimated by the metre or square metre, each estimator using whichever seems most convenient. The most common approach to estimating drywall partitions is to take the length of each different type (assembly) and thickness of wall from the plan and list them on the workup sheet. Walls that are exactly the same should be grouped together; any variation in the construction of the wall will require that it be considered separately. If the ceiling heights vary throughout the project, the lengths of walls of each height must also be kept separately.

Once the estimator makes a listing of wall lengths and heights, he has to multiply the length times height and determine the area of the partition. With this information, the amounts of material required may be estimated. Deduct all openings exceeding 2.0 m² from the area and add 8 to 10 percent for waste.

15.4 Types of Assemblies

Drywall construction may be divided into two basic types of construction: single-ply and multi-ply. *Single-ply construction* consists of a single layer of wallboard on each side of the construction, whereas *multi-ply construction* uses two or more layers of wallboard (and often, different types of board) in the various layers (Figure 15.6). The multi-ply construction may be semisolid or solid, or may have various combinations of materials.

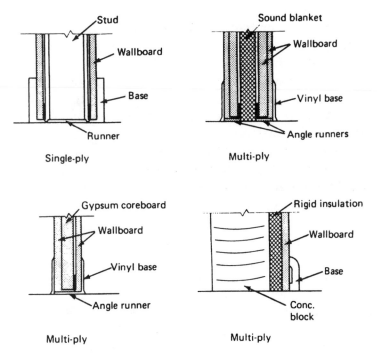

FIGURE 15.6 Typical Drywall Assemblies

Estimators analyze what is required carefully so that the takeoff and pricing may be as complete and accurate as possible.

Various types of demountable and reusable assemblies are also available. Estimators must take time to analyze the assembly specified, break it down into each piece required, and study its pieces and how they are assembled; an estimate can then be made. As questions arise about an unfamiliar assembly, they should not hesitate to call the supplier or manufacturer for clarification.

15.5 Wallboard Types

There are various types of wallboards available for use in drywall construction; among them are gypsum wallboard, tiles, wood panels, and other miscellaneous types. The various types and the special requirements of each are included in this and the following sections.

Gypsum Wallboard. Gypsum wallboard is composed of a gypsum core encased in a heavy manila-finished paper on the face side and a strong liner paper on the back. Gypsum panels are manufactured in modular sizes to suit both the metric and imperial measurement systems. The most common width is 1,200 mm, with some specialty panels at half this width. Lengths range from 2,400 mm to 4,200 mm, with the lengths generally divisible by 400 and/or 600 mm to accommodate metric framing spacing. Gypsum panel thicknesses are 9.5 mm, 12.7 mm, 15.9 mm, and 25.4 mm.

The suspension system itself is available in steel with an electro-zinc coating as well as prepainted and aluminum, and with plain, anodized, or baked enamel finishes. Special shapes—for example, steel shaped like a wood beam that is left exposed in the room—are also available.

The suspension system may be hung from the supporting construction with 18-gauge (1-mm) hanger wire spaced about 1,200 mm on centre, or it may be attached by use of furring strips and clips.

From the specifications and drawings, the type of system can be determined. The pieces required must be listed, and a complete breakdown of the number and length (m) of each piece is required. Estimators should check the drawings for a reflected ceiling plan that will show the layout for the rooms, as this will save considerable time in the estimate. They should also take note of the size of the tile to be used and how the entire system will be attached to the supporting construction. Later, this information should be broken down into the average amount of material required per square metre to serve as a reference for future estimates.

15.6 Drywall

Drywall construction consists of wallboard over supporting construction and may be installed horizontally and vertically. The types of materials used for this construction will depend on the requirements of the job with regard to appearance, sound control, fire ratings, strength requirements, and cost. Materials may be easily interchanged to meet all requirements. The supporting construction for partitions generally utilizes wood or steel studs, but often, the wallboards are applied over concrete and masonry.

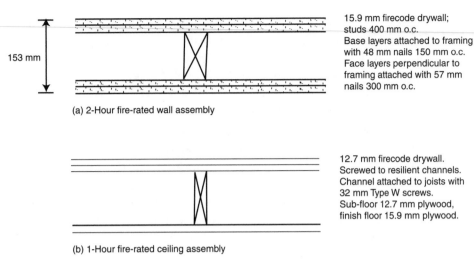

153 mm

15.9 mm firecode drywall; studs 400 mm o.c. Base layers attached to framing with 48 mm nails 150 mm o.c. Face layers perpendicular to framing attached with 57 mm nails 300 mm o.c.

(a) 2-Hour fire-rated wall assembly

12.7 mm firecode drywall. Screwed to resilient channels. Channel attached to joists with 32 mm Type W screws. Sub-floor 12.7 mm plywood, finish floor 15.9 mm plywood.

(b) 1-Hour fire-rated ceiling assembly

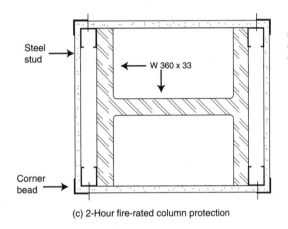

Steel stud →

← W 360 x 33

Corner bead →

12.7 firecode drywall all sides screwed to 41 mm steel studs at column corners; metal beads at corners.

(c) 2-Hour fire-rated column protection

FIGURE 15.7 Fire-Rated Drywall Assemblies

Each component of the drywall construction assembly must be taken off and estimated separately. Plain gypsum wallboard will require a finish of some type, such as painting or wallpaper.

Gypsum wallboard with a rugged vinyl film that is factory laminated to the panel is also available. The vinyl-finished panel is generally used in conjunction with adhesive fastenings and matching vinyl-covered trim; it is also available on fire-resistant gypsum wallboard where fire-rated construction is desired or required.

Fire-resistant gypsum wallboard is available generally in 12.7 mm and 15.9 mm thicknesses. These panels have cores containing special mineral materials and can be used in assemblies (Figure 15.7) that provide up to two-hour ratings in walls and three-hour ratings in ceilings and columns.

Other commonly used gypsum wallboards include insulating panels (aluminum foil on the back), water-resistant panels (for use in damp areas; they have special paper and core materials), and backing board, which may be used as a base for multi-ply construction and acoustical tile application and which may be specially formulated for a fire-resistant base for acoustical tile application.

Tiles. Tiles used in drywall construction are most commonly available in gypsum, wood and mineral fibres, vinyl, plastic, and metal. Thicknesses vary from 4.8 mm to 71.4 mm. The most common thicknesses are 12.7 mm, 15.9 mm and 19.1 mm, with the tile sizes being 300 × 300, 300 × 600, 300 × 900, 400 × 400, 400 × 800, 600 × 600, and 600 × 1,200 mm. Not all sizes are available in each type, so all manufacturers must be checked to determine what they stock.

In addition to different materials, tiles come in a variety of surface patterns and finishes. They may be acoustical or nonacoustical, and may possess varying edge conditions, light reflection, sound absorption, flame resistance, and flame spread. Tiles are even available with small slots that can be opened to provide ventilation in the space below. Plastic louvres and translucent panels are also available.

Prices may be quoted by the piece, square metre, or particular size package. An area (m²) takeoff will provide the estimating information required.

Tiles may be applied to various supporting constructions and are often used in conjunction with suspended ceilings.

Wood Panels. Available in many different wood veneers and a variety of finishes, panels are usually 1,200 mm wide, with lengths varying from 1,800 mm to 4,800 mm. The panels may be constructed of a solid piece of wood or laminated plywood with cores of veneer, flakeboard, or lumber. Either hardwood or softwood may be used.

The estimator must read the specifications carefully to determine exactly what type of panelling is required. Although inexpensive panelling is available at $10 to $25 per sheet and moderately priced panels from $40 to $80 per sheet; some panelling sells for $240 or more per sheet when special face patterns are required. The estimator should never guess at a price for materials; rather, he should always get a written quote from suppliers and manufacturers.

Panelling is taken off by the square metre or panels required. The fastening device and trim to be used must be noted in the estimate.

Miscellaneous Panels. Many types of panels are used on walls and ceilings. The majority of them will be priced by the square metre. Among the types available are vinyl-coated plywood, plastic-coated plywood, hardboards, and metal-coated panels. Each separate type has its own method of fastening, accessories, and requirements. The estimator should always check with the manufacturer for installation recommendations.

Specifications. Items that should be checked are the type, thickness, sheet size, and method of attachment required. The spacing of fasteners must also be determined either from the specifications or, if it is not given there, from the manufacturer's recommendations. The estimator should make a list of accessories, the materials from which they are fabricated, and the finish required.

Estimating. All wallboard should be taken off by the square metre, with the estimator double-checking for panel layout of the job. The first step is to determine the length (m) of each type of wall, carefully separating any wall with different sizes or types of material, fasteners, or any other variations. There are two sides to most walls, and each side requires a finish. After a complete listing has been made, the area of wallboard may be determined. Different varieties may be encountered, so each must be kept separate.

Equipment required may simply be wood or metal horses, planks, platforms, and scaffolding, as well as small electric tools and staplers. On projects with high ceilings, scaffolds on wheels are often used so the workers can work more conveniently and the scaffold may be moved easily from place to place.

Labour for drywall construction will vary depending on the type of wallboard trim, fasteners, whether the walls are straight or jogged, height of walls or ceiling, and the presence of other construction underway at the time. Many subcontractors are available with skilled workers specially trained for this type of work.

15.7 Column Fireproofing

Columns may be fireproofed by using drywall construction consisting of layers of fire-resistant gypsum wallboards held in place by a combination of wire, steel studs, screws, and metal angles. Up to a three-hour fire rating may be obtained by using gypsum board. A four-hour fire rating is available when the gypsum board is used in conjunction with gypsum tile (usually 51 mm or 76 mm thick). To receive the fire ratings, all materials must be installed in accordance with U.L. designs. A complete takeoff of materials is required. No adhesives may be used.

15.8 Accessories

Accessories for the application and installation of drywall construction include mechanical fasteners and adhesives, tape and compound for joints, fastener treatment, trim to protect exposed edges and exterior corners, and base plates and edge mouldings.

Mechanical Fasteners. Clips and staples may be used to attach the base ply in multiply construction. The clip spacing may vary from 400 to 600 mm on centre and may also vary depending on the support spacing. Staples should be 16-gauge (1.4 mm) galvanized wire with a minimum of an 11.1-mm-wide crown with legs having divergent points. Staples should be selected to provide a minimum of 15.9 mm penetration into the supporting structure. They are spaced about 178 mm on centre for ceilings and 203 mm on centre for walls.

Nails used to fasten wallboard may be bright, coated, or chemically treated; the shanks may be smooth or angularly threaded with a nail head that is generally flat or slightly concave. The angularly threaded nails are most commonly used since they provide more withdrawal resistance, require less penetration, and minimize nail popping. For a fire rating, it is usually required to have 25.4 mm or more of penetration, and in this case, the smooth shank nails are most often used. Nails with small heads should never be used for attaching wallboard to the supporting construction. The spacing of nails generally varies from 152 mm to 203 mm on centre, depending on the size and type of nail and the type of wallboard being used. Nails are bought by the kilogram. The approximate weight of the nails that would be required per 100 m^2 of gypsum board varies between 2.3 and 3.2 kg.

Screws may be used to fasten both wood and metal supporting construction and furring strips. In commercial work, the drywall screws have virtually eliminated the use of nails. Typically, these screws have self-drilling, self-tapping threads with flat recessed Phillips heads for use with a power screwdriver. The drywall screws are usually spaced about 300 mm on centre except when a fire rating is required, when the spacing is usually 200 mm on centre at vertical joists. There are three types of drywall screws: one for fastening to wood, one for sheet metal, and one for gypsum board. The approximate number of screws required per 100 m^2 of gypsum board based on single-layer application and spacing of 400 mm on centre wood framing is 1,000. If 600 mm on centre spacing is used for framing, about 850 screws are required. If the boards are applied horizontally with screws spaced 300 mm on centre, only about 820 screws are required.

Adhesives may be used to attach single-ply wallboard directly to the framing, concrete, or masonry, or to laminate the wallboard to a base layer. The base layer may be gypsum board, sound-deadening board, or rigid-foam insulation. Often, the adhesives are used in conjunction with screws and nails that provide either temporary or permanent supplemental support. Basically, the three classes of adhesives used are stud adhesives, laminating adhesives, and contact adhesives. There are also various modifications within each class. Information regarding the exact adhesives required should be obtained from the specifications of the project and cross-checked with the manufacturer. Also, determine the special preparation, application, and equipment requirements from the manufacturer. Information concerning coverage per litre and curing requirements should also be obtained.

Trim.   A wide variety of trims are available in wood and metal for use on drywall construction. The trim is generally used to provide maximum protection and neat, finished edges throughout the building. The wood trim is available unfinished and prefinished in an endless selection of sizes, shapes, and costs. The metal trim is available in an almost equal number of sizes and shapes. Finishes range from plain steel, galvanized steel, and prefinished painted to trim with permanently bonded finishes that match the wallboard; even aluminum moulding, plain and anodized, is available. Most trim is sold by the metre, so the takeoff should also be made in metres. Also to be determined is the manner in which the mouldings are to be attached to the construction.

Joint Tape and Compounds.   Joint tape and compounds are employed when a gypsum wallboard is used and it is necessary to reinforce and conceal the joints between the wallboard panels and to cover the fastener heads. These items provide a smooth, continuous surface in interior walls and ceilings.

The tape used for joint reinforcement is usually a fibre tape designed with chamfered edges feathered thin, with a cross-fibre design.

Joint compounds are classified as follows: (1) embedding compound, used to embed and bond the joint tape; (2) topping compound, used for finishing over the embedding compound (it provides final smoothing and levelling over fasteners and joints); and (3) all-purpose compound, which combines the features of both the above, providing embedding and bonding of the joint tape and giving a final smooth finish. The compounds are available premixed by the manufacturer or in a powdered form to be job mixed.

The amount of tape and compound required for any particular job will vary depending on the number of panels used with the least number of joints and the method of fastening specified. To finish 100 m² of surface area, about 120 m of tape and 39 kg of powder joint compound (or 56 kg of ready-mixed compound for the average job) will be required.

E X A M P L E   **15.2**   Commercial Building Drywall

For the commercial building in Figure 15.3, determine the required quantity of 15.9-mm-thick drywall. The information to perform this takeoff comes from the stud quantification (see Example 15.1). In that takeoff, it was found that there were roughly (108.26 m + 6.23 m) 114.49 m of interior partitions. The floor to ceiling height is 3,050 mm. In this

continued

example, 3,050-mm-length drywall will be used, but the installers will have to be careful to lift the drywall roughly 19 mm off the floor. If there is a concern about the hangers being able to do this, then 3,660-mm-length drywall should be purchased and cut. However, the latter adds material and labour costs and increases the amount of construction waste that must be hauled away from the job.

If 3,050-mm-length sheets were to be ordered and installed vertically, it would require 183 sheets.

114.49 m / 1.20 m per sheet = 96 sheets per side
Double the quantity for both sides
Add 5 percent for waste – Use 202 sheets

Taping joints is also required. Enough tape must be provided to cover the vertical joints between the gypsum board sheets. From the information above, 100 m^2 of surface area will require 120 m of tape and 56 kg of ready-mix joint compound.

Area (m^2) of surface = 114.49 m × 3.05 m ceiling height × 2 sides = 698.39 m^2
Use 700 m^2
Length (m) of tape = 7 (hundreds of m^2) × 120 m / 100 m^2
840 m of tape
Add 10 percent for waste and assume 15 m rolls
Order 62 rolls
Kilograms of joint compound = 7 (hundreds of m^2) × 56 kg/100 m^2
392 kg of joint compound
Drywall screws 1,000 per 100 m^2
7,000 drywall screws

EXAMPLE 15.3 Residential Building

Exterior walls: gypsum board, 12.7 mm × 1,200 mm × 2,400 mm
Ceiling height: 2,400 mm
44 m (from Section 12.4, Example 12.12)
Area of wall = 44 m × 2.40 m ceiling height = 106 m^2
106 m^2 / 2.88 m^2 per sheet = 37 sheets
Interior walls: 12.7 mm gypsum boards each side
Ceiling height: 2,400 mm
41 m (from Section 12.4, Example 12.19)
Area of wall = 41 m of wall × 2.40 m ceiling height × 2 sides = 197 m^2
197 m^2 of wall / 2.88 m^2 per sheet = 69 sheets
Ceiling area: 15.00 m × 7.20 m (15.9 mm gypsum wall board)
Ceiling area (m^2) = 15.00 × 7.20 = 108 m^2
108 m^2 / 2.88 m^2 per sheet = 38 sheets
Using a 5 percent waste factor, the following would be ordered:
40 sheets 15.9 mm thick 1,200 mm × 2,400 mm gypsum board
111 sheets – 12.7 mm thick 1,200 mm × 2400 gypsum board
Metal trim: corners bead: 3 pieces × 2,400 mm long

Blankets. Various types of blankets are used in conjunction with the drywall construction. The blankets are most commonly placed in the centre of the construction, between studs, or on top of the suspended ceiling assembly. The two basic types of blankets are heat insulating and sound control. The heat-insulating blankets are used to help control heat loss (winter) and heat gain (summer), while the sound-control blankets are used to improve sound transmission classification (STC) ratings of the assembly. Both types are available in a variety of thicknesses and widths. Once the area of wall or ceiling that requires the blanket has been determined, the amount of blanket required is virtually the same, but the stud spacing should be noted so that the proper width blanket will be ordered. The estimator needs to check the specifications to determine any special requirements for the blankets, such as aluminum foil on one or both sides, paper on one or both sides, and method of attachment. The most common method of attachment is with staples.

E X A M P L E **15.4** **Residential Building**

Gypsum board, nailed to studs:

$$106 \text{ m}^2 + 197 \text{ m}^2 = 303 \text{ m}^2$$
Hanger (walls), 1.2 work hours per 10 m²
Taper (walls), 1.4 work hours per 10 m²
Total 2.6 work hours per 10 m²
Hang and tape (work hours) = 30.3 (10 m²) × 2.6 work hours per 10 m²
Hang and tape = 78.78 work hours
78.78 work hours × $16.00 per work hours = $1,260.48 (walls)
Hanger (ceiling), 1.8 work hours per 10 m²
Taper (ceiling), 1.6 work hours per 10 m²
Total 3.4 work hours per 10 m²
Hang and tape = 10.8 (10 m²) × 3.4 work hours per 10 m²
Hang and tape = 36.72 work hours
36.72 work hours × $16.00 per work hour = $587.52 (ceiling)
Total cost = $1,260.48 + $587.52 = $1,848

Drywall	Work Hours per 10 m²
Gypsum Board	
Nailed to Stud:	
Walls	1.0 to 2.2
Ceiling	1.5 to 2.8
Jointing	1.0 to 1.8
Glued:	
Walls	.8 to 2.4
Ceiling	1.5 to 3.0
Jointing	.5 to 1.2
Ceilings over 2.40 m	Add 10 to 15%
Screwed to Metal Studs	Add 10%
Rigid Insulation, Glued to Wall	1.2 to 3.0

FIGURE 15.8 Work Hours Required for Drywall Installation

Labour. Subcontractors who specialize in drywall installation generally do this type of work and may price it on a unit basis (per square metre/square foot) or as a lump sum. The productivity rates for hanging drywall are shown in Figure 15.8. The hourly wage rates need to come from the local market conditions.

15.9 Wetwall Construction

Wetwall construction consists of supporting construction, lath, and plaster. The exact types and methods of assembly used for this construction will depend on the requirements of the particular job regarding appearance, sound control, fire ratings, strength requirements, and cost.

The supporting construction may be wood, steel, concrete, gypsum, tile, masonry, or lath. Certain types of lath used with the plaster are self-supporting. The plaster itself may be a two- or three-coat application, with a variety of materials available for each coat.

Proper use of plasters and bases provides the secure bond necessary to develop the required strength. A mechanical bond is formed when the plaster is pressed through the holes of the lath or mesh and forms keys on the back. A suction or chemical bond is formed when the plaster is applied over masonry and gypsum bases, with the tiny needle-like plaster crystals penetrating into the surface of the base. Both mechanical and suction bonds are developed with perforated gypsum lath.

15.10 Plaster

In its plastic state, plaster can be trowelled to form. When set, it provides a hard covering for interior surfaces, such as walls and ceilings. Plaster is the final step in wetwall construction (although other finishes may be applied over it). Together with the supporting construction and some type of lath, the plaster will complete the assembly. The type and thickness of plaster used will depend on the type of supporting construction, the lath, and the intended use. Plaster is available in one-coat, two-coat, and three-coat work and is generally classified according to the number of coats required. The last and final coat applied is called the *finish coat*, while the coat, or combination of coats, applied before the finish coat is referred to as the *base coat*.

Base Coats. Base coat plasters provide a plastic working material that conforms to the required design and serves as a base over which the finish coats are applied. Base coats are available mill mixed and job mixed. Mill-mixed base coats are available with an aggregate added to the gypsum at the mill. Aggregates used include wood fibres, sand, perlite, and vermiculite. For high-moisture conditions, a Portland cement and lime plaster base coat is available.

Three-coat plaster must be used on metal lath. The first coat (scratch coat) must be of a thickness sufficient to form keys on the back of the lath, fill it in completely, and cover the front of the lath. The thickness may vary from 3.2 mm to 6.4 mm. The second coat (brown coat) ranges from 6.4 mm to 9.5 mm thick and the finish coat 1.9 mm to 3.2 mm thick.

Two-coat plaster may be used over gypsum lath and masonry. The first coat is the base coat (scratch or brown), and the second coat is the finish coat. Base coats range from 3.2 mm to 6.4 mm in thickness and the finish coat 1.9 mm to 3.2 mm. Perforated gypsum lath will require enough material to form the mechanical keys on the back of the sheets.

Finish Coats. Finish coats serve as levelling coats and provide either a base for decorations or the required resistance to abrasion. Several types of gypsum-finish plasters are available, including those that require the addition of only water and those that blend gypsum, lime, and water (or gypsum, lime, sand, and water). The finish coat used must be compatible with the base coat. Finishing materials may be classified as prepared finishes, smooth trowel finishes, or sand float finishes. Finish coat thickness ranges from 1.9 mm to 3.2 mm.

Specialty finish coats are also available. One such specialty coat is radiant heat plaster for use with electric cable ceilings. It is a high-density plaster that allows a higher operator temperature for the heating system, as it provides more efficient heat transmission and greater resistance to heat deterioration. Applied in two coats—the first to embed the cable, the second a finish coat over the top—its total thickness is about 3.2 mm to 6.4 mm. It is usually mill prepared and requires only the addition of water.

One-coat plaster is a thin-coat, interior product used over large sheets of gypsum plaster lath in conjunction with a glass fibre tape to finish the joints. The plaster coat is 0.8 mm to 1.9 mm thick.

Keene cement plaster is used where a greater resistance to moisture and surface abrasions is required. It is available in a smooth and sand-float finish. It is a dead-burned gypsum mixed with lime putty and is difficult to apply, unless sand is added to the mixture; with sand as an additive, it is less resistant to abrasion.

Acoustical plasters, which absorb sound, are also available. Depending on the type used, they may be trowelled or machine sprayed onto the wall. Trowel applications are usually stippled, floated, or darbied to a finish. Some plasters may even be tinted various colours. Thickness ranges from 9.5 mm to 12.7 mm.

Special plasters for ornamental plastering work, such as mouldings and cornices, are also available. Moulding and casting plasters are most commonly used for such work.

Stucco. Stucco is used in its plastic state. It can be trowelled to form. When set, it provides a hard covering for exterior walls and surfaces of a building or structure. Stucco is generally manufactured with Portland cement as its base ingredient, with the addition of clean sand and, sometimes, lime. Generally applied as three-coat work, the base coats are mixed about one part Portland cement to three parts clean sand. If lime is added, no more than 2.7 to 3.6 kg per 45 kg of Portland cement should be used in the mix. The lime tends to allow the mix to spread more easily. The finish coat is usually mixed 1:2 (cement to sand), and no coat should be less than 6.4 mm thick.

Stucco is usually applied to galvanized metal lath that is furred out slightly from the wall, but it can also be applied directly to masonry. Flashing is often required and must be included in the estimate.

Per 1.6 mm Thickness of Plaster Allow	Perforated Plaster Board Lath, Add	Metal Lath, Add
0.18 m³	0.14 m³	0.27 m³

FIGURE 15.9 Approximate Plaster Quantities

E X A M P L E   **15.5**   **Plaster Quantity**

For 100 m² of wall area, a 6.4 mm thickness over metal lath will require:

0.18 m³ per 1.6 mm × 4 (for 6.4 mm) = 0.72 m³ of plaster
Add 0.27 m³ for over metal lath
Use 1 m³ of plaster

Various special finishes may also be required and will affect the cost accordingly. Finishes may be stippled, broomed, pebbled, swirled, or configured in other designs.

Specifications. The specifications should state exactly what type of plaster is required and where. The types often will vary throughout the project and each must be kept separate. The number of coats, thickness of each coat, materials used, and the proportions of the mix must all be noted. The estimator needs to check also the type of finish required, what trueness of the finish coat is required, and the room finish schedule on the drawings—since often finishes and the finish coat required are spelled out—and determine the accessories, grounds, trim, and anything else that may be required for a complete job.

Estimating. Gypsum plasters are usually packed in 45 kg bags and priced by the tonne. The estimator performs the wetwall takeoff in square metres and must then consider the number of coats, thickness of coats, mixes to be used, and the thickness and type of lath required. The amounts of materials required may be determined with the use of Figure 15.9. This table gives the cubic metre of plaster required per m² of surface.

Depending on the type of plaster being used, the approximate quantities of materials can be determined. The mix design varies from project to project and must be carefully checked. Many projects have mixes designed for a particular use included in the specifications. Read them carefully and use the specified mix to determine the quantities of materials required. Once the quantities of materials have been determined, the cost for materials may be determined and the cost per square metre calculated.

Labour time and costs for plastering are subject to variations in materials, finishes, local customs, type of job, and heights and shapes of walls and ceilings. The ability of workers to perform this type of work varies considerably from area to area. In many areas, skilled plasterers are scarce, meaning that the labour cost will be high and problems may occur with the quality of the work done. It is advisable to contact the local unions and subcontractors to determine the availability of skilled workers. In most cases, one helper will be required to work with two plasterers.

If the plastering is bid on a unit basis, the estimator needs to be certain that there is an understanding of how the area will be computed, as the methods of measuring vary in different localities. The area may be taken as the gross area, the net area, or the gross area minus the openings that are over a certain size. In addition, curved and irregular work may be charged and counted extra; it will not be done for the same costs as the flat work.

Equipment required includes a small power mixer, planks, scaffolds, mixing tools, mixing boxes, and miscellaneous hand tools. Machine-applied plaster will require special equipment and accessories, depending on the type used.

Labour. Subcontractors who specialize in wetwall installations do this type of work and they may price it on a unit basis (per square metre) or lump sum.

15.11 Lath

Lath is used as a base; the plaster is bonded to the lath. Types of lath include gypsum tile, gypsum plaster, board, metal, and wood.

The type of lath required will be specific and will vary depending on the requirements of the project. The estimator must read the specifications carefully and note on their workup sheet the type of lath required and where it is required. It is not unusual for more than one type to be used on any one job.

Gypsum Plaster Lath. In sheet form, gypsum plaster lath provides a rigid base for the application of gypsum plasters. Special gypsum cores are faced with multilayered laminated paper. The different lath types available are plain gypsum, perforated gypsum, fire-resistant, insulating, and radiant heat. Depending on the supporting construction, the lath may be nailed, stapled, or glued. The type of spacing of the attachments depends on the type of construction and the thickness of the lath. Gypsum lath may be attached to the supporting construction by use of nails, screws, staples, or clips.

Plain gypsum lath is available in thicknesses of 9.5 mm and 12.7 mm with a face size of 400 × 1,200 mm. The 9.5 mm thickness is also available in 400 × 2,400 mm sizes. When the plaster is applied to this base, a chemical bond holds the base to the gypsum lath.

Perforated gypsum lath is available 9.5 mm thick with a face size of 400 × 1,200 mm. Holes 19 mm in diameter are punched in the lath spaced 400 mm on centre. The perforated lath permits higher fire ratings because the plaster is held by mechanical as well as chemical bonding.

Fire-resistant gypsum lath has a specially formulated core of special mineral materials. It has no holes but provides additional resistance to fire exposure. It is available in a 9.5 mm thickness with a face size of 400 × 1,200 mm.

Insulating gypsum lath is plain gypsum lath with aluminum foil laminated to the back face. It serves as a plaster base, an insulator against heat and cold, and a vapour barrier. It is available in 9.5 mm and 12.7 mm thicknesses with a face size of 400 × 1,200 mm.

Radiant-heat lath is a large gypsum lath for use with plaster in electric cable ceilings. It improves the heat emission of the electric cables and increases their resistance to heat deterioration. It is available 1,200 mm wide, and in 12.7 mm and 15.9 mm thicknesses and lengths of 2,400 mm to 3,660 mm. This type of lath is used with plaster that is formulated for use with electric cable heating systems.

Estimating. Gypsum lath is sold by the sheet or 100 m². The estimator will calculate the number of square metres required and divide by the number of square metres in a sheet. Note the type and thickness required. Depending on the number of jogs and openings, about 6 percent should be allowed for waste. The materials used for attachment must be estimated and a list of accessories made.

Metal Lath. Metal lath is sheet steel that has been slit and expanded to form a multitude of small mesh openings. Ordinary, expanded metal lath (such as diamond mesh, flat-rib lath) is used in conjunction with other supporting construction. There are also metal laths that are self-supporting (such as 9.5 mm rib lath), requiring no supporting construction.

Metal lath is available painted, galvanized, or asphalt dipped; sheet sizes are generally 600 × 2,400 mm (packed 13.38 m^2 per bundle) or 675 × 2,400 mm (16.72 m^2 per bundle). Basically, the three types of metal lath available for wetwall construction are diamond, flat-rib lath, and 9.5 mm rib lath. Variations in the designs are available through different manufacturers.

The metal lath should be lapped not less than 12.7 mm at the sides and 25.4 mm at the ends. The sheets should be secured to the supports at a maximum of 150 mm on centre. The metal lath is secured to the steel studs or channels by use of 18-gauge (1 mm) tie wires about 150 mm on centre. For attachment to wood supporting construction, nails with a large head (about 12.7 mm) should be used.

Diamond Lath. Diamond lath is an all-purpose lath that is ideal as a plaster base, as reinforcement for walls and ceilings, and as fireproofing for steel columns and beams. It is easily cut, bent, and formed for curved surfaces. It is available in weights of 1.4 kg and 1.9 kg per square metre; both sizes are available in copper alloy steel, either painted or asphalt coated. Galvanized diamond lath is available only in 1.9 kg per square metre.

Flat-rib lath is a 3.2 mm lath with "flat ribs," which make a stiff type of lath. This increased stiffness generally permits wider spacing between supports than diamond lath, and the design of the mesh allows the saving of plaster. The main longitudinal ribs are spaced 38 mm apart, with the mesh set at an angle to the plane of the sheet. Available in copper alloy and steel in weights of 1.5 kg and 1.9 kg per square metre, and in galvanized steel in a weight of 1.9 kg per square metre, it is used with wood or steel supporting construction on walls and ceilings, and for fireproofing.

The 9.5 mm rib lath combines a small mesh with heavy reinforcing ribs. The ribs are 9.5 mm deep, 114 mm on centre. Used as a plaster base, it may be employed in studless wall construction and in suspended and attached ceilings. Rib lath permits wider spacing of supports than flat rib and diamond lath. This type is also used as a combination form and reinforcement for concrete floor and roof slabs. Copper alloy steel lath is available in 1.9 kg and 1.8 kg per square metre, and the galvanized is available in 1.9 kg per square metre.

Specifications. The type of lath, its weight, and finish must be checked. The spacing of the supporting construction will affect the amount of material and labour required to attach the lath. The type and spacing of attachment devices should be checked as well as a list of accessories.

Estimating. The metal lath is taken off by the square metre in the same manner as plaster. For plain surfaces, add 6 to 10 percent for waste and lapping; for beams, pilasters, and columns add 12 to 18 percent. When furring is required, it is estimated separately from the lath. Determine what accessories will be required and the quantity of each.

Gypsum Tile. Gypsum tile is a precast, kiln-dried tile used for non-load-bearing construction and fireproofing columns. Thicknesses available are 50 mm (solid), 75 mm (solid or hollow), and 100 mm and 150 mm (hollow). The 50 mm tile is used for fireproofing only, not for partitions. A face size of 300 × 750 mm (0.225 m^2) is available. Used as a plaster base, it provides excellent fire and sound resistance. Gypsum tile may be taken off as part of wetwall construction or under masonry.

Specifications. The estimator determines the type and thickness required from the specifications, makes a list of all clips and accessories, and decides how the gypsum tile is to be installed.

Estimating. The number of units required must be determined. If the square metres have been determined, their area can easily be converted to the number of units required. The thickness required must be noted as well as the accessories and the amount of each required. Resilient clips are sometimes used also. The lath required for the building will be included in the subcontractor's bid, but the estimator should check the subcontractor's proposal to be certain that it calls for the same lath as the contract documents.

15.12 Accessories

The accessories available for use with wetwall construction include various types of corner beads, control and expansion joints, screeds, partition terminals, casing beads, and a variety of metal trim to provide neat-edged cased openings. Metal ceiling and floor runners are also available, as are metal bases. Resilient channels may also be used. These accessories are sold by the metre, so the estimator makes the takeoff accordingly.

A complete selection of steel clips, nails, staples, and self-drilling screws is available to provide positive attachment of the lath. Special attachment devices are available for each particular wetwall assembly. The estimator will have to determine the number of clips or screws required on the project. The specifications will state the type of attachment required and may also give fastener spacing. The manufacturer's recommended fastener spacing may also be checked to help determine the number required.

Accessories required should be included in the subcontractor's bid, but the estimator should check the subcontractor proposal against the contract documents to be certain they are the same size, thickness of metal, and finish.

15.13 Drywall and Wetwall Checklist

Wetwall:	Drywall:
lath, metal	studs
furring	wallboards
studs	furring
channels	channels
lath, gypsum	tape
gypsum block	paste
corner beads	adhesives
accessories	staples
number of coats	clips
type of plaster	nails
tie wire	screws
moulding	
stucco	

15.14 Flooring

Flooring may be made of wood, resilient tile or sheets, carpet, clay and ceramic tiles, stone, and terrazzo. Each type has its own requirements as to types of installation, depending upon job conditions, subfloor requirements, methods of installation, and moisture conditions.

Wood Flooring

The basic wood flooring types are strip, plank, and block. The most widely used wood for flooring is oak. Other popular species are maple, southern pine and Douglas fir with beech, ash, cherry, cedar, mahogany, walnut, and teak also available. The flooring is available unfinished or factory finished.

Strip flooring is flooring up to 83 mm wide and comes in various lengths. *Plank flooring* is from 83 mm to 184 mm wide with various thicknesses and lengths. The most common thickness is 19 mm, but other thicknesses are available. It may be tongue-and-grooved, square edged, or splined. Flooring may be installed with nails, screws, or mastic. When mastic is used, the flooring used should have a mastic recess so that excess mastic will not be forced to the face of the flooring. Nailed wood flooring should be blind nailed (concealed); nail just above the tongue with the nail at a 45-degree angle. Waste on strip flooring may range from 15 to 40 percent, depending on the size of the flooring used. This estimate is based on laying the flooring straight in a rectangular room, without any pattern involved.

Strip and plank flooring may be sold either by the square metre/square foot or board foot measure. The estimator should figure the area required on the drawings, noting the size of flooring required and the type of installation and then, if required, should convert square metre/square feet to board feet, not forgetting to add waste.

Block flooring is available as parquet (pattern) floors, which consist of individual strips of wood or larger units that may be installed in decorative geometric patterns. Block sizes range from 152 × 152 mm to 762 × 762 mm, thickness from 8 mm to 19 mm. They are available tongue-and-grooved or square edged. Construction of the block varies considerably: It may be pieces of strip flooring held with metal or wood splines in the lower surface; *laminated blocks*, which are cross-laminated piles of wood; or *slat blocks*, which are slats of hardwood assembled in basic squares and factory assembled into various designs. Block flooring is estimated by the square metre/square foot, with an allowance added for waste (2 to 5 percent). The type of flooring, pattern required, and method of installation must be noted.

Wood flooring may be unfinished or factory finished. Unfinished floors must be sanded with a sanding machine on the job and then finished with a penetrating sealer, which leaves virtually no film on the surface, or with a heavy solid type finish, which provides a high luster and protective film. The penetrating sealer also will usually require a coat of wax. The sanding of the floors will require from three to five passes with the machine. On especially fine work, hand sanding may be required. The labour required will vary, depending on the size of the space and the number of sanding operations required. The surface finish may require two or three coats to complete the finishing process. Factory-finished wood flooring requires no finishing on the job, but care must be taken during and after installation to avoid damaging the finish.

In connection with the wood floor, various types of supporting systems may be used. Among the more common are treated wood sleepers, a combination of 3.2 mm hot asphalt fill and treated sleepers, steel splines, and cork underlayment.

Resilient Flooring

Resilient flooring may be made of asphalt, vinyl, rubber, or cork. Resilient sheets are available in vinyl and linoleum. The flooring may be placed over wood or concrete subfloors by use of the appropriate adhesive. The location of the subfloor (below grade, on grade, or suspended above grade) will affect the selection of resilient flooring, since moisture will adversely affect some types. All types may also be used on suspended wood subfloors and on concrete as long as it is sufficiently cured. Where moisture is present below grade and on grade, the materials may be used *except* cork and rag felt-backed vinyl.

Tile sizes range from 225 × 225 mm to 300 × 300 mm, except for rubber tile, which is available up to 900 × 900 mm and vinyl accent strips, which come in various sizes. Thicknesses range from 1.3 mm to 3.2 mm except for rubber and cork tiles, which are available in greater thicknesses. Sheet sizes most commonly used are 1,800 mm and 3,600 mm wide, while 1,350 mm is also available.

The subfloor may require an underlayment on it to provide a smooth, level, hard, and clean surface for the placement of the tile or sheets. Over wood subfloors, panel underlayments of plywood, hardboard, or particleboard may be used, while the concrete subfloors generally receive an underlayment of mastic. The panel underlayment may be nailed or stapled to the subfloor. The mastic underlayments may be latex, asphalt, polyvinyl-acetate resins, or Portland and gypsum cements.

Adhesives used for the installation of resilient flooring may be trowelled on with a notched trowel or brushed on. Because there are so many types of adhesives, it is important that the proper adhesive be selected for each application. Check the project specifications and the manufacturers' recommendations to be certain they are compatible.

The wide range of colours and design variations are, in part, responsible for the wide use of resilient flooring. For the estimator, this means taking care to bid the colour, design, thickness (gauge), size, and finish that is specified.

Accessories include wall base, stair tread, stair nosings, thresholds, feature strips, and reducing strips. The colour and design variations are more limited in the accessories than in the flooring materials.

Specifications. From the specifications, the colour, design, gauge (thickness), size of tile or sheet, adhesives, subfloor preparation, and any particular pattern requirements may be found. Fancy patterns may be shown on the drawings and, in most cases, will have a higher percentage of waste.

Note which areas will require the various types of resilient flooring, since it is unusual for one type, colour, size, design, and so forth to be used throughout the project. The wall base may vary in height from area to area; this can be determined from the specifications, room finish schedule, and details.

Estimating. Resilient tile is estimated by the square metre/square foot: the actual area of the surface to be covered plus an allowance for waste. The allowance will depend on the area and shape of the room. When designs and patterns are made of tile or a combination of feature strips and tile, a sketch of the floor and an itemized breakdown of required materials should be made. The cost of laying tile will vary with the size and shape of the floor, size of the tile, type of subfloor and underlayment, and the design. Allowable waste percentages are shown in Figure 15.10.

EXAMPLE **15.6**   **Commercial Building**

Assume that the entire commercial building in Figure 9.15 was to be covered with resilient flooring (main level only).

Net floor area (Example 9.7) = 547 − 34 − 30 = 483 m²
Area taken by interior walls (see Example 15.1)
108.26 m × 0.12 m = 12.99 m²
6.23 m × 0.18 m = 1.12 m²
Area under walls = 12.99 m² + 1.12 m² = 14.11 m²
Net area of floor (m²) = 483 m² − 14 m² = 469 m²
Add 5 percent for waste
Resilient flooring required 492 m²

Litres of Adhesive:

Required adhesive: coverage 4 m² per litre
Litres of adhesive = 469 m² resilient flooring / 4 m² per litre
Use 118 litres

Length (m) of Vinyl Base:

From Example 15.1, the interior walls measure 114.49 m.
From the drawings, there are 44 m of exterior walls.
Gross metre of base = (2 × 114.49) + 44
Gross metre of base = 273 m
Deductions for doors:
1 at 1,800 mm (1 side) = 1.80 m
2 at 1,200 mm (2 sides) = 2.40 m
10 at 900 mm (1 at 1 side, 9 at 2 sides) = 17.10 m
Net base = 273 m − 21 m = 252 m
Add 5 percent for waste – Use 265 m
Area of base = 265 m × 0.10 m = 27 m²
With an adhesive coverage of 4 m² per litre – Order 7 litres

Area m²	Percent Waste
Up to 10	10 – 12
10 to 15	7 – 10
15 to 30	6 – 7
30 to 100	4 – 6
100 to 500	3 – 4
500 and up	2 – 3

FIGURE 15.10 Approximate Waste for Resilient Tile

Feature strips must be taken off in metres and, if they are to be used as part of the floor pattern, the area of feature strips must be subtracted from the floor area of tile or sheets required.

For floors that need sheet flooring, the estimator must do a rough layout of the floors involved to determine the widths required, the location where the roll will be cut, and ways to keep waste to a minimum. Waste can amount to between 30 and 40 per-cent if the flooring is not well laid out or if small amounts of different types are required.

Each area requiring different sizes, designs, patterns, types of adhesives, or any-thing else that may be different must be kept separately if the differences will affect the cost of material and amount of labour required for installation (including sub-floor preparation).

Wall base (also referred to as *cove base*) is taken off by the metre. It is available in vinyl and rubber, with heights of 64 mm, 100 mm, and 150 mm and in lengths of 1,067 mm, and 15 m and 30 m. Corners are preformed, so the number of interior and exterior corners must be noted.

Adhesives are estimated by the number of litres required to install the flooring. To determine the number of litres, divide the total area of flooring by the coverage of the adhesive per litre (in square metres). The coverage usually ranges from 4 to 5 m^2 per litre but will vary depending on the type used and the subfloor conditions.

Labour. Subcontractors who specialize in resilient floor installation generally do this type of work, and they may price it on a unit basis (per square metre/square foot) or on a lump sum. The time required for resilient floor installation is shown in Figure 15.11.

Carpeting

Carpeting is selected and specified by the type of construction and the type of pile fibres. The types of construction (how they are made) are tufted, woven, and knitted; punched and flocked have become available more recently. In comparing carpeting of similar construction, such factors as pile density, pile thickness, and the number of tufts per square millimetres are evaluated in different carpets. Carpeting is generally available in widths of 2,700 mm, 3,600 mm, 4,500 mm, and 5,400 mm, but not all types come in all widths.

Pile fibres used include wool, nylon, acrylics, modacrylics, and polypropylene for long-term use. Acetate, rayon, and polyester fibres are also used. The type of pile used will depend on the type of use intended and the service required. The installed per-formance of the carpet is not dependent on any single factor but on all the variables involved in the construction of the carpet and the pile characteristics.

The cushion over which the carpeting is installed may be manufactured from ani-mal hair, rubberized fibres, or cellular rubber. The cushion increases the resilience and durability of the installation. The cushion may be bonded to the underside of the car-pet, but it is more common to have separate cushions. The type of cushion used will depend on the intended usage of the space, and a variety of designs are available for each

Tile	Work Hours per 10 m^2
Resilient Squares	
225 x 225 mm	1.5 to 2.5
300 x 300 mm	1.0 to 2.2
Seamless Sheets	.8 to 2.4
Add for felt underlayment	Add 10%
Less than 50 m^2	Add 15%

FIGURE 15.11 Work Hours Required for Resilient Floor Installation

type of material. The various cushions within each group are rated by weight in grams per square metre (ounces per square yard). The heavier the cushion, the better and more expensive it will be. Cushioning is generally available in widths of 686 mm, 914 mm, and 1,372 mm; and 1,800, 2,700 mm, and 3,600 mm.

Specifications. The specifications should state the type of carpeting required, the pile yarn weight and thickness, number of tufts per inch (25.4 mm), construction, backing, rows, and other factors relating to the manufacture of the carpet. Many specifications will state a particular product or "equal," which means the estimator will either use the product mentioned or ask other suppliers (or manufacturers) to price a carpet that is equal in quality. In the latter case, the estimator should compare the construction specifications of the carpeting to be certain that the one chosen is equal. Different types of carpeting may be used throughout the project. Take note of what types are used and where they are used.

The cushion type required and its material, design, and weight must be noted. If variations in the type of cushion required throughout the project are evident, they should be noted.

E X A M P L E   **15.7**   **Residential Building**

Using 3,600-mm-wide carpet, the following length of carpet will be required.

Bedroom 1	2,400 mm
Bedroom 2	2,450 mm
Bedroom 3	3,425 mm
Bedroom 4	3,000 mm
Living/Dining room	9,000 mm
Hall (0.90 m × 9.00 m = 27.00 m²)	2,700 mm
Closets	1,800 mm
Total carpet	24.78 m

Area of carpet = 24.78 × 3.60 = 89 m² of carpet

Estimating. Carpeting is estimated by the square metre/square yard with special attention given to the layout of the space for the most economical use of the materials. Waste and excess material may be large without sufficient planning. Each space requiring different types of carpeting, cushion, or colour must be figured separately. If the specifications call for the colour to be selected by the consultant at a later date, it may be necessary to call and try to determine how many different colours may be required. In this manner, a more accurate estimate of waste may be made.

Certain types of carpeting can be bought by the roll only, and it may be necessary to purchase an entire roll for a small space. In this case, waste may be high, since the cost of the entire roll must be charged to the project.

The cushion required is also taken off in square metres/square yards, with the type of material, design, and weight noted. Since cushions are available in a wider range of widths, it may be possible to reduce the amount of waste and excess material.

Labour. Subcontractors who specialize in carpet installations do this type of work, and they may price it on a unit basis (per square metre/square yard) or lump sum. The time required for the installation is shown in Figure 15.12.

Carpet	m² per Work Hour
Carpet and pad, wall to wall	0.8–2.0
Carpet, pad backing, wall to wall	1.0–2.2
Deduct for gluing to concrete slab	10%
Less than 100 m²	Add 15 to 20%

FIGURE 15.12 Work Hours Required for Carpet Installation

Tile

Tile may be used on floors and walls. The tile used for floors is usually ceramic or quarry tile, whereas the tile used for walls and wainscots may be ceramic, plastic, or metal.

Ceramic tile is available in exterior or interior grades, glazed or unglazed. Individual tile size may range from 9.5 × 9.5 mm to 216 × 108 mm. Tile may come in individual pieces or sheets of 0.05 to 0.20 m² per sheet. Tile mounted in sheets will be much less expensive to install than unmounted tile. Ceramic tiles come in various shapes and a wide range of sizes and colours. The tile may be installed by use of Portland cement mortar, dry-set mortar, organic adhesives, and epoxy mortars. The Portland cement mortar is used where levelling or slopes are required in the subfloor; the thickness of this mortar ranges from 19 mm to 32 mm, and it requires damp curing. The mortar will receive a coat of neat grout cement coating and the tile will be installed over the neat cement. The other methods are primarily thin-set (1.6 mm to 6.4 mm) one-coat operations. After the tile has been installed, the joints must be grouted. The grouts may be Portland cement based, epoxies, resins, and latex.

Specifications. The type of tile (material) should be determined for each space for which it is specified. The type of tile and the finish often vary considerably throughout the job. The specifications usually provide a group from which a tile will be selected and a price range (e.g., American Olean, price range A). The groupings and price ranges vary among manufacturers, so care must be taken in the use of specifications written in this manner. Other specifications will spell out precisely what is required in each area, which makes it easier to make an accurate estimate. Each area requiring different types, sizes, or shapes of tile must be taken off separately.

The methods of installation must be noted, and if the methods vary throughout the job, they must be kept separate. The type of grout required in each area must also be noted.

The types of trimmers are also included in the specifications. The number of trimmers required is kept separate from the rest of the tile takeoff because it is more expensive. Note exactly what is required because some trimmer shapes are much more expensive than others. Estimators should bid what is specified, and if the specifications are not clear, they should contact the consultant.

Estimating. Floor and wall tiles are estimated by the square metre/square foot. Each area must be kept separate, according to the size and type being used. It is common to have one type of tile on the floor and a different type on the walls. The different colours also vary in cost even if the size of the tile is the same, so caution is advised. The trim pieces should be taken off by the length (m) of each type required. Because of the large variety of sizes and shapes at varying costs, the specifications must be checked carefully and the bid must reflect what is required. If Portland cement mortar is used as a base, the tile contractor installs it. This requires the purchase of cement, sand, tile backer board, and, sometimes, wire mesh. Tile available in sheets is much more quickly installed than individual tiles. Adhesives are sold by the litres or sack, and approximate coverage is obtained from the manufacturer. The amount of grout used depends on the size of the tile.

When figuring wall tile, estimators note the size of the room, number of internal and external corners, height of wainscot, and types of trim. Small rooms require more labour than large ones. A tile setter can set more tile in a large room than in several smaller rooms in a given time period.

Accessories are also available and, if specified, should be included in the estimate. The type and style are in the specifications. They may include soap holders, tumbler holders, toothbrush holders, grab rails, paper holders, towel bars and posts, doorstops, hooks, shelf supports, or combinations of these. These accessories are sold individually, so the number required of each type must be taken off. The accessories may be recessed, flush, or flanged, and this must also be noted.

15.15 Painting

The variables that affect the cost of painting include the material painted, the shape and location of the surface painted, the type of paint used, and the number of coats required. Each of the variables must be considered, and the takeoff must list the different conditions separately.

Although painting is one of the items commonly subcontracted, the estimator should still take off the quantities so that the subcontractor's quotation can be checked. In taking off the quantities, the areas of surface are taken off the drawings, and all surfaces that have different variables must be listed separately. With this information, the amount of materials can be determined by use of the manufacturer's information on coverage per litre.

The following methods for taking off the painting areas are suggested, with interior and exterior work listed separately.

Interior

Walls—actual area in square metres
Ceiling—actual area in square metres
Floor—actual area in square metres
Trim—length (m)(note width); amount of door and window trim
Stairs—square metres
Windows—size and number of each type, square metres
Doors—size and number of each type, square metres
Baseboard radiation covers—length (m) (note height)
Columns, beams—square metres

Exterior

Siding—actual area in square metres
Trim—length (m) (note width)
Doors—square metres of each type
Windows—square metres of each type
Masonry—square metres (deduct openings over 1 m^2)

Painting, Brushes	m² or m per Hour
Interior	
Primer and 1 Coats	20–26 m²
Primer and 2 Coats	15–20 m²
Stain, 2 Coats	15–20 m²
Trim	
Primer and 1 Coats	40–60 m
Primer and 2 Coats	30–50 m
Stain, 2 Coats	30–50 m
Exterior	
Primer and 1 Coats	16–22 m²
Primer and 2 Coats	12–18 m²
Stain, 2 Coats	20–28 m²

FIGURE 15.13 Work Hours Required for Painting and Staining

The specifications should list the type of coating, number of coats, and finish required on the various surfaces throughout the project. Interiors receive different treatment from exteriors; different material surfaces require different applications and coatings—all this should be in the specifications. Paints may be applied by brush, roller, or spray gun. The method to be used is also included in the specifications.

Sometimes the specifications call for prefinished and factory-finished materials to be job finished also. Except for possible touchups, this is usually due to an oversight in the consultant's office, and the estimator should seek clarification. The most common items to be factory finished are doors, floorings, windows, baseboard, radiation covers, and grilles. The estimator should keep a sharp eye out to see that each item of work is figured only once.

Structural steel work often requires painting also. It usually comes to the job primed, with only touchup of the prime coat required. Sometimes, it is delivered unprimed with the priming done on the job. Touchup painting is impossible to figure accurately and depends on the type of structural system being used, but an average of 5 to 10 percent of the area is usually calculated as the touchup required.

The structural steel work is taken off by the tonnage of steel required with the types and sizes of the various members required. It must be noted which type of steel is to be painted: steel joists, rectangular or round tubes, H-sections, or any other type. The shape of the member will influence the cost considerably. The area to be painted per tonne of steel may vary from 15 for large members to 50 m² for trusses and other light framing methods.

Labour. Subcontractors who specialize in painting and staining often do this type of work and they almost always price it on a lump-sum basis. The time required for painting and staining is shown in Figure 15.13.

15.16 Floors and Painting Checklist

Floors: spikes
    type of material   nails
    type of fastener   adhesives

screws	**Painting:**
finish	filler
thickness	primer
size, shape	paint, type
accent strips	number of coats
pattern	shellac
cushion	varnish
base	stain
corners	check specifications for all areas requiring paint

Web Resources

www.cgcinc.com
www.gypsum.org
www.americangypsum.com
www.usg.com
www.gp.com
www.tblp.org
www.eifsalliance.com

www.woodfloors.org
www.forestdirectory.com
www.nofma.org
www.maplefloor.org
www.wfca.org
www.flooryou.com
www.floorfacts.com
www.paint.org

Review Questions

1. What is the difference between drywall and wetwall?

2. Why should walls of various heights, thicknesses, and finishes be listed separately?

3. What procedure is used to estimate the steel studs and runners used?

4. List the types of lath used for wetwall and the unit of measure for each.

5. What are the advantages and disadvantages of using subcontractors for drywall and wetwall construction?

6. What unit of measure is used for wood block flooring, and what information should be noted on the workup sheets?

7. What unit of measure is used for resilient flooring, and what information should be noted on the workup sheets?

8. What unit of measure is used to estimate carpet, and what can be done to minimize waste?

9. Determine the quantities of the drywall for the residential building shown in Appendix A.

10. Determine the quantities of the drywall for the commercial building shown in Appendix B.

11. Determine the quantities of the interior and exterior areas to be painted in the residential building shown in Appendix A.

12. Determine the quantities of the interior areas to be painted in the commercial building shown in Appendix B.

13. Determine the quantities of the floor finish for the residential building shown in Appendix A.

14. Determine the quantities of the floor finish for the commercial building shown in Appendix B.

CHAPTER 16
Mechanical

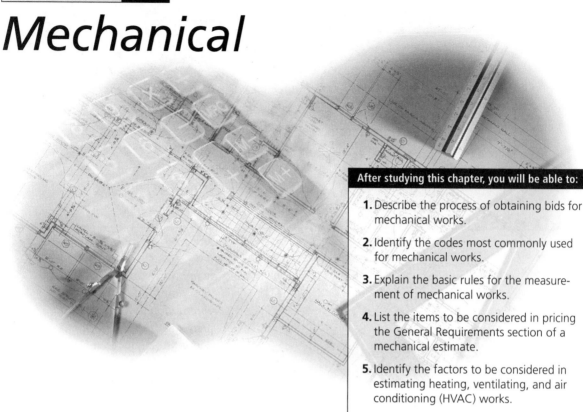

After studying this chapter, you will be able to:

1. Describe the process of obtaining bids for mechanical works.

2. Identify the codes most commonly used for mechanical works.

3. Explain the basic rules for the measurement of mechanical works.

4. List the items to be considered in pricing the General Requirements section of a mechanical estimate.

5. Identify the factors to be considered in estimating heating, ventilating, and air conditioning (HVAC) works.

16.1 Mechanical Work

To estimate mechanical works, the mechanical estimator must have a complete understanding of the scope of work specified in this division. Depending on the type and use of the building, the mechanical portion of the bid may be approximately 15 to 30 percent of the total estimate. The use of the bid depository (Section 4.8) by bid-calling authorities for mechanical bids is a common practice. However, because of responsibility and coordination issues, some bid-calling authorities prefer that general contractors be allowed to carry a mechanical subcontractor of their choice. Additionally, the general contractor's estimator must ensure that bids received do not duplicate the work of other trades. General contractors usually carry the lowest mechanical subtrade bid received through the bid depository, unless the lowest bidder did not submit a bid to the selected general contractor.

The mechanical work must be installed in accordance with all applicable building codes—national, provincial, and municipal. The codes most commonly used for mechanical works are the *National Plumbing Code of Canada*, *National Fire Code of Canada*, *Model National Energy Code of Canada for Houses*, and *Model National Energy Code of Canada for Buildings* and applicable provincial codes. These codes contain information relating to water supply, pipe sizes of all types, sewage drainage systems, heating and ventilation systems, storm water drainage, and various design data.

Field experience, familiarity with plumbing and HVAC principles, and an understanding of the drawings and specifications are key points to the successful bidding of mechanical work. Typical symbols and abbreviations are shown in Figures 16.1 and 16.2.

Piping symbols:

Vent — — — — — — — — — —

Cold water —— · —— · —— ·

Hot water —— ·· —— ·· —— ··

Hot water return —— — —— — ——

Gas |—— G —— G ——

Soil, waste or leader ——————
(above grade)

Soil, waste or leader — — — — —
(below grade)

Fixture symbols:

Baths

Water closet (with tank)

Water closet (flush valve)

Shower

Lavatory

Dishwasher   DW

Service sink   SS

Hot water heater   HWH

Hot water tank   HWT

HWT

Drinking fountain   DF

Meter   M

Hose bib   HB

Cleanouts   C/O   CO

Floor drain   FD

Roof drain   RD

A.F.D.	area floor drain
B.W.V.	backwater valve
CODP.	deck plate cleanout
C.W.	cold water
C.W.R.	cold water return
DEG.	degree
D.F.	drinking fountain
D.H.W.	domestic hot water
DR.	drain
D.W.	dishwasher
F.	fahrenheit
FDR.	feeder
FIXT.	fixture
F.D.	floor drain
F.H.	fire hose
F.E.	fire extinguisher unit
H.W.	hot water
H.W.C.	hot water circulating line
H.W.R.	hot water reserve
H.W.S.	hot water supply
H.W.P.	hot water pump
I.D.	inside diameter
LAV.	lavatory
LDR.	leader
O.D.	outside diameter
(R)	roughing only
R.D.	roof drain
S.C.	sill cock
S.S.	service sink
TOIL.	toilet
UR.	urinal
V.	vent
W.C.	water closet
W.H.	wall hydrant

FIGURE 16.1 Plumbing Symbols and Abbreviations

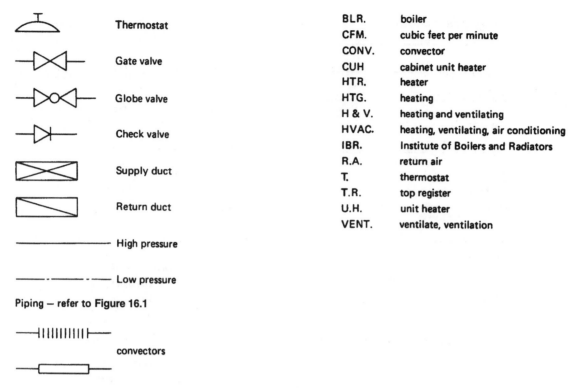

FIGURE 16.2 HVAC Symbols and Abbreviations

16.2 Method of Measurement

The *Method of Measurement of Construction Works* published by the Canadian Institute of Quantity Surveyors (CIQS) contains basic rules for the measurement of mechanical works. Pipework is measured in metres with no deductions made to pipe lengths for fittings not exceeding 225 mm in diameter; fittings and valves are enumerated as well as all other items, with the exception of ductwork and mechanical insulation.

16.3 General Requirements

Itemized prices for the following categories (as per the CIQS *Method of Measurement of Construction Works*) are required in this section of the mechanical bid:

Certificates and Fees. Charges for obtaining permit and licences for plumbing, welding tests and other similar regulatory requirements as specified.

Liability and Protection. Allowance for making good errors and protection of finishing trades' works.

Record Drawings. Cost of producing as-built drawings, incorporating changes made to the original drawings.

Shop Drawings. Cost of processing and checking shop drawings.

Operating Instructions and Maintenance Manuals. Associated costs of providing written operating instructions and maintenance manual including spare parts, if required.

Temporary Facilities. Mobilization and demobilization costs for site offices and storage sheds, including all necessary temporary services and facilities.

Insurances. Cost of required insurances, including public liability, all risk, and Workers' Compensation.

Warranty. Cost of providing a guarantee to cover defects in materials and workmanship for the specified period of time.

Preparation for Painting. Cost for cleaning surfaces ready for painting.

Valve Tags and Charts. Cost for providing valve tags and charts.

Wall Plates and Access Doors. Cost for providing wall plates and access doors or hatches.

Bases and Supports. The cost of any steel bases or stands for mechanical equipment, if not to be provided by the miscellaneous metals subtrade.

Temporary Plumbing. Provision of temporary services, including costs of permit and connection fees.

Testing and Commissioning. Cost for testing each part of the system as required during the progress of the work and final testing upon completion.

Cleanup. Cost of cleaning up and disposing rubbish created by this trade.

Electrical Wiring and Starters. Cost of cables, wiring, and accessories required to be supplied and installed by the mechanical subtrade.

Personnel. Salaries and wages for managerial, technical, and administrative personnel to be assigned to the project, including benefits, travel, and other reimbursable expenses.

Layout. Costs of labour, material, and instruments required for laying out the work.

Hoisting and Scaffolding. Cost of hoisting major equipment that is beyond the capabilities of the general contractor's hoisting equipment, and provision of any special scaffolding.

Plant and Equipment. Cost for providing, installing, and operating all tools and equipment, including fuel, maintenance, repairs, and spare parts.

Premium Time. Additional costs in respect of work involving overtime or shift work.

Bonds. The cost of any required bid bond, performance bond, and labour and material payment bond.

Lost Time. Allowance for time lost in reporting to site when work cannot be performed due to inclement weather as stated in the trade agreement.

Cutting, Patching, and Making Good. Cost of any cutting, patching, coring, and making good the work of other trades.

16.4 ■ Mechanical Insulation

The scope of work under mechanical insulation includes duct, piping, and equipment insulation. Piping insulation is measured in metres, with an allowance of 0.50 m for each fitting or valve. Ductwork insulation is measured in square metres. Equipment insulation is measured in square metres. Prefabricated insulated panels are measured in square metres, with the number of panels indicated.

16.5 ■ Building Services Piping

Using the working drawings and specifications, estimators prepare a complete list of everything that will be required. First, they determine exactly where the responsibility for the plumbing begins. It may begin, or terminate, 1 m from the outside face of the structure, at the structure, or somewhere else; they should always check to have a clear picture of the project in mind.

The takeoff for plumbing includes the costs of domestic water pipes, gas pipes, sewer pipes, drains, soil and vent stacks, soil, waste, vent pipes, and all rough work required for fixtures. Information required includes size, length, pipe material, weight, and fittings (unions, hangers, Ts, Ys, elbows, and so on). Also included are kinds and sizes of cleanouts, grease traps, plugs, traps, valves, and the like. The trenches required for plumbing pipe, especially from the road to the building, may be the responsibility of the mechanical subcontractor or the general contractor. If the general contractor is responsible for trenching, it should be noted as such on the estimate. Pipe materials include cast iron, copper, lead, galvanized, and plastic.

16.6 ■ Plumbing Fixtures and Equipment

The finished plumbing list should include a list of all items and fixtures furnished and installed by the contractor. All different types and sizes of fixtures are kept separate. In pricing the fixtures, the estimator must also consider the accessories and fittings required for each fixture. The listing requires care in counting the fixtures on the working drawings and making note of all specification requirements. When listing the prices for the fixtures, estimators note whether the fixtures and accessories were priced separately. The make, kind, type, and quality will greatly affect the price of the fixtures and accessories. Again, there should be no guesswork, only accurate prices. Items that should be included in this takeoff include lavatory, water closet, shower with stall, shower without stall, bathtub, bathtub with shower, slop sink, water heater, dishwasher, garbage disposal, laundry tubs, and water cooler. Another situation that must be noted is that which occurs when the owner or others supply an item to the mechanical subcontractor for installation. It is important that estimators know exactly the limits of responsibility in this case, since this is often an area of dispute.

16.7 ■ HVAC Work

Heating, ventilating, and air conditioning equipment includes heat exchangers, air handling units, air conditioning equipment, heat pumps, humidity control equipment, terminal heating and cooling units, floor-heating and snow-melting equipment, and energy recovery equipment.

Air distribution is measured under the categories of ducts, duct accessories, fans, air terminal units, air outlets and inlets, and air cleaning devices. Spiral ductwork is measured in metres with a count of all fittings. Rectangular and circular ductwork is measured in metres and the resultant length converted into weight. For sheet metal ducts add 20 percent to cover seams, joints, cleats, hangars, sealants, and waste. For fibreglass ducts, add 15 percent to cover waste plus 7 kg per 10 m^2 to cover hangars and supports. For duct liners add 15 percent to cover waste. For welded black iron ducts, add 30 percent to cover seams, joints, cleats, hangars, sealants, and waste.

16.8 Coordination Requirements

Figure 16.3 lists the major areas of coordination required between the mechanical and general contractors for plumbing works, and Figure 16.4 lists the major areas of coordination required between the mechanical and general contractors for HVAC work.

Coordination among the mechanical contractors is also important in the understanding of who is responsible for what, why, and when.

Item	Coordination Requirements
1. Underground utilities	Location, size, excavation by whom, from where?
2. Building entrance	Floor sleeves, supports.
3. Mechanical room equipment	Supports required, location, anchors, by whom?
4. Distribution	Wall sleeves, hangers, chases, in wall, roof vents, access doors.
5. Fixtures	Method of support, feed, outlets, built-in, floor drains, vents.
6. Finishes	Factory or field.
7. Specialty equipment	Field provisions, storage.
8. Scheduling	Work to be done, when required? Job to be completed on time – who – why – when?

FIGURE 16.3 Coordination Issues for Plumbing Works

Item	Coordination Requirements
1. Underground utilities	Location, size, excavation by whom, from where?
2. Equipment	Method of support, location, by whom, anchors, access for receiving and installing, size limitations, flues, roof curbs.
3. Piping	Wall sleeves, size limitations, chases, in walls, under floors, floor sleeves, expansion compensators, access doors.
4. Ductwork	Sizes, support, access doors, drops, outlet sizes, chases, outside air louvres, roof curbs, lintels.
5. Terminal equipment	Recesses for CUH, RC, FC, etc., size of radiation, method of concealing, grille fastenings
6. Mechanical-electrical responsibility	Who is doing what?
7. Finishes	Field or factory?
8. Scheduling	Who? Why? When?

FIGURE 16.4 Coordination Issues for HVAC Works

16.9 Checklist

Rough:
- permits
- excavation and backfill
- water, gas, and sewage lines
- required pipes and fittings
- cleanouts
- valves
- tanks
- sleeves

Finish:
- water closets
- bath tubs
- lavatories

Heating (hot water):
- boiler
- stoker
- oil tanks
- gauges
- fuel
- piping
- insulation
- circulating pumps
- piping accessories
- radiators
- fin tubes
- enclosures
- clocks
- hangers
- unit heaters

Heating (warm air):
- boiler
- fuel
- oil tanks
- gauges
- accessories
- thermostats
- wiring
- chimney
- ducts
- diffusers

- fans
- drinking fountains
- showers
- tubs
- service sinks
- water heater
- urinals
- washers and dryers
- dishwashers

Miscellaneous:
- hookup to equipment supplied by owner or other contractors may be required
- valves
- accessories
- chimney
- thermostats
- wiring
- filters
- humidifiers
- dehumidifiers
- insulation
- baffles

Air Conditioning:
- central
- units
- coolant
- fans
- piping
- diffusers
- registers
- wiring
- ducts
- filters
- humidifiers
- dehumidifiers
- inlets
- returns
- fresh air
- louvres
- thermostats

Review Questions

1. How do the various codes affect the installation of the mechanical portions of the project?

2. When and under what circumstances would a mechanical estimator allow for *premium time* and *lost time* in a subtrade bid?

3. What type of work is most generally included under plumbing?

4. Why should the general contractor's estimator review the mechanical portion of the project prior to receiving bids from the mechanical subtrades?

5. How do the various types of construction affect the cost of the plumbing work?

6. How do the various types of construction affect the cost of the heating work?

7. Under which of the mechanicals would electric heat most likely be placed?

CHAPTER **17**

Electrical

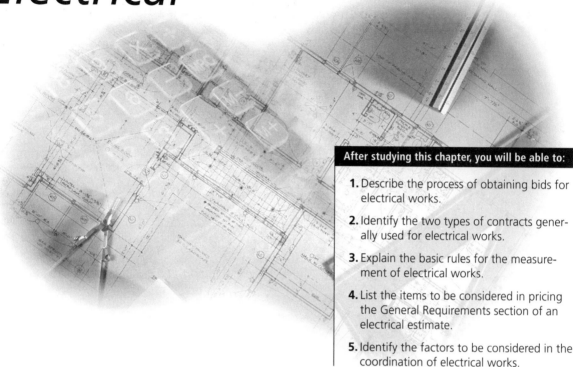

After studying this chapter, you will be able to:

1. Describe the process of obtaining bids for electrical works.

2. Identify the two types of contracts generally used for electrical works.

3. Explain the basic rules for the measurement of electrical works.

4. List the items to be considered in pricing the General Requirements section of an electrical estimate.

5. Identify the factors to be considered in the coordination of electrical works.

17.1 Electrical Work

Under stipulated lump-sum contracts, the electrical work is the responsibility of the general contractor. In most cases, this means that electrical subcontractors will submit prices on the work to be completed to the general contractor. Bids may be received through the bid depository (see Section 4.8) or if the bid-calling authorities did not specify this process, bids are solicited from electrical subcontractors who have previously worked with the general contractor. The general contractor will include an electrical contractor price plus overhead and profit in the bid price to the owner. In this case, the general contractor is directly responsible for the work to the owner and must coordinate all the parties involved in the project.

With construction management contracts, the electrical contractor will bid the electrical work directly to the owner/construction manager; the owner/construction manager will select the contractor and sign a contract. The responsibility for the electrical work is the electrical contractor's, and although there shall be certain mutual responsibilities and coordination, the general contractor's responsibilities are not as great as they would be under stipulated-sum contracts.

All electrical work must be installed in accordance with the *Canadian Electrical Code*, published by the Canadian Standards Association. Provincial and municipal regulations must also be considered. Before beginning the takeoff, the electrical estimator should review the plans and carefully read the specifications. Often, the specifications will require that "all work and installations be in conformance with all applicable provincial electrical safety codes and building codes." This statement means that contractors are responsible for compliance with the laws; if they are responsible for them, then they had better be familiar with them. The codes contain information regarding wiring design and protection, wiring methods and materials, equipment, special occupancies, and other information.

Field experience in construction will be helpful in understanding the problems involved in electrical work and how the electrical aspect should be integrated into the rest of the construction. Without field experience, an understanding of the fundamentals of electrical work, and an ability to read and understand the electrical drawings and specifications, it will be difficult to do a meaningful takeoff on these items.

17.2 Stipulated Lump-Sum Contracts

Because an electrical subcontractor will undoubtedly do the electrical work, that contractor will do the bidding. Learning to estimate electrical work is a special skill requiring extensive knowledge of the properties and behaviour of electricity. The complexity of this topic would fill a book by itself. What are presented here are some issues that contractors face when dealing with electrical subcontractors. However, by using some common sense and complete files from past jobs, it is possible to obtain approximate estimates for checking whether the bids submitted are reasonable. With experience and complete files, it is possible to figure an estimate close to the low bid.

The selection of an electrical subcontractor should not be based on price alone, although price is an important consideration. Other factors, such as the speed with which the subcontractors complete their work and the cooperation they show in dealing with the prime contractor and other subcontractors, are also important. General contractors are responsible to the owners for all work; it is in their own best interests to consider all factors when selecting an electrical subcontractor and ensure that there is no duplication in the bids received from the various trades.

The major areas of coordination required between the electrical contractor and the general contractor are outlined in Section 17.7, while Figure 17.1 shows typical electrical symbols and abbreviations.

When electrical subcontractors bid a project, they are often asked to include required temporary wiring and lighting. By having the electrical subcontractor include these costs in their bid, the contractor has a negotiating advantage.

17.3 Construction Management Contracts

The electrical contractor does the takeoff and bidding, but this does not mean that the estimator for the construction manager should not review the drawings and specifications for this work. Often, when projects are being bid under separate contracts, the contractors for each phase receive the drawings only for that phase (or portion of work) on which they will be bidding. (The bidders for general construction may receive no mechanical.) In this case, a trip to the plans room or to the office of the mechanical contractor who is bidding the project should be made so that the drawings and specifications may be investigated.

Even under construction management contracts, there are many areas of mutual responsibility and coordination. Keep in mind that the entire building must fit together and operate as one unit. The major areas of coordination for stipulated-sum contracts and construction management contracts are outlined in Section 17.7. If construction management contracts are being used, the contractor needs to be aware of who is providing the temporary wiring and lighting.

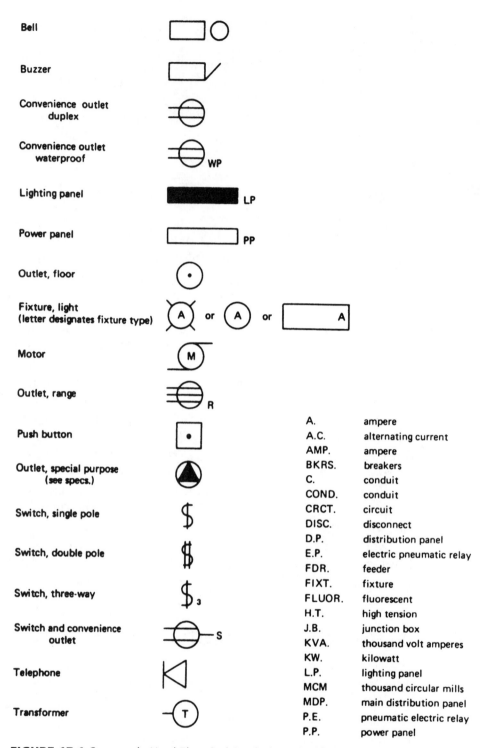

FIGURE 17.1 Commonly Used Electrical Symbols and Abbreviations

17.4 Method of Measurement

The *Method of Measurement of Construction Works* published by the Canadian Institute of Quantity Surveyors (CIQS) contains basic rules for the measurement of electrical works. Cables, wire, ducts, raceways, and conduits are measured in metres; bends, fittings, supports, fastenings, and all other items are enumerated.

17.5 General Requirements

Itemized prices for the following categories (as per the CIQS *Method of Measurement of Construction Works*) are required in this section of the mechanical bid:

Certificates and Fees. Charges for obtaining permit and licences for electrical tests and other similar regulatory requirements, as specified.

Liability and Protection. Allowance for making good errors and protection of finishing trades' works.

Record Drawings. Cost of producing as-built drawings, incorporating changes made to the original drawings.

Shop Drawings. Cost of processing and checking shop drawings.

Operating Instructions and Maintenance Manuals. Associated costs of providing written operating instructions and maintenance manual.

Temporary Facilities. Mobilization and demobilization costs for site offices and storage sheds, including all necessary temporary services and facilities.

Insurances. Cost of required insurances including public liability, all risk, and workmers' Compensation.

Warranty. Cost of providing a guarantee to cover defects in materials and workmanship for the specified period of time.

Identification Tags. Cost for providing identification tags.

Bases and Supports. The cost of any steel bases or stands for electrical equipment, if not to be provided by the miscellaneous metals subtrade.

Testing. Cost for testing each part of the system, as required during the progress of the work and final testing upon completion.

Cleanup. Cost of cleaning up and disposing rubbish created by this trade.

Personnel. Salaries and wages for managerial, technical, and administrative personnel to be assigned to the project, including benefits, travel, and other reimbursable expenses.

Layout. Cost of labour, material, and instruments required for laying out the work.

Hoisting and Scaffolding. Cost of hoisting major equipment that is beyond the capabilities of the general contractor's hoisting equipment and provision of any special scaffolding.

Plant and Equipment. Cost for providing, installing, and operating all tools and equipment, including fuel, maintenance, repairs, and spare parts.

Premium Time. Additional costs in respect of work involving overtime or shift work.

Bonds. The cost of any required bid bond, performance bond, and labour and material payment bond.

Lost Time. Allowance for time lost in reporting to site when work cannot be performed due to inclement weather, as stated in the trade agreement.

Cutting, Patching, and Making Good. Cost of any cutting, patching, coring, and making good the work of other trades.

17.6 Electrical Wiring and Fixtures

The wiring is considered the *rough* work, and the fixtures are considered the *finish* work. The wiring will usually be concealed in a conduit, which is installed throughout the building as it is erected. The wiring is pulled through the conduit much later in the job. Cables are also used extensively. The fixtures are usually the last items to go into the building, often after the interior finish work is complete.

To work up an estimate, estimators go through the plans and specifications in a systematic manner, taking each different item and counting the number of each. Every item must be kept separate. For example, floor outlets are different from wall outlets. The estimators must not hesitate to check off (lightly) each item as they count it to reduce the possibility of estimating the same item twice. Included on the list are all outlets, floor plugs, distribution panels, junction boxes, lighting panels, telephone boxes, switches, television receptacles, fixtures, and any other items, such as snow-melting mats.

It is the estimator's job to determine exactly where the responsibility begins for the wiring. Does it begin at the property line, at the structure, or 1 m from the structure? If a transformer is required, who pays for it, who installs it, and who provides the base on which it will be set?

Different types of construction affect the installation of rough and finish work. When using steel joists, there is usually ample space through which to run conduits easily. Cast-in-place concrete requires that there be closer cooperation between the general contractor and the electrical contractor because the conduit (as well as sleeves) often must be cast in the concrete and fixture hangers cast in. The use of hollow-core, pre-stressed, precast concrete causes other problems, such as where to run the conduit and how to hang fixtures properly. The conduit can be run in the holes (or joints) that are in the direction of the span, but care must be taken to run them in other directions unless the conduit can be exposed in the room. These problems greatly increase the amount of conduit required as well as the cost of the installation. If installation is difficult, it also becomes more expensive. Similar problems occur when using precast double tees, except in that case, no holes are available in the spanning direction. The problem can be alleviated to some extent by pouring a 50 mm to 75 mm concrete floor over the slabs in which to run the conduit.

One last point: Estimators must not guess the price of fixtures. Prices vary considerably. What seems to be an inexpensive fixture may turn out to be very expensive. They should never trust guesswork but, rather, check prices with the light fixture manufacturers and request a firm price.

17.7 Coordination Requirements

Figure 17.2 lists the major areas of coordination required between the electrical and general contractors.

Coordination of work among the mechanical contractors themselves is also important, since often all three of the mechanical contractors have work to do on a particular piece of equipment. For example, the heating contractor may install the boiler unit in place, the electrical contractor may make all power connections, and the plumbing contractor may connect the water lines. There are many instances of several trades connecting to one item. Coordination and an understanding of the work to be performed by each contractor are important for the smooth running of the job.

Item	Coordination Requirements
1. Underground utilities	Location, size, excavation by whom, from where?
2. Equipment	Recessed depth, size of access openings, method of feed, supports, by whom, size limitations.
3. Distribution	Outlet locations, materials, method of feed, chases, in walls, under floor, overhead, special considerations.
4. Terminal fixtures and devices	Location, method of support, finish, colour, and material.
5. Mounting surfaces	Mounting surface? Can it work?
6. Specialty equipment	Field provisions storage.
7. Scheduling	Work to be done? When required? Job to be completed on time – who, why, when?

FIGURE 17.2 Coordination Requirements

17.8 Checklist

Rough:

> conduit (sizes and lengths)
> wire (type, sizes, and lengths)
> outlets (floor, wall, overhead)
> switches (2-, 3-, and 4-way)
> panel boards
> breakers (size, number of each)
> outlets, weatherproof
> control panels
> power requirements

Finish:

> fixtures (floor, wall, ceiling, etc.)
> mounting requirements

> clocks
> time clocks
> bells (buzzers)
> alarm systems
> TV outlets
> heaters

Miscellaneous:

> hook-up for various items from the other trades (motors, boilers, etc.)

Review Questions

1. What is the difference between stipulated lump-sum and construction management contracts?

2. Explain the procedure to be followed by the electrical contractor in submitting a bid through the bid depository system.

3. Why should the general contractor estimator review the electrical portions of the job in a stipulated lump-sum contract?

4. Why is cooperation and coordination so important between the various contractors on a project?

5. How will the various types of construction affect the cost of the electrical work?

CHAPTER 18
Bid Closing

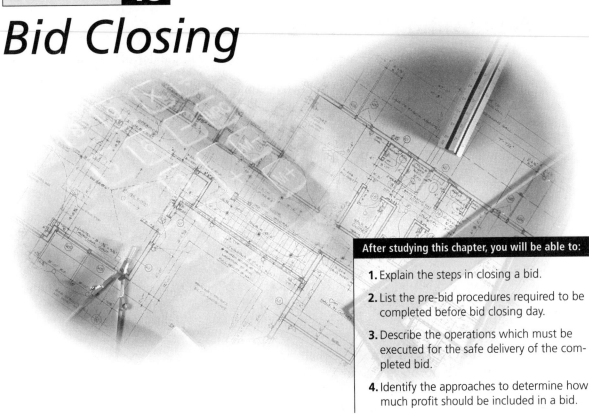

After studying this chapter, you will be able to:

1. Explain the steps in closing a bid.

2. List the pre-bid procedures required to be completed before bid closing day.

3. Describe the operations which must be executed for the safe delivery of the completed bid.

4. Identify the approaches to determine how much profit should be included in a bid.

18.1 Steps to Closing a Bid

Contractors should expect to close, not just drop off the bid at the stipulated time and location. Bid closing procedures for the subtrade estimator are substantially different from those which the general contractor's estimator must follow. Approximately 80 percent of a general contractor's bid price is derived from subtrade quotations, and this adds to the anxiety and panic in a general contractor's office on bid closing day. Other than quotations from specialist suppliers, such as those for washroom accessories and toilet partitions, which may be submitted one or two days prior to closing, most subtrades submit their bids to the general contractor in the last hour prior to the time of closing. Generally, trades that will influence the final bid prices, such as the mechanical and electrical subtrades, will wait for approximately half an hour before bid closing to submit their prices to the general contractor. Subcontractors are reluctant to divulge their prices and will wait to the last 20 minutes, just prior to the closing, thereby giving the general contractors little or no opportunity to peddle their bids. Estimators are required to work under these stressful circumstances, and so, bid closing day calls for creative actions to deal with the usual snags. To minimize the problems and avoid costly mistakes, estimators must have established procedures for closing a bid.

18.2 Bid Closing Procedures

The following processes should be completed prior to the day of the bid closing:

Partial Completion of the Summary Sheet. An example of a bid summary sheet is shown in Figure 4.8. It is a huge advantage for an estimator to complete a contractor's own forces quantity takeoff and pricing prior to bid closing day. The general expenses items should also be estimated, except for those items that will be priced on a percentage or dependent on the final value of the bid. Obviously, the subcontractor prices cannot be entered at this stage.

Bid Form Requirements. There are significant items on the Bid Form that can be addressed in advance of bid closing day. Attached to the Bid Form (see Figure 3.4) are appendices that must be completed. This may include unit prices and alternative or separate prices. Calculation of these for the contractor's own work should be done ahead of time and steps taken to obtain those of other trades. The number of addenda issued can be confirmed on the Bid Form, as stipulated. The signing and/or sealing of the Bid Form by a duly authorized person can also be done ahead of time.

Insurances and Bonds. Obtain and complete insurance documents, bonds, and any other documents accompanying the bid.

Duplicate Bid Documents. Make copies of bid documents.

Prepare Envelope for Bid Documents. Follow the instructions to bidders with respect to the information required on the envelope.

Prepare for Bid Delivery. Brief the person who is to deliver the bid. Designate a route to be taken and what means of communication will be used, including backup, and the information that will be relayed before the envelope is sealed and delivered.

Staff Selection. Select the staff for bid closing day. Instruct them on priorities for receiving bids via fax machines and use of a standard form for logging telephone bids.

Check Office Equipment. Check fax machines to ensure an adequate supply of ink and paper. Perform a routine check on all computers and telephones.

18.3 Closing Day Procedures

The process of closing a bid is exciting but can also be very stressful. Some estimators may enjoy the thrill of a bid closing, while others approach the day with anxiety and a sense of panic. The bid closing staff must be able to work under extreme pressure to ensure the following operations are executed for the safe delivery of the completed bid:

Bid Review. The level of profit (as discussed in Section 18.4) and overhead allowances must be discussed and determined with the owner/general manager and changes made, if required.

Subtrade Analysis. As discussed in Section 4.7, subtrade bids received must be tabulated (see Figure 4.4), analyzed, and included on the summary sheet.

Final Check for Information. Check the mail, telephone voice messages, and fax machines for last-minute information.

Relay Information to Bid Runner. The person filling out the final price on the Bid Form must be given the final bid price in time to insert the sum both in number and written words and get the bid in in spite of any conditions that may be obstructive. Driving to the bid-closing location on an empty gas tank is an estimator's horror story too often told.

Bid Delivery. The bid is delivered according to instructions in the location and stipulated time.

18.4 Profit

The amount of profit to be added to the bid price is one of the last things considered and must be dealt with separately. Keep in mind that construction is among the top in the percentage of business failures. Would you believe that in the haste to arrive at the final bid price, some estimators even forget to add profit?

First, let us understand that by *profit* we mean the amount of money added to the total estimated cost of the project; this amount of money should be clear profit. All costs relating to the project, including project and office overhead and salaries, are included in the estimated cost of the building.

There are probably more approaches to determining how much profit should be included than could be listed. Each contractor and estimator seems to have a different approach. A few typical approaches are listed as follows:

1. Add a percentage of profit to each item as it is estimated, allowing varying amounts for the different items; for example, 8 to 15 percent for concrete work, but only 3 to 5 percent for work subcontracted out.

2. Add a percentage of profit to the total price tabulated for materials, labour, overhead, and equipment. The percent would vary from small jobs to larger jobs (perhaps 20 to 25 percent on a small job and 5 to 10 percent on a larger one), taking into account the accuracy of the takeoff and pricing procedures used in the estimate.

3. Various methods of selecting a figure are employed that will make a bid low, but not too low, by trying to analyze all the variables and other contractors who are bidding.

4. There are "strategies of bidding" that some contractors (and estimators) apply to bidding. Most of the strategies require bidding experience that has to be accumulated over the years and competitive patterns from past biddings to be used as patterns for future biddings. This will also lend itself nicely to computerized job-costing systems.

5. Superstition sometimes plays a part. And why not, since superstition is prevalent in our lives? Many contractors and estimators will use only certain numbers to end their bids; for example, some always end with a seven or add 50 cents to a million-dollar bid.

One approach is to include all costs of the project before profit is considered. Then, make a review of the documents to find if the drawings and specifications were clear, if you understood the project you were bidding, and how accurate a takeoff was made (it should always be as accurate as possible). The other factors to be considered are the consultants' reputations, and how they handled work on previous jobs.

After reviewing the factors, you, the contractor, must decide how much money (profit, over and above project and office overheads) you want to make on this project. This amount should be added to the cost of construction to give the amount of the bid (after it is adjusted slightly to take into account superstitions and strategy types of bids). Exactly what is done at this point, slightly up or down, is an individual matter, but you should definitely know your competitors, keep track of their past bidding practices, and use those against them, whenever possible.

Since profit is added at the end of the estimate, the estimator has a pretty good idea of the risks and problems that may be encountered. Discuss these risks thoroughly with other members of the firm. It is far better to bid what you feel is high enough to cover the risks than to neglect the risks, bid low, and lose money. There is sometimes a tendency to "need" or "want" a job so badly that risks are completely ignored. Avoid this sort of foolishness—it only invites disaster. If a project entails substantial risk and it is questionable that a profit can be made, consider not even bidding it and let someone else have the heartaches and the loss. Always remember, construction is a business in which you are supposed to make a fair, reasonable profit. In a period of recession, a contractor may have very little choice but to bid with a very low profit margin with hopes to keep his best employees gainfully employed as better times are eagerly awaited.

Review Questions

1. Why do trades that will influence the final bid prices wait for approximately half an hour before bid closing to submit their prices to the general contractors?

2. What are the significant items on a Bid Form that can be addressed in advance of bid-closing day?

3. What are the objectives of a bid review?

4. What factors determine how much profit should be included to the total estimated cost of the project?

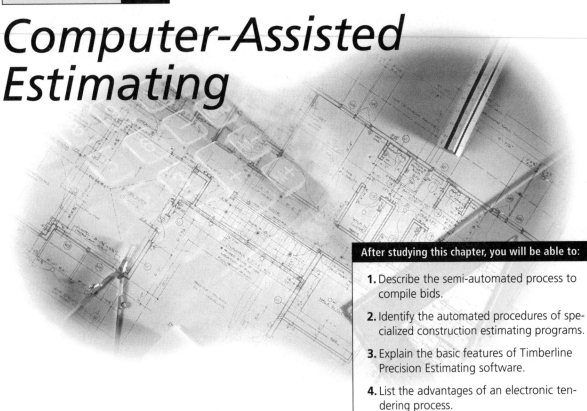

CHAPTER **19**

Computer-Assisted Estimating

After studying this chapter, you will be able to:

1. Describe the semi-automated process to compile bids.

2. Identify the automated procedures of specialized construction estimating programs.

3. Explain the basic features of Timberline Precision Estimating software.

4. List the advantages of an electronic tendering process.

19.1 Use of Computers in Estimating

The use of computers in all phases of construction has grown rapidly in the last decade. The increased speed and storage capacities of the personal computers, combined with the sophisticated software packages and lower prices, have made the computer an integral part of the construction process. Computers are used with CAD (Computer-Aided Design) software from the beginning of a project at the design concept stage, through the development of the contract documents and the bidding stages of a project, to management of actual construction. The extent to which computers are used in estimating building costs varies considerably. Most computerized estimating is either performed using spreadsheet programs (semi-automated) or specialized construction estimating programs (automated).

Semi-automated. This process is a combination of manual quantity takeoff and computer-assisted preparation of cost estimates using electronic spreadsheet software, such as Microsoft Excel, Lotus 1-2-3, and Corel's Quattro Pro, which require manual input of data. The advantages of this process are the low initial investment in equipment and training, less duplication of time and effort, ability to modify previous estimates to a certain extent, and the suitability for small projects. The disadvantages are that manual takeoff and manual input of data are comparatively tedious and costly and there is very little flexibility in the presentation of estimate reports.

Automated. In this process, the computer hardware and software are designed and configured in order that the process of quantification, the selection of item description from a database, and performing the measurements of area, length or counting by a digitizer (electronic tablet) are all interlinked. The estimator executes the manipulation and selection of data to achieve the required results. The advantages of this process are the increased speed and improved accuracy with preparation of estimates; less time is spent on takeoff and more time is available for the estimator to devote to assessment of project costs and execution; a fully computerized estimating system for recording, analyzing, and storing of cost data from previous projects to provide a database of historical pricing information; and enhanced sophisticated report capabilities. The disadvantages are the high initial cost of purchasing hardware and software, high maintenance costs, databases requiring continual updating, and the software being dependent on good input. As the saying goes, "Good data input, good data output."

19.2 Windows-Based Estimating Software

Using computer construction estimating software for Windows to estimate makes the estimator's job easier in many ways. Employing electronic calculations or formulas helps reduce calculation errors. Using the CSC/CSI format (see Figure 3.1) makes it less likely that any major areas of construction work will be missed. Although the computer provides all the tools to do a better job, the estimator will need to become familiar with new ways of doing business. He will still need to know all about quantity takeoffs; yet more time will be spent developing databases and work crews, running audit trails, and analyzing category costs. The estimator needs to know work processing, formulate setups, work categories and development, and manipulation of databases.

Accompanying this text is an estimating compact disk (CD) that will allow the user to gain some experience estimating with windows-based computer software. *Timberline Software*, a pioneer in computerized construction management techniques, developed this software. This basic edition will allow the user to estimate quantities and labour for the textbook problems and use databases for labour and material costs. The instructions for installing the software and its manual are found on the last page of this book.

19.3 Overview of Timberline Precision Estimating

Precision Estimating is driven by a number of databases, a versatile estimating spreadsheet that can be customized to a format that is familiar to the estimator, several takeoff functions, sophisticated reporting capabilities, and interfaces to other construction management software applications. Of greater significance, Precision Estimating creates a Windows environment that most estimators will recognize.

Databases. The database stores all the information required to produce an estimate including item descriptions, labour, material and equipment prices, subcontractor information, and takeoff formulas. It is the price book that contains the construction estimator's knowledge. You can assemble a database to reflect your techniques and methods, or you can purchase one that can be customized to suit your needs.

The development of a customized database incorporating previous projects is essential to the successful implementation of a computerized estimating system. The input based on the MasterFormat coding system is quite common to most systems. Formulas, rate tables, and productivity factors can all be integrated. Computers cannot replace an intelligent estimator, they just do things faster than him or her. They simply number crunch accurately.

Precision databases typically consist of a three-tiered hierarchy:

Group Phases (Divisions). These form the highest and most general level of a database. This should follow the "MasterFormat" numbering system:

E.g., 04000.00 MASONRY

Phases (Subdivisions). These organize items into groups of work-related tasks, again following the "MasterFormat" numbering system:

E.g., 04200.00 Unit Masonry

Items. These are the lowest level of the database. A good database contains item descriptions representative of work items commonly used in your office:

E.g., 10 240 mm bond beam

Estimates and the Estimating Spreadsheet. The Estimating spreadsheet in Precision Estimating is similar to the spreadsheets used in the manual and semi-automated estimating processes. Columns are provided for divisions, subdivisions, item numbers, takeoff quantities, and provisions for several unit pricing categories. Because the Estimating spreadsheet is disk-based, estimates are saved automatically.

Customizing the Spreadsheet. The spreadsheet can be customized to suit the particular needs of most estimating practices. Features include functions to display in another order (takeoff, assembly, or by location), the flexibility to change the size of the columns and turn them on or off, and commands to change the fonts and colours used on the spreadsheet. Spreadsheet details can be collapsed or expanded, and typical spreadsheets can be saved for future use.

Creating the Estimate. When an estimate is created, information must be copied from the database into the spreadsheet during the takeoff process, entering dimensions and specifications as required. The database and estimates are stored in separate files. Changes made to the database do not affect existing estimates.

Takeoff. Precision Estimating provides several ways to take off items during estimate preparation. The basic version only allows you to perform a *quick takeoff*. This is a fast, easy way to get items into the estimate spreadsheet. It involves clicking an icon in the toolbar to open the Quick Takeoff window and double-clicking on the mouse (or dragging and dropping) the items you want to take off.

```
01000.00  GENERAL REQUIREMENTS
02000.00  SITE WORK
03000.00  CONCRETE
04000.00  MASONRY
   04150.00  Masonry Accessories
   04200.00  Unit Masonry
      05  250 mm concrete block in foundation wall
      10  240 mm bond beam
05000.00  METALS
06000.00  WOOD AND PLASTICS
```

FIGURE 19.1 Typical Database Hierarchy

Group	Phase	Item	Description	Takeoff Quantity	Other Cost/Unit	Total Amount
02000.00			SITE WORK			
	02200.00		Earthwork			
		20	Crushed stone bed	53.75 CM	27.50	1,478
03000.00			CONCRETE			
	03100.00		Concrete Formwork			
		05	Formwork to foundation wall footing; 200 mm wide	138.00 LM	10.00	1,380
	03200.00		Concrete Reinforcement			
		05	Rebar; 15M	238.33 KG	1.50	357
		10	W.W.M; 152 x 152 - MW 18.7 x 18.7 MW	295.63 SM	3.05	902
	03300.00		Cast-in-Place Concrete			
		05	25 MPa concrete in foundation footing	6.21 CM	130.00	807
		10	25 MPa concrete slab-on-grade	26.88 CM	132.00	3,548
		15	Broom finish to concrete slab-on- grade	268.75 SM	3.00	806
04000.00			MASONRY			
	04150.00		Masonry Accessories			
		05	Rebar; 15M	238.33 LM	3.15	751
	04200.00		Unit Masonry			
		05	250 mm concrete block in foundation wall	87.63 SM	80.00	7,010
		10	240 mm bond beam	69.00 LM	25.00	1,725

\Phase/Item ⟨ Takeoff Order ⟨ Assembly /

FIGURE 19.2 Typical Estimating Spreadsheet

The standard and extended version of the program allows for more versatile operations in the takeoff process including assembly takeoff, which is a convenient way to take off groups of items and quickly calculate their takeoff quantities. Items can be reviewed and modified for use in the current estimates before being generated into the spreadsheet.

Reports. Precision Estimating offers a wide variety of standard database and estimate reports. *Database reports* are used to document the database setup. These reports can also be used to check or review the information stored in the database. Reports can be printed for divisions, items, subcontractors, formulas, and so on. *Estimate reports* include standard estimate details as well as the customized spreadsheet details. *Customized reports* are reports where information is selected for inclusion in a report, by the order in which to sort the information, and the level of detail to be included. Reports can be printed on paper, previewed on the screen, and exported to a text file that can be inserted into another program, such as a word processor.

Interfaces with Other Applications. Precision Estimating can be linked to other Precision applications, including Bid Analysis, Buyout, Cad Integrator, Paydirt, and Digitizer and to scheduling software.

19.4 Electronic Tendering

Electronic tendering is a process whereby information is exchanged between participants of the bidding process in digital format. Currently, several systems are in place for processing tender documents in Canada, including systems that display descriptions of projects for construction which may be viewed via access to a specific website on the Internet and a system that in addition to displaying descriptions of projects, allows bidders to research potential projects, and review and download tender documents including specifications in electronic form upon payment of a fee.

Bid calling authorities, in particular the federal and various provincial governments, regional municipalities, and city councils, have moved toward the use of systems which involve some form of electronic tendering. The requirement that bids be advertised, that tender information and documentation be reviewed, assumes that within the industry, general contractors, subcontractors, and suppliers are sufficiently aware of and equipped to deal with the processing of information electronically. For those that are not, this technological advance carries with it the risk that they will be at a disadvantage.

The costs to contractors include computer equipment/hardware necessary to access the various electronic tendering systems, the cost of software available from the various websites necessary for the electronic processing of tender information, the fees associated with registration to the websites, and the costs of downloading documents electronically.

Presently, electronic tendering is used in varying forms in Canada as follows:

1. The Ontario Public Buyers Association at website **www.vaxxine.com/opba**. The site provides purchasing services to those who purchase public goods. A fee is payable to obtain membership. Tender documents are available only from the tendering entity directly.

2. Electronic Tendering Network (ETN) at website **www.etnbids.com/**. The site provides information to potential bidders with respect to available projects and goods and services from several hundred public sector owners in Ontario. In Ontario, the regional municipalities, cities, and some of the organizations currently using ETN ask bidders to provide an electronic copy of their tenders in disk form, together with the traditional paper copy of the form of tender.

3. Merx at website **www.merx.cebra.com**. Merx is an Internet-based system used as a bulletin board by buyers to describe construction and/or supply opportunities at projects and used by vendors to locate available projects by searching, viewing, and then ordering tender documents electronically. Merx is a Canada-wide electronic tendering system used for tenders of goods with a value in excess of $25,000 and for tenders of services with a value in excess of $100,000. The federal government, seven of the provincial governments, various municipalities, universities, schools, and hospitals use it. Merx allows users to review tender documents, select and place orders for specific documents, and download tender documents which may also be obtained by way of courier or facsimile transmission.

Review Questions

1. In its simplest form, what can computer estimating software do? How will that help the estimator?

2. In its most elaborate form, what can computer software do?

3. What factors have made computers increasingly popular for estimating?

4. What is the advantage of using the CSC/CSI format?

5. How does the estimator's job emphasis change when using computer estimating?

6. (Computer homework). Install Timberline's Precision Estimating–basic version CD.

APPENDIX A

Common Terms Used in the Building Industry

Addenda Statements or drawings that modify the basic contract documents after the latter have been issued to the bidders, but prior to the closing of the bids.

Alternatives Proposals required of bidders reflecting amounts to be added to or subtracted from the base bid price in the event that specific changes in the work are ordered.

Anchor Bolts Bolts used to anchor structural members to concrete or the foundation.

Approved Equal The term used to indicate that material or product finally supplied or installed must be equal to that specified and as approved by the consultant.

As-Built Drawings Drawings made during the progress of construction, or subsequent thereto, illustrating how various elements of the project was actually installed.

Astragal A closure between the two leafs of a double-swing or double-slide door to close the joint. This can also be a piece of moulding.

Axial Anything situated around, in the direction of, or along an axis.

Baseplate A plate attached to the base of a column that rests on a concrete or masonry footing.

Bay The space between column centre lines or primary supporting members, lengthwise in a building. Usually, the crosswise dimension is considered the span or width module, and the lengthwise dimension is considered the bay spacing.

Beam A structural member that is normally subjected to bending loads and is usually a horizontal member carrying vertical loads. (An exception to this is a purlin.) There are three types of beams:

1. Continuous Beam: A beam that has more than two points of support.

2. Cantilevered Beam: A beam that is supported at only one end and is restrained against excessive rotation.

3. Simple Beam: A beam that is freely supported at both ends, theoretically with no restraint.

Beam and Column A primary structural system consisting of a series of beams and columns; usually arranged as a continuous beam supported on several columns with or without continuity that is subjected to both bending and axial forces.

Beam-Bearing Plate Steel plate with attached anchors that are set on top of a masonry wall so that a purlin or a beam can rest on it.

Bearing The condition that exists when one member or component transmits load or stress to another by direct contact in compression.

Benchmark A fixed point used for construction purposes as a reference point in determining the various levels of floor, grade, and so on.

Bid Proposal prepared by a prospective contractor specifying the price for doing the work in accordance with the contract documents.

Bid Bond A surety bond guaranteeing that a bidder will sign a contract, if accepted, in accordance with their bid.

Bid Security A bid bond, certified cheque, or other forfeitable security guaranteeing that a bidder will sign a contract, if accepted, in accordance with the bid.

Bill of Materials A list of items or components used for fabrication, ordering, shipping, receiving, and accounting purposes.

Bird Screen Wire mesh used to prevent birds from entering the building through ventilators or louvres.

Bond Masonry units interlocked in the face of a wall by overlapping the units in such a manner as to break the continuity of vertical joints.

Bonus and Penalty Clause A provision in the bid form for payment of a bonus for each day the project is completed prior to the time stated, and for a charge against the contractor for each day the project remains uncompleted after the time stipulated.

Brace Rods Rods used in roofs and walls to transfer wind loads and/or seismic forces to the foundation (often used to plumb building but not designed to replace erection cables when required).

Bridging The structural member used to give lateral support to the weak plane of a truss, joist, or purlin; provides sufficient stability to support the design loads, sag channels, or sag rods.

Builtup Roofing Roofing consisting of layers of rag felt or jute saturated with coal tar pitch, with each layer set in a mopping of hot tar or asphalt; ply designation as to the number of layers.

Camber A permanent curvature designed into a structural member in a direction opposite to the deflection anticipated when loads are applied.

Canopy Any overhanging or projecting structure with the extreme end unsupported. It may also be supported at the outer end.

Cantilever A projecting beam supported and restrained only at one end.

Cap Plate A horizontal plate located at the top of a column.

Cash Allowances Stipulated sums that the contractor is requested to include in their bid which are to be expended on specific work items on the project.

Caulk To seal and make weathertight the joints, seams, or voids by filling with a waterproofing compound or material.

Certificate of Occupancy Statement issued by the building officials granting permission to occupy a project for a specific use.

Certificate of Payment Statement by an architect informing the owner of the amount due a contractor on account of work accomplished and/or materials suitably stored.

Change Order A contract instruction, usually prepared by the consultant and signed by the owner or the owner's representative, authorizing a change in the scope of the work and a change in the cost of the project.

Channel A steel member whose formation is similar to that of a C-section without return lips; may be used singularly or back to back.

Clip A plate or angle used to fasten two or more members together.

Clip Angle An angle used for fastening various members together.

Collateral Load A load, in addition to normal live, wind, or dead loads, intended to cover loads that are either unknown or uncertain (sprinklers, lighting, and so on).

Column A main structural member used in a vertical position on a building to transfer loads from main roof beams, trusses, or rafters to the foundation.

Contract Documents Working drawings, specifications, General Conditions, Supplementary General Conditions, the Owner-Contractor Agreement, and all addenda (if issued).

Curb A raised edge on a concrete floor slab.

Curtain Wall Perimeter walls that carry only their own weight and wind load.

Datum Any level surface to which elevations are referred (see Benchmark).

Dead Load The weight of the structure itself, such as floor, roof, framing, and covering members, plus any permanent loads.

Deflection   The displacement of a loaded structural member or system in any direction, measured from its no-load position, after loads have been applied.

Design Loads   Those loads specified by building codes, or owner's or architect's specifications to be used in the design of the structural frame of a building. They are suited to local conditions and building use.

Door Guide   An angle or channel guide used to stabilize and keep plumb a sliding or rolling door during its operation.

Downspout   A hollow section, such as a pipe used to carry water from the roof or gutter of a building to the ground or sewer connection.

Drain   Any pipe, channel, or trench for which wastewater or other liquids are carried off, that is, to a sewer pipe.

Equal   See *Approved Equal*.

Erection   The assembly of components to form the completed portion of a job.

Expansion Joint   A connection used to allow for temperature-induced expansion and contraction of material.

Fabrication   The manufacturing process performed in the plant to convert raw material into finished metal building components. The main operations are cold forming, cutting, punching, welding, cleaning, and painting.

Fascia   A flat, broad trim projecting from the face of a wall, which may be part of the rake or the eave of the building.

Field   The jobsite or building site.

Field Fabrication   Fabrication performed by the erection crew or others in the field.

Field Welding   Welding performed at the jobsite, usually with gasoline-powered machines.

Filler Strip   Preformed neoprene material, resilient rubber, or plastic used to close the ribs or corrugations of a panel.

Final Acceptance   The owner's acceptance of a completed project from a contractor.

Fixed Joint   A connection between two members in such a manner as to cause them to act as a single continuous member; provides for transmission of forces from one member to the other without any movement in the connection itself.

Flange   That portion of a structural member normally projecting from the edges of the web of a member.

Flashing   A sheet-metal closure that functions primarily to provide weather tightness in a structure and secondarily to enhance appearance; the metalwork that prevents leakage over windows, doors, around chimneys, and at other roof details.

Footing   That bottom portion at the base of a wall or column used to distribute the load into the supporting soil.

Foundation   The substructure that supports a building or other structure.

Framing   The structural steel members (columns, rafters, girts, purlins, brace rods, and so on) that go together to comprise the skeleton of a structure ready for covering to be applied.

Furring   Levelling up or building out of a part of wall or ceiling by wood, metal, or strips.

Glaze (Glazing)   The process of installing glass in window and door frames.

Grade   The term used when referring to the ground elevation around a building or other structure.

Grout   A mixture of cement, sand, and water used to solidly fill cracks and cavities; generally used under setting places to obtain a solid, uniform, full bearing surface.

Gutter   A channel member installed at the eave of the roof for the purpose of carrying water from the roof to the drains or downspouts.

Head   The top of a door, window, or frame.

Holdback   A sum withheld from each payment to the contractor in accordance with the terms of the Owner-Contractor Agreement.

Impact Load   The assumed load resulting from the motion of machinery, elevators, cranes, vehicles, and other similar moving equipment.

Instructions to Bidders   A document stating the procedures to be followed by bidders.

Insulation   Any material used in building construction for the protection from heat or cold.

Invitation to Bid   An invitation to contractors furnishing information on the submission of bids for a project.

Jamb   The side of a door, window, or frame.

Joist   Closely spaced beams supporting a floor or ceiling. They may be wood, steel, or concrete.

Lavatory   A bathroom-type sink.

Liens   Legal claims against an owner for amounts due to those engaged in or supplying materials for the construction of the building.

Lintel   The horizontal member placed over an opening to support the loads (weight) above it.

Liquidated Damages   An agreed-to sum chargeable against the contractor as reimbursement for damages suffered by the owner because of the contractor's failure to fulfill contractual obligations.

Live Load   The load exerted on a member or structure due to all imposed loads except dead, wind, and seismic loads. Examples include snow, people, and movable equipment. This type of load is movable and does not necessarily exist on a given member of structure.

Loads   Anything that causes an external force to be exerted on a structural member. Examples of different types are:

1. *Dead Load:* in a building, the weight of all permanent constructions, such as floor, roof, framing, and covering members.
2. *Impact Load:* the assumed load resulting from the motion of machinery, elevators, craneways, vehicles, and other similar kinetic forces.
3. *Roof Live Load:* all loads exerted on a roof (except dead, wind, and lateral loads) and applied to the horizontal projection of the building.
4. *Seismic Load:* the assumed lateral load due to the action of earthquakes and acting in any horizontal direction on the structural frame.
5. *Wind Load:* the load caused by wind blowing from any horizontal direction.

Louvre   An opening provided with one or more slanted, fixed, or movable fins to allow flow of air, but to exclude rain and sun or to provide privacy.

Mullion   The large vertical piece between windows. (It holds the window in place along the edge with which it makes contact.)

Nonbearing Partition   A partition that supports no weight except its own.

Parapet   That portion of the vertical wall of a building that extends above the roof line at the intersection of the wall and roof.

Partition   A material or combination of materials used to divide a space into smaller spaces.

Pave   The line along the sidewall formed by the intersection of the inside faces of the roof and wall panels; the projecting lower edges of a roof, overhanging the walls of a building.

Performance Bond   A bond that guarantees to the owner, within specified limits, that the contractor will perform the work in accordance with the contract documents.

Pier   A structure of masonry (concrete) used to support the bases of columns and bents. It carries the vertical load to a footing at the desired load-bearing soil.

Pilaster   A flat, rectangular column attached to or built into a wall masonry or pier; structurally, a pier, but treated architecturally as a column with a capital, shaft, and base. It is used to provide strength for roof loads or support for the wall against lateral forces.

Precast Concrete   Concrete that is poured and cast in some position other than the one it will finally occupy; cast either on the jobsite and then put into place, or away from the site to be transported to the site and erected.

Prestressed Concrete   Concrete in which the reinforcing cables, wires, or rods are tensioned before there is load on the member.

Progress Payments   Payments made during progress of the work, on account, for work completed and/or suitably stored.

Progress Schedule   A diagram showing proposed and actual times of starting and completion of the various operations in the project.

Punch List   A list prepared by the architect or engineer of the contractor's uncompleted work or work to be corrected.

Purlin   Secondary horizontal structural members located on the roof extending between rafters, used as (light) beams for supporting the roof covering.

Rafter   A primary roof support beam usually in an inclined position, running from the tops of the structural columns at the eave to the ridge or highest portion of the roof. It is used to support the purlins.

Recess   A notch or cutout, usually referring to the blockout formed at the outside edge of a foundation, providing support and serving as a closure at the bottom edge of wall panels.

Reinforcing Steel (Rebar)   The steel placed in concrete to carry the tension, compression, and shear stresses.

Rolling Doors   Doors that are supported on wheels that run on a track.

Roof Overhang   A roof extension beyond the end or the sidewalls of a building.

Roof Pitch   The angle or degree of slope of a roof from the eave to the ridge. The pitch can be found by dividing the height, or rise, by the span; for example, if the height is 2 m and the span is 4 m, the pitch is 2/4 or 1/2 and the angle of pitch is 45 degrees (see Roof Slope).

Roof Slope   The angle that a roof surface makes with the horizontal, usually expressed as a certain rise in millimetres of run.

Sandwich Panel   An integrated structural covering and insulating component consisting of a core material with inner and outer metal or wood skins.

Schedule of Values   A statement furnished to the architect by the contractor reflecting the amounts to be allotted for the principal divisions of the work. It is to serve as a guide for reviewing the contractor's periodic application for payment.

Sealant   Any material that is used to close cracks or joints.

Separate Contract   A contract between the owner and a specialist contractor other than the general contractor for the construction of a portion of a project.

Sheathing   Rough boarding (usually plywood) on the outside of a wall or roof over which is placed siding or shingles.

Shim   A piece of steel used to level or square beams or column base plates.

Shipping List   A list that enumerates by part, number, or description each piece of material to be shipped.

Shop Drawings   Drawings that illustrate how specific portions of the work shall be fabricated and/or installed.

Sill   The lowest member beneath an opening, such as a window or door; also, the horizontal framing members at floor level, such as sill girts or sill angles; the member at the bottom of a door or window opening.

Sill Lug   A sill that projects into the masonry at each end of the sill. It must be installed as the building is being erected.

Sill Slip   A sill that is the same width as the opening—it will slip into place.

Skylight   An opening in a roof or ceiling for admitting daylight; also, the reinforced plastic panel or window fitted into such an opening.

Snow Load In locations subject to snow loads, as indicated by the average snow depth in the reports of the Meteorological Service of Canada (MSC), the design loads shall be modified accordingly.

Soffit The underside of any subordinate member of a building, such as the undersurface of a roof overhang or canopy.

Soil Borings A boring made on the site in the general location of the proposed building to determine soil type, depth of the various types of soils, and water table level.

Soil Pressure The allowable soil pressure is the load per unit area a structure can safely exert on the substructure (soil) without exceeding reasonable values of footing settlements.

Spall A chip or fragment of concrete that has chipped, weathered, or otherwise broken from the main mass of concrete.

Span The clear distance between supports of beams, girders, or trusses.

Spandrel Beam A beam from column to column carrying an exterior wall and/or the outermost edge of an upper floor.

Specifications A statement of particulars of a given job as to size of building, quality and performance of workers and materials to be used, and the terms of the contract. A set of specifications generally indicates the design loads and design criteria.

Square One hundred square feet (9.29 m^2).

Stock A unit that is standard to its manufacturer; it is not custom made.

Stool A shelf across the inside bottom of a window.

Stud A vertical wall member to which exterior or interior covering or collateral material may be attached. Load-bearing studs are those that carry a portion of the loads from the floor, roof, or ceiling as well as the collateral material on one or both sides. Non-load-bearing studs are used to support only the attached collateral materials and carry no load from the floor, roof, or ceiling.

Subcontractor A separate contractor for a portion of the work (hired by the general contractor).

Substantial Completion For a project or specified area of a project, the date when the construction is sufficiently completed in accordance with the contract documents, as modified by any change orders agreed to by the parties so that the owner can occupy the project or specified area of the project for the use for which it was intended.

Supplementary General Conditions One of the contract documents, prepared by the architect, that may modify provisions of the General Conditions of the contract.

Temperature Reinforcing Lightweight deformed steel rods or wire mesh placed in concrete to resist possible cracks from expansion or contraction due to temperature changes.

Time of Completion The number of days (calendar or working) or the actual date by which completion of the work is required.

Truss A structure made up of three or more members, with each member designed to carry basically a tension or a compression force. The entire structure, in turn, acts as a beam.

Veneer A thin covering of valuable material over a less expensive body; for example, brick on a wood frame building.

Wainscot Protective or decorative covering applied or built into the lower portion of a wall.

Wall Bearing In cases where a floor, roof, or ceiling rests on a wall, the wall is designed to carry the load exerted. These types of walls are also referred to as load-bearing walls.

Wall Covering The exterior wall skin consisting of panels or sheets and including their attachment, trim, fascia, and weather sealants.

Wall Nonbearing Wall not relied upon to support a structural system.

Water Closet More commonly known as a toilet.

Working Drawing The actual plans (drawings and illustrations) from which the building will be built. They show how the building is to be built and are included in the contract documents.

APPENDIX B

Conversions and Formulas

International System of Units (SI)

Units

length	metre	m
mass	kilogram	kg
time	second	s
electric current	ampere	A
thermodynamic temperature	Kelvin	K
amount of substance	mole	mol
luminous intensity	candela	cd
area	square metres	m^2
temperature	degree Celsius	°C
volume	cubic metres	m^3

Conversion and Rounding Values

When converting imperial units to SI units, an exact or "soft" metric equivalent is utilized or a "hard" or rounded metric number convenient to work with is adopted.

In rounding values, when the first digit discarded is less than 5, the last digit retained is not changed. When the first digit discarded is greater than 5 or is a 5 followed by at least one digit other than 0, add 1 to the last digit retained.

Example:

7.2456 rounded to four digits would be 7.246

7.2456 rounded to three digits would be 7.25

7.2456 rounded to two digits would be 7.2

Conversions: Imperial to SI

Distance

From	To	Multiply By:
inches	millimeters	25.4
feet	metres	0.3048
yards	metres	0.9144
miles	kilometres	1.6093

Areas

From	To	Multiply By:
square feet	square metres (m²)	0.0929
square yards	square metres (m²)	0.836
square miles	square kilometres (km²)	2.59

Volume

From	To	Multiply By:
cubic feet	cubic metres (m³)	0.0283
cubic yards	cubic metres (m³)	0.7646
imperial gallons	litres	4.5460
American gallons	litres	3.7853

Mass

From	To	Multiply By:
pounds	kilograms (kg)	0.4536
short ton	kilograms (kg)	907.1847
long ton	kilograms (kg)	1016.047
short ton	tonne (t)	0.9072
long ton	tonne (t)	1.0160

SI Lengths

1 metre = 100 cm

1 kilometre = 1,000 metres

Imperial Lengths

12 inches = 1 foot

3 feet = 1 yard

SI Areas

1 square kilometre (km²) = 100 hectares (ha)

1 hectare (ha) = 10,000 square metres (m²)

Imperial Areas

144 s.i. = 1 s.f.

9 s.f. = 1 s.y.

100 s.f. = 1 square

SI Volume

1 cubic metre (m³) = 999.972 litres

Imperial Volume

1,728 c.i. = 1 c.f.
27 c.f. = 1 c.y.
1 c.f. = 7.4850 gallons
1 gallon = 231 c.i.
1 gallon = 8.33 pounds of water
1 c.f. = 62.3 pounds of water

SI Weight

1 tonne (t) = 1,000 kilograms (kg)

Imperial Weight

1 short tonne = 2,000 pounds
1 long tonne = 2,240 pounds

Inches Reduced to Decimals of a Foot

Inches	Decimal
$\frac{1}{2}$	0.041
1	0.083
$\frac{1}{2}$	0.125
2	0.167
$2\frac{1}{2}$	0.209
3	0.250
$3\frac{1}{2}$	0.292
4	0.333
$4\frac{1}{2}$	0.375
5	0.417
$5\frac{1}{2}$	0.458
6	0.500
$6\frac{1}{2}$	0.542
7	0.583
$7\frac{1}{2}$	0.625
8	0.667
$8\frac{1}{2}$	0.708
9	0.750
$9\frac{1}{2}$	0.792

Formulas for Area and Volume Calculations

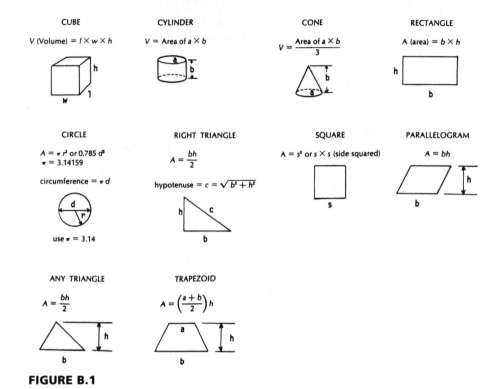

CUBE

V (Volume) $= l \times w \times h$

CYLINDER

$V =$ Area of $a \times b$

CONE

$V = \dfrac{\text{Area of } a \times b}{3}$

RECTANGLE

A (area) $= b \times h$

CIRCLE

$A = \pi r^2$ or $0.785\ d^2$
$\pi = 3.14159$

circumference $= \pi d$

use $\pi = 3.14$

RIGHT TRIANGLE

$A = \dfrac{bh}{2}$

hypotenuse $= c = \sqrt{b^2 + h^2}$

SQUARE

$A = s^2$ or $s \times s$ (side squared)

PARALLELOGRAM

$A = bh$

ANY TRIANGLE

$A = \dfrac{bh}{2}$

TRAPEZOID

$A = \left(\dfrac{a + b}{2}\right) h$

FIGURE B.1

APPENDIX C

Residential Project Drawings

Listed below are the Residential Project drawings that you will find in an insert at the end of the book.

Proposed Bungalow

SP-1 Specification Sheet
SS-1 Site Servicing
A1 Basement Plan
A2 First Floor Plan
A3 Front Elevation
A4 Front Elevation - B
A5 Rear Elevation
A6 Right Elevation
A7 Left Elevation
A8 Cross Section

The above drawings should be used to answer the Review Questions at the end of Chapters 8, 9, 10, 12, 13, 14, and 15.

Students are required to manually take off quantities and to create an estimate using Timberline Precision Estimating for the trades in the following divisions:

Chapter 8:	Division 2 - Excavation
Chapter 9:	Division 3 - Formwork, Reinforcement, Cast-in-Place Concrete
Chapter 10:	Division 4 - Masonry
Chapter 12:	Division 6 - Wood Framing
Chapter 13:	Division 7 - Insulation
Chapter 14:	Division 8 - Doors and Windows
Chapter 15:	Division 9 - Finishes

APPENDIX D

Commercial Project Drawings

Listed below are the Commercial Project drawings that you will find in an insert at the end of the book.

Proposed Two-Storey Building

SS1	Site Statistics and Details
SS2	Siting Plan
A1	First and Second Floor Plan
A2	Elevations and Roof Plan
A3	North/East Elevation Cross-sections
A4	Wall Sections
A5	Construction Details
A6	Stair Detail/Section
A7	Schedules
SP-1	Structural Notes
S1	Structural Floor Plans
S2	Wall Sections
E1	Electrical Plans, Lighting Fixture Schedule, and Voice/Data/Power Outlet Detail
M1	Mechanical Plans and Specs

The above drawings should be used to answer the Review Questions at the end of Chapters 5, 8, 9, 10, 11, 12, 13, 14, and 15.

Students are required to manually take off quantities and to create an estimate using Timberline Precision Estimating for the trades in the following divisions:

Chapter 8:	Division 2 - Excavation
Chapter 9:	Division 3 - Formwork, Reinforcement, Cast-in-Place Concrete
Chapter 10:	Division 4 - Masonry
Chapter 11:	Division 5 - Structural Steel
Chapter 12:	Division 6 - Wood Framing
Chapter 13:	Division 7 - Insulation
Chapter 14:	Division 8 - Doors and Windows
Chapter 15:	Division 9 - Finishes

Electronic Spreadsheets

The CD enclosed in this book contains the Basic Edition of "Timberline Precision Estimating software," as well as several spreadsheets that were used in the previous chapters. The spreadsheets were generated using Microsoft Excel. Since these spreadsheets are on a CD, they are designated as read only. To access the Excel spreadsheets, do the following:

1. Load the CD-ROM in the drive
2. When the Precision Estimating installation screen comes up, click EXIT
3. From the desktop, open My Computer
4. Right click on the CD-ROM icon and choose OPEN

You can then access the spreadsheet files found in the *Spreadsheets* folder. These spreadsheets can be copied to a computer's hard drive. The properties can be changed so that they are *read–write*, or they can be opened in Excel. If the latter option is used, they will need to be saved to a computer's hard drive. The table below contains the file name, description, and figure reference. These spreadsheets are provided as guides and may need to be revised to meet specific needs. Set out below is a list (and visual depiction) of each spreadsheet included on the CD.

File Name	Description	Figure
Bid Summary	Bid Summary Sheet	Figure E-1
Cut-Fill	Cut and Fill Worksheet	Figure E-2
Recap	Own Forces Estimate Summary Sheet	Figure E-3
Slab Rebar	Slab Reinforcing Steel Worksheet	Figure E-4
Steel	Structural Steel Worksheet	Figure E-5
Estimate Form	Standard Form of Estimate	Figure E-6

		ESTIMATE SUMMARY			

Project: _____ Estimate No. _____
Location _____ Date _____
Architect _____ By _____ Checked _____

DIV	DESCRIPTION	LABOUR $	MATERIAL $	EQUIPMENT $	SUBCONTRACT $	TOTAL $
	TRADE COSTS					
	2 SITEWORK					
	3 CONCRETE					
	4 MASONRY					
	5 METALS					
	6 WOOD & PLASTICS					
	7 THERMAL AND MOISTURE PROTECTION.					
	8 DOORS AND WINDOWS					
	9 FINISHES					
	10 SPECIALTIES					
	11 EQUIPMENT					
	12 FURNISHINGS					
	13 SPECIAL CONSTRUCTION					
	14 CONVEYING SYSTEMS					
	15 MECHANICAL					
	16 ELECTRICAL					
	TOTAL TRADE COSTS					
	GENERAL EXPENSES					
	SITE STAFF					
	TEMPORARY OFFICES					
	TEMPORARY FACILITIES					
	TEMPORARY UTILITIES					
	PROTECTION					
	CLEANING					
	MISC. EQUIPMENT					
	PERMITS					
	BONDS					
	INSURANCES					
	LABOUR BURDENS					
	RETAIL SALES TAX					
	TOTAL GENERAL EXPENSES					
	HEAD OFICE COSTS					
	LAST MINUTE CHANGES					
	TOTAL LAST MINUTE CHANGES					
	PROFIT					
	TOTAL PROJECT COSTS					

COMMENTS _____
_____
_____
_____
_____

FIGURE E-1 Bid Summary Sheet

ESTIMATE WORK SHEET

Project: _____
Location _____
Architect _____
Items _____

CUT & FILL WORK SHEET

Estimate No. _____
Sheet No. _____
Date _____
By _____ Checked _____

Grid	Fill									Cut								
	Fill At Intersections					Points	Average	Area	Total	Cut At Intersections					Points	Average	Area	Total
	1	2	3	4	5					1	2	3	4	5				
1																		
2																		
3																		
4																		
5																		
6																		
7																		
8																		
9																		
10																		
11																		
12																		
13																		
14																		
15																		
16																		
17																		
18																		
19																		
20																		
21																		
22																		
23																		
24																		
25																		
26																		
27																		
28																		
29																		
30																		
31																		
32																		
33																		
34																		
35																		
36																		
37																		
38																		
39																		
40																		
41																		
42																		

TOTAL FILL - Cubic metres
Shrinkage Factor
Required Cubic Metres of Fill
Net Cubic Metres to Import

TOTAL CUT   Cubic metres
Swell Factor
Cubic Metres of Cut to Haul

FIGURE E-2 Cut and Fill Workshop Sheet

ESTIMATE SUMMARY SHEET

Project _____
Location _____
Architect _____
Items _____

Estimate No. _____
Sheet No. _____
Date _____
By _____ Checked _____

Cost Code	Description	Q.T.O.	Waste Factor %	Purch. Quan.	Unit	Crew	Prod Rate	Wage Rate	Work Hours	Unit Cost			Labour	Material	Equipment	TOTAL
										Labour	Material	Equipment				
	TOTALS															

FIGURE E-3 Own Forces Estimate Summary Sheet

ESTIMATE WORK SHEET
REINFORCING STEEL

Project _____
Location _____
Architect _____
Items _____

Estimate No. _____
Sheet No. _____
Date _____
By _____ Checked _____

Cost Code	Description	Length/ Width m	Bar Spacing mm	Pcs.	Bar Length m	Cover- age mm	Bar Length		Bar Size				Bar Weights				Quantity	Unit
							Ea	Total	10 M	15 M	20 M	25 M	10 M 0.785	15 M 1.570	20 M 2.355	25 M 3.925		
	Total Weight																	

FIGURE E-4 Slab Reinforcing Steel Worksheet

ESTIMATE WORK SHEET

Project _____
Location _____
Architect _____
Items _____

STRUCTURAL STEEL

Estimate No. _____
Sheet No. _____
Date _____
By _____ Checked _____

Cost Code	Description	Designation	Kilogram/ Metre		Count			Quantity	Unit
								0	kg
								0	kg
								0	kg
								0	kg
								0	kg
								0	kg
								0	kg
								0	kg
								0	kg

FIGURE E-5 Structural Steel Worksheet

ESTIMATE

Project: ___ Project No: ___ Estimate No: ___ Page: ___ of ___
Element/Trade: ___ Measured: ___ Estimate Type: ___ Date: ___
Element/UCI Reference: ___ Extended: ___ Priced: ___ Checked: ___

Description	No.	Dimensions			Extensions	Quantity	Unit	Unit Price	Cost ($)
					Brought Forward from Page:				
					Carried Forward to Page:				

FIGURE E-6
Standard Form of
Estimate

Index

$250

Timberline's Estimating Basic software that comes with this book provides the essentials you need to create a construction estimate. You start with an easy-to-use spreadsheet that is backed by a "price book" database, which you can easily customize to include your own unique formulas, productivity and conversions.

Pull-down menus give you instant access to everything you need to prepare an estimate, whether you use items from another estimate, do a takeoff, modify your pricing database, or print a report. Point-and-click, drag-and-drop, and other easy-to-use features are all a part of Timberline's estimating software.

How to install this software

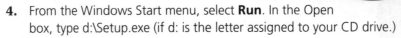

You must install the software using the Setup command:

1. If you use Microsoft Windows NT Workstation 4.0, Windows 2000 Professional, or Windows XP Professional, log on as the administrator (not as an operator with administration rights).

2. If you use antivirus software, turn it off for the duration of the installation. **Shut down** other unnecessary programs or services for the duration of the installation.

3. Insert the CD-ROM into your CD drive.

4. From the Windows Start menu, select **Run**. In the Open box, type d:\Setup.exe (if d: is the letter assigned to your CD drive.)

5. In the Welcome window, click **Next**.

6. After you read the Timberline Software Corporation end user license agreement, click **Accept**.

7. To install the Pervasive Work Group Engine, click **Yes**. After installation, click **Continue** to restart your computer. Log in again, if necessary. Estimating Basic installation will resume.

8. Click **Next** to accept the default destination folder, or click **Browse** to select a different folder.

9. In the Select Components window, select **Estimating-Basic** and **Sample Database-Standard**. Click **Next**. In the Select Program Folder window, accept the default choices and click **Next**.

10. In the Start Copying Files window, review the list of components and destination folders and click **Next** to begin the installation.

11. Click **Yes** if you would like to view the release notice now. Click **No** to continue.

12. Restart your antivirus software and any other programs that you closed in Step 1.

LICENSEE AGREES THAT THE FOREGOING LIMITED WARRANTY IS IN LIEU OF ALL OTHER WARRANTIES OF TIMBERLINE AND TIMBERLINE DISCLAIMS ALL OTHER WARRANTIES , EXPRESS OR IMPLIED, INCLUDING, BUT NOT LIMITED TO, ANY IMPLIED WARRANTY OF MERCHANTABILITY, FITNESS OR ADEQUACY FOR ANY PARTICULAR PURPOSE OR USE, QUALITY OR PRODUC-TIVENESS, OR CAPACITY.

REQUIRED OPERATING ENVIRONMENT. It is Licensee's responsibility to conform to Timberline's stated hardware and operating system requirements as set forth in the user documentation.

PATENT AND COPYRIGHT INDEMNIFICATION. Timberline will defend, at its expense, any action brought against Licensee to the extent that it is based on a claim that the Software supplied to Licensee constitute direct infringement of copyright filed in the United States on or before the date of this Agreement. In the event any Software furnished hereunder is, in Timberline's opin-ion, likely to, or does, become the subject of a claim of infringement of a copyright or a patent, Timberline may, at its option and expense, procure for Licensee the right to continue using the Software or modify the Software to make them non-infringing or replace them with non-infringing software, which may, at Timberline's option, come under this same Agreement. If, in Timberline's opinion, none of the foregoing alternatives is reasonably available to Timberline, then Timberline may refund the purchase price for the Software and terminate this Agreement. THE FOREGOING STATES THE ENTIRE LIABILITY OF TIMBERLINE WITH RESPECT TO INFRINGEMENT OF ANY COPYRIGHTS, PATENTS, OR OTHER INTELLECTUAL PROPERTY RIGHTS BY THE SOFTWARE OR ANY PARTS THEREOF AND IS IN LIEU OF ALL WARRANTIES OR CONDITIONS, EXPRESS OR IMPLIED.

GENERAL

This License may not be assigned or otherwise transferred by Licensee without the prior written consent of Timberline.

The parties agree that no action, regardless of form, arising hereunder, may be instituted by either party more than one (1) year after the cause of action arose, except that the above limitations shall not apply to the enforcement of any of Timberline's pro-prietary rights.

It is agreed that this Agreement shall be governed by and construed in accordance with the laws of the State of Oregon. Any and all legal transactions must be transacted or brought in the courts in the State of Oregon. This Agreement will not be gov-erned by the United Nations Convention on Contracts for the International Sales of Goods, the application of which is expressly excluded. Licensee agrees that the Software will not be shipped, transferred or exported into any country or used in any man-ner prohibited by the United States Export Administration Act, or any other export laws, restrictions, or regulations.

LIMITATION OF REMEDIES AND LIABILITY

A. LICENSEE ACKNOWLEDGES THIS LICENSE EXPRESSLY STATES THE ENTIRE RIGHTS OF AND REMEDIES AVAILABLE TO THE LICENSEE AS AGAINST LICENSOR ARISING IN ANY WAY OUT OF THIS AGREEMENT, AND LICENSOR SHALL NOT BE UNDER ANY LIABILITY TO LICENSEE WHATSOEVER, WHETHER SUCH LIABILITY IS FOR LOSS OR DAMAGE (INCLUDING CONSEQUENTIAL LOSS OR DAMAGE) OR OTHERWISE AND WHETHER SUCH LIABILITY AROSE IN CONNECTION WITH TIMBERLINE SOFTWARE OR SERVICES OR OTHERWISE.

B. SUBJECT TO THE FOLLOWING PARAGRAPH C, ALL CONDITIONS, WARRANTIES, REPRESENTATIONS, PROMISES, UNDERTAK-INGS, TERMS, GUARANTEES AND COLLATERAL AGREEMENTS IN ANY WAY RELATING TO THE SUBJECT MATTER OF THIS AGREEMENT WHICH ARE NOT EXPRESSLY STATED IN THIS AGREEMENT ARE EXCLUDED, INCLUDING WITHOUT LIMITATION ANY IMPLIED OR STATUTORY WARRANTIES, SUCH AS MERCHANTABILITY OR FITNESS FOR A PARTICULAR PURPOSE.

C. IF ANY ACT OF PARLIAMENT OR OTHER APPLICABLE LAW OR REGULATION RESTRICTS LICENSOR'S ABILITY TO EXCLUDE ITS LIABILITY IN THE MANNER REFERRED TO IN PARAGRAPHS A AND B, BUT PERMITS LICENSOR TO LIMIT ITS LIABILITY TO ANY ONE OR MORE OF SEVERAL OPTIONS SPECIFIED IN THAT LAW, LICENSOR'S LIABILITY TO LICENSEE, IN LICENSOR'S ABSOLUTE DIS-CRETION, SHALL BE LIMITED TO ANY ONE OR MORE OF THOSE OPTIONS.

D. LICENSEE WARRANTS THAT IT HAS NOT RELIED UPON ANY REPRESENTATION MADE BY LICENSOR OR UPON ANY DESCRIP-TIONS OR ILLUSTRATIONS OR SPECIFICATIONS CONTAINED IN ANY DOCUMENT, INCLUDING ANY CATALOGUES OR PUBLICITY MATERIAL, PRODUCED BY LICENSOR.

INTEGRATION. LICENSEE AND TIMBERLINE AGREE THAT THE TERMS OF THIS LICENSE AGREEMENT ALLOCATE THE RISKS ASSO-CIATED WITH THE USE OF THE SOFTWARE BETWEEN TIMBERLINE AND LICENSEE. TIMBERLINE SOFTWARE PRICING REFLECTS THIS ALLOCATION OF RISK IN THE LIMITED WARRANTY AND IN THE LIMITATION OF REMEDIES AND LIABILITY. LICENSEE ACKNOWLEDGES THAT LICENSEE HAS READ THE SIGNED LICENSE AGREEMENT, WHICH INCLUDES ALL ADDENDUMS, UNDER-STANDS EACH AND EVERY TERM AND CONDITION, AND AGREES TO BE BOUND BY ITS TERMS AND CONDITIONS. IN THE EVENT THE TERMS OF THIS LICENSE AGREEMENT CONFLICT WITH THE TERMS OF A SIGNED LICENSE AGREEMENT FOR THE USE OF THE SOFTWARE, THE TERMS OF THE SIGNED LICENSE AGREEMENT SUPERSEDE THIS LICENSE AGREEMENT.